MW01435821

First Mile Access Networks and Enabling Technologies

Ashwin Gumaste, Ph.D.

Tony Antony

Cisco Press

Cisco Press
800 East 96th Street
Indianapolis, IN 46240 USA

First Mile Access Networks and Enabling Technologies

Ashwin Gumaste, Ph.D.

Tony Antony

Copyright© 2004 Cisco Systems, Inc.

Published by:
Cisco Press
800 East 96th Street
Indianapolis, IN 46240 USA

All rights reserved. No part of this book may be reproduced or transmitted in any form or by any means, electronic or mechanical, including photocopying, recording, or by any information storage and retrieval system, without written permission from the publisher, except for the inclusion of brief quotations in a review.

Library of Congress Cataloging-in-Publication Number: 2002115402

ISBN: 1-58705-129-x

Printed in the United States of America 1 2 3 4 5 6 7 8 9 0

First Printing January 2004

Warning and Disclaimer

This book is designed to provide information about first mile access networks and enabling technologies. Every effort has been made to make this book as complete and as accurate as possible, but no warranty or fitness is implied.

The information is provided on an "as is" basis. The authors, Cisco Press, and Cisco Systems, Inc., shall have neither liability nor responsibility to any person or entity with respect to any loss or damages arising from the information contained in this book or from the use of the discs or programs that may accompany it.

The opinions expressed in this book belong to the author and are not necessarily those of Cisco Systems, Inc.

Trademark Acknowledgments

All terms mentioned in this book that are known to be trademarks or service marks have been appropriately capitalized. Cisco Press or Cisco Systems, Inc. cannot attest to the accuracy of this information. Use of a term in this book should not be regarded as affecting the validity of any trademark or service mark.

Corporate and Government Sales

Cisco Press offers excellent discounts on this book when ordered in quantity for bulk purchases or special sales.

For more information please contact: U.S. Corporate and Government Sales 1-800-382-3419
corpsales@pearsontechgroup.com

For sales outside the U.S. please contact: International Sales international@pearsontechgroup.com

Feedback Information

At Cisco Press, our goal is to create in-depth technical books of the highest quality and value. Each book is crafted with care and precision, undergoing rigorous development that involves the unique expertise of members from the professional technical community.

Readers' feedback is a natural continuation of this process. If you have any comments regarding how we could improve the quality of this book, or otherwise alter it to better suit your needs, you can contact us through e-mail at feedback@ciscopress.com. Please make sure to include the book title and ISBN in your message.

We greatly appreciate your assistance.

Publisher	John Wait
Editor-in-Chief	John Kane
Executive Editor	Jim Schachterle
Cisco Representative	Anthony Wolfenden
Cisco Press Program Manager	Nannette M. Noble
Production Manager	Patrick Kanouse
Acquisitions Editor	Amy Moss
Development Editor	Andrew Cupp
Copy Editor	Keith Cline
Technical Editors	Chiara Regale
	Priya D. Shetty
Team Coordinator	Tammi Barnett
Cover Designer	Louisa Adair
Composition	ContentWorks
Proofreader	Missy Pluta
Indexer	Tim Wright

CISCO SYSTEMS

Corporate Headquarters
Cisco Systems, Inc.
170 West Tasman Drive
San Jose, CA 95134-1706
USA
www.cisco.com
Tel: 408 526-4000
　　　800 553-NETS (6387)
Fax: 408 526-4100

European Headquarters
Cisco Systems International BV
Haarlerbergpark
Haarlerbergweg 13-19
1101 CH Amsterdam
The Netherlands
www-europe.cisco.com
Tel: 31 0 20 357 1000
Fax: 31 0 20 357 1100

Americas Headquarters
Cisco Systems, Inc.
170 West Tasman Drive
San Jose, CA 95134-1706
USA
www.cisco.com
Tel: 408 526-7660
Fax: 408 527-0883

Asia Pacific Headquarters
Cisco Systems, Inc.
Capital Tower
168 Robinson Road
#22-01 to #29-01
Singapore 068912
www.cisco.com
Tel: +65 6317 7777
Fax: +65 6317 7799

Cisco Systems has more than 200 offices in the following countries and regions. Addresses, phone numbers, and fax numbers are listed on the **Cisco.com Web site at www.cisco.com/go/offices.**

Argentina • Australia • Austria • Belgium • Brazil • Bulgaria • Canada • Chile • China PRC • Colombia • Costa Rica • Croatia • Czech Republic Denmark • Dubai, UAE • Finland • France • Germany • Greece • Hong Kong SAR • Hungary • India • Indonesia • Ireland • Israel • Italy Japan • Korea • Luxembourg • Malaysia • Mexico • The Netherlands • New Zealand • Norway • Peru • Philippines • Poland • Portugal Puerto Rico • Romania • Russia • Saudi Arabia • Scotland • Singapore • Slovakia • Slovenia • South Africa • Spain • Sweden Switzerland • Taiwan • Thailand • Turkey • Ukraine • United Kingdom • United States • Venezuela • Vietnam • Zimbabwe

Copyright © 2003 Cisco Systems, Inc. All rights reserved. CCIP, CCSP, the Cisco Arrow logo, the Cisco *Powered* Network mark, the Cisco Systems Verified logo, Cisco Unity, Follow Me Browsing, FormShare, iQ Net Readiness Scorecard, Networking Academy, and ScriptShare are trademarks of Cisco Systems, Inc.; Changing the Way We Work, Live, Play, and Learn, The Fastest Way to Increase Your Internet Quotient, and iQuick Study are service marks of Cisco Systems, Inc.; and Aironet, ASIST, BPX, Catalyst, CCDA, CCDP, CCIE, CCNA, CCNP, Cisco, the Cisco Certified Internetwork Expert logo, Cisco IOS, the Cisco IOS logo, Cisco Press, Cisco Systems, Cisco Systems Capital, the Cisco Systems logo, Empowering the Internet Generation, Enterprise/Solver, EtherChannel, EtherSwitch, Fast Step, GigaStack, Internet Quotient, IOS, IP/TV, iQ Expertise, the iQ logo, LightStream, MGX, MICA, the Networkers logo, Network Registrar, *Packet*, PIX, Post-Routing, Pre-Routing, RateMUX, Registrar, SlideCast, SMARTnet, StrataView Plus, Stratm, SwitchProbe, TeleRouter, TransPath, and VCO are registered trademarks of Cisco Systems, Inc. and/or its affiliates in the U.S. and certain other countries.

All other trademarks mentioned in this document or Web site are the property of their respective owners. The use of the word partner does not imply a partnership relationship between Cisco and any other company. (0303R)

Printed in the USA

About the Authors

Ashwin Gumaste received a Ph.D. in electrical engineering from the University of Texas at Dallas. Currently, Ashwin is with Fujitsu Laboratories in the Photonics Networking Laboratory (PNL) group in Richardson, Texas, where his research includes network development and design. Previously, Ashwin worked with Cisco Systems under the Optical Networking Group. He has written numerous papers and has more than two dozen pending or approved U.S. and EU patents, and during 1991, Ashwin was awarded the National Talent Search Scholarship in India. His research interests include optical and wireless networking and self-similar phenomenon in social and networking environments. He proposed the first architecture to implement optical burst transport and dynamic lightpath provisioning, called Light-trails, and also proposed the Light-frame framework—a conceptual model for future packet mode optical communication. He is the coauthor of *DWDM Network Designs and Engineering Solutions* published by Cisco Press.

Tony Antony has more than 14 years of telecommunications/data-networking experience and is currently working at Cisco Systems in the Technical Marketing Group. He received a master's degree in telecommunications from Southern Methodist University (SMU), Dallas and also holds CCNP and CCIP (Optical) certifications along with multiple other certifications such as FC-SAN. Tony's previous experience includes engineering positions at Texas Instruments, Raytheon, Knowledge Laboratories, TTI, and KPMG. He has authored numerous technical papers at international conferences in the networking area. His research interests include optical Internet and network simulations. He coauthored the Cisco Press book *DWDM Network Designs and Engineering Solutions*. He has served as General Chair for Workshop on Global SAN (ONSAN). Tony can be reached at tantony@cisco.com.

About the Technical Reviewers

Chiara Regale is a technical marketing engineer for Cisco Metro Ethernet Solutions. She joined Cisco in September 1998 as a software engineer. In this role she implemented a Layer 2 protocol to cut off the convergence time of a network in case of failures. After a master's degree in telecommunications engineering (Politecnico of Turin, Italy) in March 1999, she joined the Spanning Tree Protocol development team for the Catalyst 6000 family of switches. In this role, she participated in the design and implementation of features and customer requirements to improve Spanning Tree performance (Cisco Multi-Instance Spanning Tree, IEEE 802.1w/1s, Spanning Tree BPDU Guard, and Spanning Tree High Availability features). In December 2001, she changed career paths and joined the Cisco Service Provider Business Unit as technical marketing engineer, where she is focusing on solution design and validation in the metro Ethernet arena.

Priya D. Shetty received a master's degree in computer science from the University of Texas at Dallas. Priya's research interests are business modeling of broadband networks and interfaces between heterogeneous mediums, in particular first mile access interfaces in optical and wireless networking. She is with Advance Networks Laboratory at the University of Texas at Dallas where her research interest focuses on scheduling mechanisms for gigabit speed switching fabrics. She has worked for multiple companies such as Nucleus Software Offshore, Ltd., and Agicta Laboratories.

Dedication

Dedicated to the loving memory of Zanjeer and Dr. Anil Gumaste.

—Ashwin

I dedicate this book to my parents, C.P. Antony and Ritha Antony; my wife, Sheela; and my daughters, Chelsey and Melanie.

—Tony

Acknowledgments

Ashwin Gumaste: A very special thanks to my advisor and the distinguished chair professor of telecommunications, Dr. Imrich Chlamtac, at the University of Texas at Dallas, for the encouragement he bestowed upon me, as well as for being a strong source of motivation in the optical networking field. A special thanks to Csaba Szabo my employer in BCN. A special thanks to Susumu Kinoshita from Fujitsu Labs Japan for introducing intricacies in the technology. I wish to thank Jeff Norman, CEO of Main.net, for useful information on power-line communication. I wish to thank the two reviewers, Chiara and Priya, for comments and Amy, Jim, and Drew for coordination and development of this book. Also would like to acknowledge support of Dana Bisaro (Zero dB), Mike DiMauro (FTTH Council), and Russ Gyurek (Cisco Systems) for their comments on FTTH and first mile in general.

Tony Antony: I would like to thank my parents, friends, and family for supporting me during the entire process of this book. A special thanks to my team members and managers, especially Faraz Aladin and Russ Tarpey at Cisco Systems. I would also like to thank Artie Thomas, Chris Harvey, and Nicolas Brenton and all my other colleagues for the continuous encouragement they offered to me. I extend my thanks to Amy Moss for coordinating the efforts, Drew Cupp for developing the book, and Chiara Regal and Priya Shetty for their excellent comments.

Contents at a Glance

Introduction xvii

Chapter 1 Introduction to First Mile Access Technologies 3

Chapter 2 Passive Optical Networks in the First Mile 27

Chapter 3 Enterprise Fiber Solution 59

Chapter 4 Data Wireless Communication 87

Chapter 5 Broadband Wireless Access (IEEE 802.16: WirelessMAN) 113

Chapter 6 Free Space Optics Solutions in the First Mile 137

Chapter 7 DSL Technologies 161

Chapter 8 Power-Line Communication 183

Chapter 9 First Mile Access Management and Business Model 203

Chapter 10 Business Case for First Mile Access Networks 233

Chapter 11 The Complete First Mile Network: Integration, Interfacing, and Management 267

Appendix Unit of Optical Power Measurement: Decibel 277

Index 279

Contents

Introduction xvii

Chapter 1 Introduction to First Mile Access Technologies 3

The Network Today 3

Bandwidth Management in Worldwide Networks: Impact on First Mile Access 5

The First Mile Issue 7

Business Justification for the First Mile Problem 7

Technological Aspirations Forecast for First Mile Access Networks 10

Technologies Deployed in Access Networks 11
- DSL Technology 11
- Fiber 12
  - Attenuation Windows in a Fiber 14
  - Passive Optical Networks 15
- Cable Networks 17
- The PSTN 19
- Wireless Networks 20
  - Fixed Wireless 21
  - WiFi: Wireless Fidelity 22
  - Infrared and Free Space Optics 22
- Comparison of First Mile Access Technologies 22

Summary 23

Review Questions 23

References 24

Chapter 2 Passive Optical Networks in the First Mile 27

Network Profiling for Passive Optical Networks 29
- The Network Topology 29

Passive Optical Networking: Problem Formulation 31

PON Classification—APON and EPON 36
- Advantages of BPON 37
- Gigabit PON 38

Emergence of EPON—An Effective PON Solution, Particularly for IP-Centric Communication 39
- Ethernet as a Disruptive Technology 40

Standardization Efforts in EPON—The EFM Push 42
    Multipoint Application 43
    Transmission upon Reception 44
        System Operation 47
    Interleaved Polling with Adaptive Cycle Time 48

Numeric Evaluation of EPONs 49

Comparison with Contemporary Schemes 53

Summary 54

Review Questions 55

References 56

Chapter 3 Enterprise Fiber Solution 59

Coarse Wavelength-Division Multiplexing 60

Business Case for CWDM and Access Networks 61

CWDM Network Configuration 61

Technicalities of CWDM 63

CWDM Network Elements 63
    Vertical Cavity Surface Emitting Laser 64
        Line Width of a Laser 65
    Thin Film Filters 65

Standards for CWDM Deployment 66

CWDM Benefits 67
    CWDM Services 67

CWDM Network Protection 68
    Protection in Ethernet over CWDM Networks 71
        Protection Using STP and UDLD 71
        Protection Using EtherChannel and UDLD 71
    CWDM Network Architecture 72
    Mux/Demux and OADM Characterization 72

Design Rules for CWDM Networks in Metro Access 74
    Power-Budget Calculations 75
        Maximum Distance Transmitted in a Point-To-Point System 75
        Maximum Distance Transmitted in a Ring Using OADMs 76

CWDM Design Examples 76
    Example 1 77
        Example 1 Solution 77

          Example 2   79
                 Example 2 Solution   79
          Exercise   82

     Summary   82

     Review Questions   83

     References   83

Chapter 4   Data Wireless Communication   87

     The Wireless Medium and the Three Wireless Technologies for Communication   87
          Time-Division Multiplexing   88
          Frequency-Division Multiplexing   89

     Code-Division Multiple Access   90
          Classification of CDMA: Frequency Hopping and Direct Sequence   91
               Frequency-Hopping Spread Spectrum   91
               Direct-Sequence Spread Spectrum   92

     Frequency-Hopping Spread Spectrum Theory   93

     The Advantage of Spread Spectrum   94

     Antenna Theory: Gain and Half-Power Beamwidth   95
          Effective Isotropic Radiated Power (EIRP) AND G/T   96
          Smart Antennas   97

     Other Wireless Effects: Fading and Delay Spread   100

     Wireless Data: The Evolution of the 802.11 Series of Standards   100
          IP Mobility   101
          The 802.11b High-Rate Wireless LAN   101
               Infrastructure Method   102
               Ad Hoc Mode   103
          The 802.11b Physical Layer   103
               Note on QPSK   103
               Complementary Code Keying   104
          Dynamic Rate Shifting in 802.11b Standards   105
          802.11b Physical Layer Functioning   105
          802.11b MAC Sublayer   106
          Accessing the 802.11b Medium and Interframe Space   106
          Management Frames for 802.11b   106

     WLAN Device   107

     Operation of the IEEE 802.11 Standard   107

Performance of 802.11 Standards in First Mile Access   108

802.11g   109

Wireless Data Applications: Hotspot Concept   110

Summary   110

Review Questions   110

References   111

Chapter 5   Broadband Wireless Access (IEEE 802.16: WirelessMAN)   113

IEEE 802.16—The WirelessMAN Standard   113

Reference Model   114
    Physical Layer   115
        Burst FDD and TDD Operations   117
    Map Information Elements   119
    Physical Layer Support for Subframing   119

Modulation Schemes   121

Privacy Sublayer   122
    Packet Data Encryption   123
    Key Management Protocol   123
    Security Association   123

MAC Common Part Sublayer   124
    MAC Addressing   124

Scheduling Services   126
    Request: Grant per Subscriber Mode and Grant per Connection Mode   127

Contention Resolution   128

Initialization   128

Propagation Conditions   131

Design Considerations   131
    Design Issues   131

Services Offered and Relation to the First Mile Problem   132

Summary   132

Review Questions   133

References   134

Chapter 6 Free Space Optics Solutions in the First Mile 137

    Free Space Optics: Complement to Radio-Frequency Wireless 137

    What Is Free Space Optics? 138

    FSO Operation 138

    Transmitters: Lasers 139
        Distributed Feedback (DFB) Lasers 141
        Distributed Bragg Reflector Lasers 142

    Receivers: Photodetectors 142
        The PIN Photodetector 143
        Avalanche Photodiodes 145
        Optical Receivers 146
            Receiver Noise 147

    Receiver Performance 148
        Bit Error Rate 148
        Signal to Noise Ratio 150

    Lenses and Mirrors in FSO 151

    Tracking and Acquisition 152

    Link Margin and Design Considerations 152

    Factors Affecting FSO Operations 153
        Mie Scattering 153
        Rayleigh Scattering 154
        Beam Spreading 154
        Impact Due to Rain, Fog, and Snow 154
        Link Design Example 155

    Laser Safety for FSO Networks 155

    FSO Communication in First Mile Networks 156

    Summary 157

    Review Questions 157

    References 158

Chapter 7 DSL Technologies 161

    What Is DSL? 163
        Symmetrical Services 164
        Asymmetrical Services 164

    DSL Technologies 165

ISDN-DSL   165
High-Data-Rate DSL   166
Symmetrical DSL   168
HDSL2: The Next-Generation HDSL   169
Asymmetrical DSL   170
Rate-Adaptive DSL   172
Very-High-Data-Rate DSL   173
    Symmetrical VDSL   173

Long-Reach Ethernet   174

Line-Encoding Standards   175
    Discrete Multi-Tone   175
    Carrierless Amplitude/Phase Modulation   176

A Typical xDSL Network   176

Issues with xDSL   177

DSL in the First Mile   179

Summary   179

Review Questions   180

Chapter 8 Power-Line Communication   183

Introduction to Power-Line Communication   183

A Typical Indoor Power-Line Communication Network   184

Channel Characteristics That Affect PLC   185

Noise in PLC Systems   186

Modulation Requirements in PLC   186

Media Access Control Issues in PLC   187
    Orthogonal Frequency-Division Multiplexing   188

Standards and Regulatory Aspects (From "Power Line Communications: State of the Art and Future Trends")   189

A Case Study of Power-Line Communications (Courtesy of Main.net)   189

Adaptations to PLC to Meet RBOC and ILEC Requirements   192

An Ideal First Mile Network: Coalescing PLC with EPON   193

Comparison of PLC with Other Contemporary Solutions   195

HomePlug 1.0 Specifications for PLC   195
    Signal Processing   196

HomePlug 1.0 MAC   196
Carrier Sense and Collision Detection over Power Lines   197

Market Analysis and Trends   197

Hybrid Fiber Coaxial Plant as a Possible First Mile Access Solution   198

Summary   199

Review Questions   199

References   200

Chapter 9   First Mile Access Management and Business Model   203

Unified Management for Access Networks   203

The Emergence of Alternative Models—Intelligent Cities   204

Intelligent City/Corporation Networks   205
Nontechnological Challenges   207
Intelligent Optical Layer   208
Economic Implications   209

Community Network Management   210

Mobility Management   212

Potential of Effective Bandwidth Management   213

Directions for Network Management   215

QoS in the First Mile   216
Toolkit for QoS and Traffic Engineering   217
Using MPLS to Queue Traffic   218
Failure of OSPF and BGP for Providing QoS   219

Constraint-Based Routing   219
Attributes of CBR   219

Differentiated Services (DiffServ)   219

Integrated Services (IntServ)   220
Reservation Protocol   221

802.1Q VLAN   221

QoS in ATM   222

Implementation of Different Network Architectures   222

Directions for Access Networking   223
Comparison of ATM Solutions with MPLS for Core Networks   223

ATM in Core Networks   224
Multiprotocol Label Switching   225
Comparison of ATM and MPLS: ATM Perspective   226
MPLS Versus ATM: MPLS Perspective   226
ATM and MPLS in Conjunction   227

ATM and MPLS Implementation Costs   227

Summary   228

Review Questions   229

References   229

Chapter 10   Business Case for First Mile Access Networks   233

The Application Business Case   234
Security   234
Healthcare   235
Networking Economy   235
Data-Centric Applications and QoS-Demanding Applications   236

Product and Services Portfolio for the First Mile Business Case   236
Voice   236
Data Services   238
Video   238
Videoconferencing   239
Video Distribution and Programming   239
Telemedicine   240
Visualization   240

The Technology Business Case   240
The Ethernet Solution   244
The TDM Solution   244

Business Case: Characterizing a First Mile Access Network   247

Case Study: Distributed and Unified Technology Solutions in a Community Network   247
Conventional Distributed Model   248
Cumulative Comprehensive Model   249

Planning the First Mile Access Network: CCM Perspective   250
Solution to Designing the CCM Scheme for Access Networks   252

Case Study Distributed Pricing Versus Cumulative Pricing in a First Mile Network   259

Market Analysis for CCM Method in the Access Area   262

Summary 264

Review Questions 264

References 265

Chapter 11 The Complete First Mile Network: Integration, Interfacing, and Management 267

Interfacing Requirements in the First Mile Access Space 267

CWDM and PON 268

PLC and LRE 269

EPON and WiFi 270

EPON and ADSL 270

BPON and PLC 271

BPON and DSL 271

X to FSO 272

Multimode Fiber 272

Management of Multiple Technologies in the Network 272

Billing Issues 274

Future Applications: First Mile Networks 274

Summary 275

Review Questions 275

Appendix Units of Optical Power Measurement: Decibel 277

Index 279

Introduction

The massive growth of Internet traffic and the simultaneous increase in computing speeds has led to a dual-state of high-speed networks in the core of the Internet, as well as high-speed desktop computers at end user premises—thus leaving a void in the access area that connects these two superfast phenomenon. This access region, popularly known as the first mile area (sometimes also called the last mile) is a market segment witnessing phenomenal growth even in bearish times. This book discusses the multiple technologies in the first mile area and showcases network solutions that help solve the first mile bottleneck. In addition to the diverse technological solutions, this book deals with business case analysis of the first mile and justifies the tremendous interest shown in this area. The book covers multiple technologies, protocols, and business methods useful for the understanding and future growth of first mile access networks.

Goals and Methods

The book aims at providing a platform for showcasing first mile access technologies and network solutions. This book is a first effort to highlight the bandwidth bottleneck in the first mile area and also serves as a tool for helping the audience understand business success in this area. By covering multiple technologies, this book shows the diversity in the solution options and then discusses management methods for bringing these solutions under the same umbrella. The book addresses the business case with practical viewpoint and highlights the application and technology business cases in the first/last mile area. This book showcases one solid method of pragmatic implementation of modern networking to better the standard of living and thereby remove anomalies such as the digital divide. Finally, this book brings to the fore a method to foster growth in the networking segment that can lead to a fast recovery of the telecommunication industry.

Some of the most important details covered include the following:

- Study the motivations for the increased interest and need in the first mile access area.
- Examine why the first mile is such an important market segment.
- Explain end users as consumers as opposed to customers and technologies as opposed to solutions.
- Learn how to design passive optical networks (PON). Build EPON solutions and differentiate from BPON and GPON.
- Enhance bandwidth capacity and network planning in the access area by using CWDM technology. Design CWDM networks.
- Discover the use of WiFi in the first mile and study the subsystems involved in a WiFi solution. Learn about CDMA.

- Understand the importance of WMAN solution—namely 802.16 and free space optics (FSO) in the first mile.
- Discover how electric lines can be doubled as data carriers by use of power-line communication (PLC).
- Find out how DSL works and learn about issues that affect DSL networks.
- Learn the importance of management in first mile access networks.
- Learn about intelligent cities and communities as case studies for first mile access networks.
- Discover how multiple protocols interact in a distributed environment.
- Study business cases and learn to formulate methods to build business case for first mile access networks.
- Learn pricing and distribution strategies in first mile networks.

Who Should Read This Book?

In general terms, network engineers, network architects, network design engineers, customer support engineers, salespeople, system engineers, and consultants who design, deploy, operate, and troubleshoot first mile access networks and who want to provide new-world enabling services on their networks should read this book. Academicians and researchers will find new technology solutions to hone their innovative skills. Business planners and economists will find new methods to quantify access networks and help understand a business model that is different from legacy networking and one where returns are subtle yet guaranteed.

How This Book Is Organized

The book is divided into 11 chapters, and each chapter deals with an important facet of first mile access networking technologies. It is structured in a way that the reader is initially acquainted with premier first mile technologies.

The first chapter deals with the motivations in the first mile and the reason why this area is becoming so popular. It talks about solutions that are viable in this market space. Chapters 2 through 8 discuss different first mile technologies used in access networking. Chapter 2 deals with PON, Chapter 3 talks about CWDM, and hence, these two chapters showcase fiber-based solutions. Chapter 4 discusses WiFi and the important 802.11b implementation, while Chapter 5 extends wireless to a larger periphery by discussing

802.16 wireless MAN standard. Chapter 6 then covers FSO. Thus, Chapters 4 through 6 sum up the wireless implementation in the first mile. Chapter 7 discusses DSL and long reach Ethernet (LRE) implementations in the first mile. Chapter 8 demonstrates how the electric wires can be used for data transfer. Chapter 9 talks about management issues in the first mile and how protocols interact. Chapter 10 deals with business aspects of the first mile and helps formulate a business case as well as pricing strategies in the first mile. Finally Chapter 11 shows how these technologies can be integrated and the interface requirements for successful first mile network implementation.

- **Chapter 1, "Introduction to First Mile Access Technologies"**—In this chapter, we showcase multiple technologies and solutions that can be used for first mile access. We define the basic problem faced in first mile networks and justify the need for multiple solutions in this area. Based on this diversity of needs and applications, there is a generic evolution rather than revolution of new technologies from core and LANs that solve the first mile access area problem. First mile access networks are thus an eloquent solution to solve business aspects for future of telecommunication networks.

  In this regard, we briefly discuss the multiple technologies that underline first mile advanced access networks. Among others we outline PON, fiber (CWDM), DSL, wireless, and PSTN as discrete contenders for first mile solutions. In the following chapters we deal with each of these technologies and design first mile access networks to fit the required space created from the void emerged due to high-speed core and high-speed personal computers.

- **Chapter 2, "Passive Optical Networks in the First Mile"**—In this chapter, we discuss the multiple PON technologies and elements that affect first mile access networks. We discuss the multiple PON technologies among which are APON, GPON, and EPON. We show the benefits of EPON as compared to APON and GPON. We focus on the aspects of cost, performance, and protocol issues. In essence outlining EPON networks as the key technology of the future in the first mile. We study the disruptive aspects and understand the protocol behavior. We study an academically proposed protocol named TUR in detail.

- **Chapter 3, "Enterprise Fiber Solution"**—We show CWDM as a technology for enterprise access networks. The advantage of CWDM is its ability to exploit the massive bandwidth offered by the fiber yet keep network equipment costs low. We look at CWDM from deployment perspective rather than from a technology perspective. We see how CWDM networks are beneficial to access area networking—especially considering the fact that these networks are built from low cost components. With the help of design guidelines, we introduce the reader into CWDM network deployment and show the intricacies of network designs.

- **Chapter 4, "Data Wireless Communication"**—We outline the effects of wireless networking in the access area. We define wireless networking for Wireless LANs (WLANs) and introduce TDM, FDM, and spread-spectrum technologies. We look at QPSK modulation format for spread spectrum and discuss frequency hopping and direct sequence spread-spectrum technology. We look at antennas, the mathematics behind the antennas, and finally WLAN behavior. The chapter ends by placing WLANs in the framework for first mile access networks.

- **Chapter 5, "Broadband Wireless Access (IEEE 802.16: WirelessMAN)"**—In this chapter, we talk about a substitute technology for the fiber in the access area. We also talk about the WirelessMAN standard called 802.16. We then discuss enabling technologies for implementation of the WirelessMAN standard.

- **Chapter 6, "Free Space Optics Solutions in the First Mile"**—In this chapter, we study FSO as a method to solving the first mile advance access bottleneck. This is an amazing technology that needs a lot of work before effective deployment. FSO design is an intriguing subject, consisting of aspects of wireless and optical design and taking into consideration multiple issues such as tracking. This chapter aims to cover these issues.

- **Chapter 7, "DSL Technologies"**—In this chapter, we discuss DSL in the first mile. DSL technology is an evolutionary technology built on the basis of the existence of POTS. Over the past few years multiple variants of DSL have emerged—namely ADSL, VDSL, and HDSL. We study these variants in this chapter and discuss implementation as well as management issues in these technologies. We also discuss LRE—a new paradigm for point-to-point communication over copper. This provides a good view of the final outlook of the first mile: an all Ethernet approach. The chapter discusses some of the key limitations of DSL implementation that need to be solved for achieving a complete broadband network.

- **Chapter 8, "Power-Line Communication"**—In this chapter, we discuss the possible dual use of electric utility lines for data transport. We see that this solution us particularly important for residential usage in the first mile. In this chapter, we showcase the technologies required in the first mile for implementation of a PLC solution. We also study PLC through a case study: HomePlug 1.0.

- **Chapter 9, "First Mile Access Management and Business Model"**—We analyze the implementation of ATM and MPLS solutions in core networks. Benefits of MPLS can be seen in terms of costs as well as implementation. Benefits in ATM can be seen for stringent QoS requirements and service parameters (TOS). ATM has several drawbacks, such as an extra AAL for further processing and route dissemination using VP/VCs. MPLS is questionable for large networks due to management issues as well as the large stacking of labels. From a future perspective, Ethernet and MPLS appear to offer low cost survivable methods for core network communication and provide good quantity (throughput) in addition to QoS.

- **Chapter 10, "Business Case for First Mile Access Networks"**—In this chapter we showcase the business case analysis of first mile networks. The chapter illustrates a method to build a business case for first mile networks. We show how to build the network, taking into account the network economics. We, through a formulation and a case study, integrate the various aspects such as technology and the economics of a first mile network.

  The chapter treats four levels of business case analysis for the first mile. In the first level, we study the applications of the first mile—the bandwidth killer applications that are the key motivators of the first mile network. In the second level, we discuss the products and services portfolio in the first mile—that is, how these applications can be met through real offerings-products through a first mile solution. In the third level, we discuss the implementation of the network taking into consideration the first two levels of the business case: applications and products. In the last level, we study the financial business case and validate the motivation and intuitive correctness for the first mile. This level also gives ideas on billing, and we discuss one popular billing mechanism while proposing an alternative billing scheme that is reliable and pragmatic and that yields a more efficient business case for first mile advance access networks.

- **Chapter 11, "The Complete First Mile Network: Integration, Interfacing, and Management"**—In this chapter, we study how multiple technologies fit together and form the complete first mile access solution. We study the multiple requirements such as technologies and interfaces that are needed to culminate into a first mile solution.

CHAPTER 1

Introduction to First Mile Access Technologies

The Network Today

Since the deployment of ARPANET, networking has changed how we live our day-to-day lives. The tremendous reliance on networks as a method to enhance our quality of living has created a multibillion-dollar industry that has experienced near-exponential growth over almost three decades. From a conglomeration of semiconductor and telecommunication (telco) services to application-specific services such as optical and wireless networking, the telco industry has evolved dramatically over this time, adding newer products to the generic telco portfolio and emerging as a huge industry spread over large geographic areas and facilitating the transfer of both small and large amounts of information. This transfer of information between end users was propelled by the need to communicate over long distances and created the need for a telco infrastructure. This transfer of information also drove the information technology (IT) revolution.

The World Wide Web, based on the initial ideas of ARPANET, blossomed into a distributed architecture that allowed multiple users to communicate and created a parallel electronic economy. In both developed and developing countries, this growth led to the rapid deployment of network infrastructures. The emergence of data networking as opposed to voice transfer was a paradigm shift in network behavior and deployment in the 1990s. The significant surge in bandwidth—a result of data networking—created a requirement for high-speed core networks that were easy to install and that incorporated already existing fiber infrastructure and supporting network equipment.

Soon, nationwide backbones were built that could handle the transfer of several terabits of information in a few seconds. Two technologies that prompted the deployment of these high-speed core links were IP and wavelength-division multiplexing (WDM). IP emerged as the protocol of choice for data transfer traffic between hosts, evolving as a best-effort service that created a universally acceptable methodology. On the other hand, WDM showcased a way to maximize the usage of the near-infinite (30 terahertz [THz]) bandwidth fiber offered. With WDM, multiple wavelengths could be used to provide independent optical circuits (lightpaths), thereby alleviating the basic bandwidth bottleneck problem so evident because of the technology mismatch created from optical and electronic interfaces (optoelectronic bandwidth constraints). Another development that occurred for routing IP traffic over high-speed links was Multiprotocol Label Switching (MPLS), which served as a premier switching (Layer 2.5) technology that enabled the very fast gigabit-rate switching of IP packet trains at router interfaces.

On the end-user side, there was a massive increase in the use of personal computers, which enabled end users to create applications and compute software for personal applications at near-gigabit speeds.

These developments created a high-speed core and a high-speed end-user apparatus. However, this resulted in a void, often referred to as the *first mile access problem*, which emerged as a byproduct of the high-speed core and high-speed end computers; the high-speed core and high-speed end computers required the transfer of high-volume applications from the core to the end user and vice versa.

The first mile access problem can be summarized as the bottleneck created by the absence of technologies and solutions to accommodate the high capacity needs of core networks to reach end users. After all, end users want and need to access applications spread across network domains in real time and to access applications that require significant bandwidth. The development of bandwidth-killer applications, such as video on demand, e-learning, and so on, intensified the need for high-speed dedicated end-user connections.

Despite the exponential growth in the number of Internet end users, serious concerns slowed investment in (and, therefore, technology maturation of) first mile products. One concern was end-user dynamic bandwidth requirements, or *bandwidth on demand*. Bandwidth on demand meant that needs would vary, and this uncertainty implied that there was not enough justification and motivation for investment in this area of networking. However, as bandwidth-critical applications became more and more prevalent, there surfaced once again a desperate need to solve the first mile access issue. An investment and technological plateau in the networking core segment shifted the investment focus to the first mile access area, enabling the development of technological solutions for first mile access problems.

Two approaches were proposed for first mile access solutions: wired networks and wireless networks. Wired networks, as the name implies, were created out of a wired infrastructure, and the material of the wire often dictated the bandwidth that could be provided to the end user. For example, a coaxial and copper solution provided no more than 8 to 10 Mbps of bandwidth to the end user. Power-line-based communication yielded 4 to 6 Mbps of line rates to the end user. Optical fiber-to-the-user (FTTU) yielded a wider bandwidth of about 100 to 1000 Mbps, creating an absolute abundance of bandwidth for the end user. In the wireless approach, wireless fidelity (WiFi) is gradually emerging as a standard for first mile applications, yielding a range of bandwidth availabilities depending on the WiFi quality and end-user distance from a wireless base station.

Bandwidth Management in Worldwide Networks: Impact on First Mile Access

Figure 1-1 describes a typical end user–to–end user scenario in which geographically dispersed users communicate with each other over multiple network segments. Typically we observe multiple users, both on wired and wireless access networks, connected to the network core. The sheer diversity of network protocols and user requirements gives rise to a bandwidth-management problem: the real-time challenge to provision networks to provide the pool of network users their multiple quality requirements.

Figure 1-1 *A Generic Network Showcasing Multiple-Tier Networks*

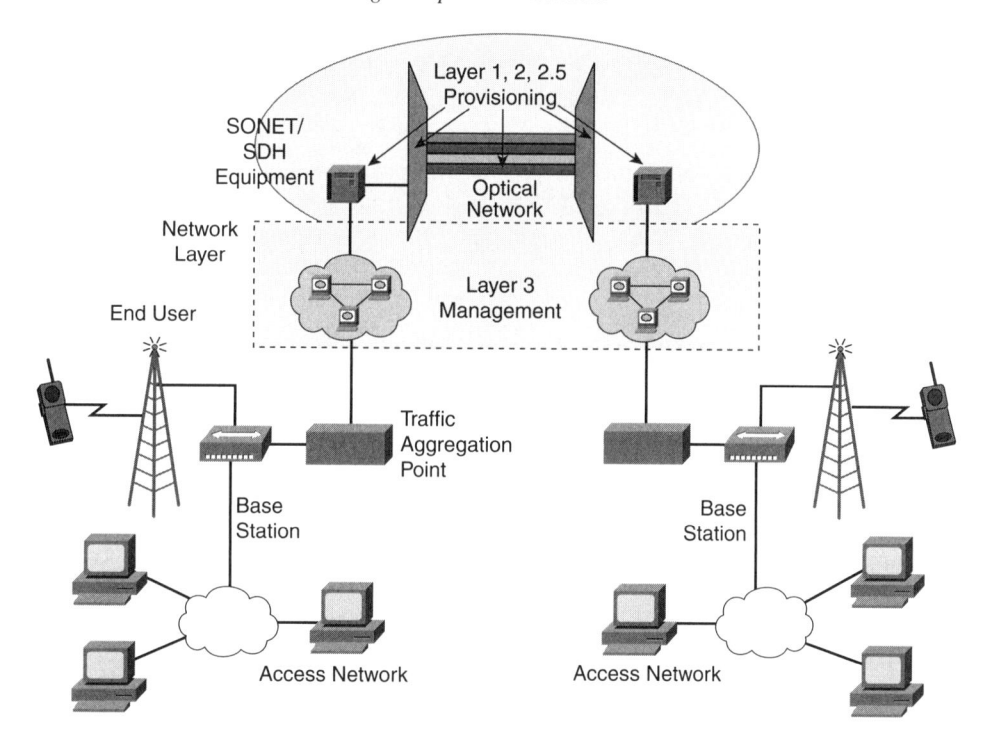

We can view the network as hierarchically layered based on core, metro, and access philosophies. The core network is often laid on fiber, such that the multichannel-based WDM system does transports raw bits through the high-speed connection. In this case, the bandwidth-management issue is to allocate wavelengths (upon demand) on a real-time basis and create these all-optical lightpaths between the source and destination. A migration from circuit-based lightpaths has yielded a more significant solution called *light-trails*, which is an optical burst solution using existing hardware and creating an out-of-band protocol that enables fast provisioning of the bandwidth to yield sublambda-level communication in the core. The absence of a protocol or framing procedure does not allow the legitimate marriage of IP directly over WDM—hence the need for a framing procedure. Synchronous Optical Network/Synchronous Digital Hierarchy (SONET/SDH) and Gigabit Ethernet are two conventional methods for transporting IP over WDM. Resilient packet rings (RPRs) are a more recent innovation to cater to bursty IP traffic transport over the optical layer. SONET/SDH has matured over the years and currently provides a synchronous transport mechanism in the core at speeds as high as 10 Gbps (OC-192/STM 64) and OC-768 or 40 Gbps under experimental setups. The drawback of such synchronous technology is the massive investment cost to cater to expensive high-speed electronics. Second, SONET/SDH or even RPR is not a very efficient protocol for IP-centric communication. At Layer 3, we have gigabit-capacity routers often working on a Layer 2.5 principle (that of MPLS), enabling high-speed switching of IP packet streams without the actual address resolution of the IP packets.

The significant influence of the enterprise segment of the industry has led to a surge in the need for metropolitan-area networks (MANs). Typically, metro rings, built on pure SONET/SDH or WDM technology, were built to cater to enterprise as well as aggregated access traffic. Metro networks have produced a steady and positive gradient in revenue, and hence there is strong justification for new technology and innovation (and investment) in this area.

As shown in Figure 1-1, the access area is a multitechnology zone. The primary bifurcation of technology yields wired and wireless segments, and it can be assumed that all end users are connected to the network using either of the two connection methods.

This book covers both wired and wireless ideologies, but delves deeper than just these approaches to focus on stronger and clearer philosophies such as passive optical networks (PONs), digital subscriber line (DSL), and power-line communication (among others). Before considering the individual technology stacks, we must look at the bigger picture—that is, how these various technologies work together, amalgamated, to provide a single network. It is at the bigger-picture level that the basic bandwidth-management problem emerges from the plethora of showcased access technologies.

The bandwidth-management problem can be defined as a network set up with a fixed number of resources (such as equipment and interfaces) and connected to a fixed number of end users whose demands vary over time. It is desired to provision services (bandwidth, quality of service [QoS], and so on) to these end users on a per-demand basis, maximizing the network use. This problem is a resource-provisioning problem and for most networks requires complex mathematical time computations (nondeterministic polynomial time complete, also called NP complete) for provisioning. In practice, the solution may be a best effort, or a pre-assigned scheme depending on the QoS a user desires at initialization. Such a solution may not be optimal but may be near optimal with the added time computation benefit. End-user bandwidth on demand also requires network resource management to ensure, in the most optimal way, the network functioning (including the network's capacity to deal with bandwidth flows through the network). The presence of multiple technologies in the same network creates a greater challenge involving the interoperability of these technologies at network peers. End users, whose requirements vary over time, also use multiple technologies, creating a potential management disaster—one that has multiple solutions at any given instant, with the optimum one being the hardest to find!

The First Mile Issue

The first mile problem can be thought of as an aspect of the phenomenon called the *digital divide*, which refers to an increasing gap between those who have access to useful online information and opportunities and those who do not. The expanded notion of digital divide holds that those who are underserved are at risk of more rapidly falling behind economically and in their quality of life to the extent they are unable to use IT effectively. This evolving notion of digital divide has come to refer to organizations and businesses as well as communities worldwide. One of the most optimal solutions to the digital divide is to provide broadband services to end users on a massive worldwide scale.

The term *broadband* is referred to under the Telecommunications Act of 1996 as advanced services (that is, data transmission rates) significantly higher than those that can be sent through ordinary high-quality voice circuits (that is, more than 56 kbps). The industry definition of broadband services is the capability to provide end users (consumers as opposed to customers) a dedicated connection greater than 200 kbps.

Business Justification for the First Mile Problem

Technically, and as explained previously, the first mile access problem concerns providing high-speed access to end users. Two rapid developments caused the first mile problem:

- High-speed desktops
- High-speed network cores

The products and services in these industry segments have expanded (and matured) significantly, driven by developments in the telco and computer industries. Gigahertz-rate-capable computers have become an easily obtained commodity and are increasingly seen on the desks of end users. The rapid surge in Internet traffic led to service providers and carriers buying and installing in the core of networks high-speed (Gbps rate) equipment manufactured by vendors, such as Cisco, Nortel, and so on. The Internet soon became a high-speed backbone, a network of networks, capable of gigabit-rate throughputs on high-speed connections (typically on optical media). This resulted in a bottleneck between the high-speed network core and high-speed computers sitting on the desks of the end users. Figure 1-2 shows a schematic of the current situation. We can observe that between the high-speed core and high-speed desktops exists the access networks, which are typically low-cost, low-speed access links often on coaxial cable and copper that cannot support data transfers of more than a few megabits per second.

Figure 1-2 *Access Bottleneck*

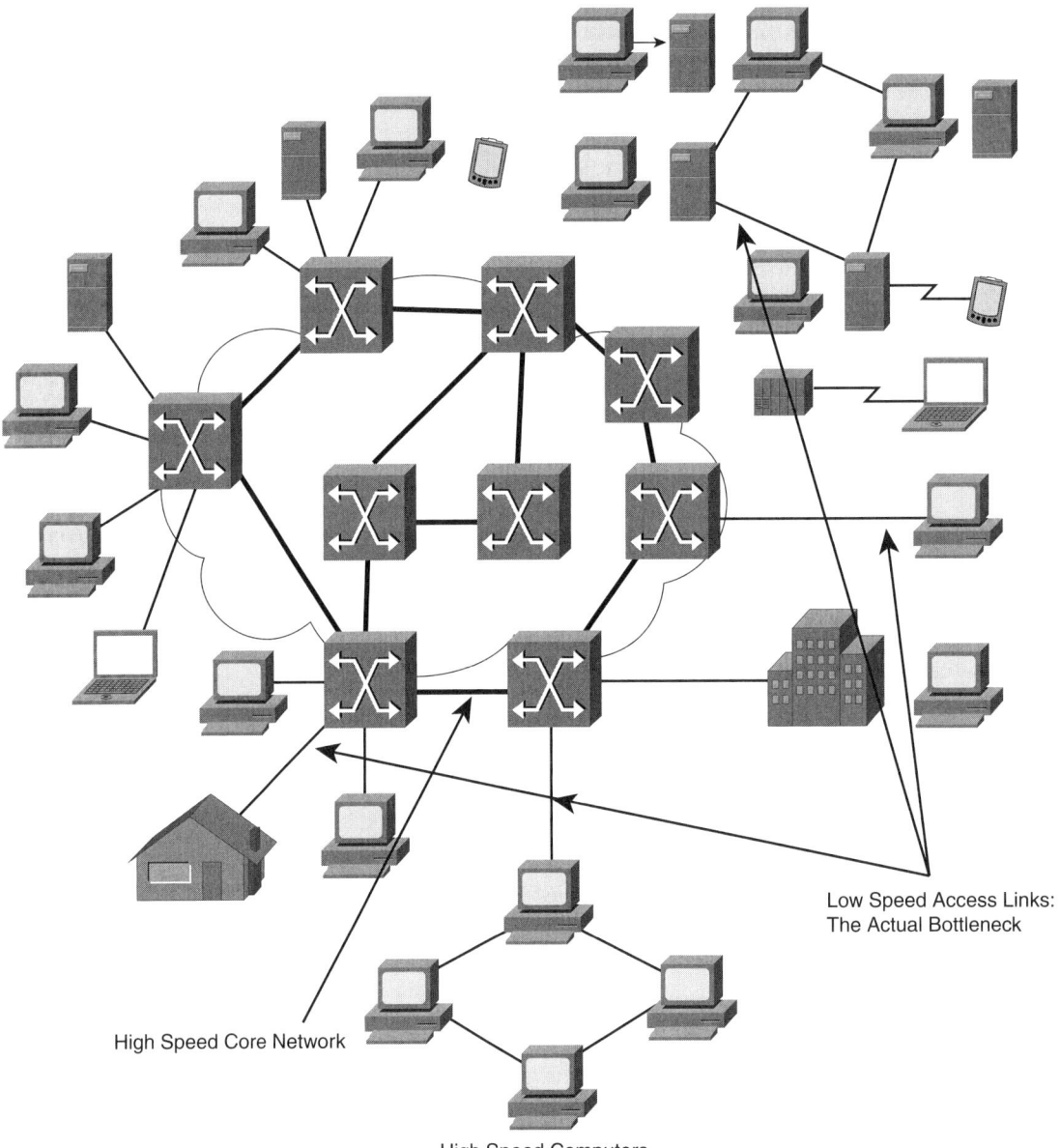

This tremendous growth in information transfer rates (both network core and end-user transfers) created what is today commonly referred to as the first mile access problem. Although it was necessary (and desirable) to upgrade technology at both the ends (core and user), the IT industry, often hampered by regulatory laws, failed to upgrade the technologies that carried data to end users.

In addition, because investments in coaxial cable and shielded twisted pair (STP; phone lines) to facilitate data transfers were so high initially, telcos did not to invest to rejuvenate this area. STP emerged as a technology (and solution) that was far more resilient and useful than for just the transport of plain voice signals (at 64 kbps). By using the same spectrum either completely, as in dial-up access, or in shared-spectrum format (coexistence of the two signals), as in DSL, data transport could be clubbed on to the same STP wire. By using a modulation technique, a baseband data signal could be modulated to provide a high-frequency signal that was transmitted over the STP medium. Modulation here meant the variation of a high-speed carrier with the baseband signal to produce a modulated high-frequency (frequency modulation) or variable-amplitude (amplitude modulation) or phase-shifted (phase modulation) waveform. Devices that can modulate and demodulate the baseband signals are called *modems*, and modem modulation speeds were critical in determining the data rate that could be extracted from STP media. Therefore, access networks typically relied on STP as the medium by which to provide a reasonable data flow to end computers. Of course, the data flow provided even after maturation of technology was typically less than 56 Kbps in direct modulation and about 1.5 Mbps using variants of DSL.

In no way was this data rate comparable to the speeds obtained by the two ends of the network, namely the high-speed core and the super-fast personal computers. As mentioned previously, this speed discrepancy highlighted the basic access bottleneck problem faced today. However, carriers and providers could still justify their lack of investment in the access area. First, as explained earlier, STP was a very efficient medium for data transfer. Initial Internet applications were restricted to e-mail and simple web use. Second, the amount of existing infrastructure and the investments already made for STP-based Public Switched Telephone Networks (PSTNs) were very large. Most developed countries took six to seven decades to lay this massive PSTN infrastructure, but this perseverance resulted in an excellent medium to reach end users and homes. Because they had a high telephone density, developed countries were able to absorb data traffic seamlessly in their PSTNs.

However, as web-based applications grew from plain e-mail and browser applications to more attractive multimedia and video applications, the need for bandwidth to the end users increased phenomenally. Supporting such high-bandwidth killer applications required an upgrade to these large and cumbersome (geographically diverse) telephone networks. However, it was observed that the need to provide high bandwidth to end users did not justify the level of investment required to upgrade these networks. Other factors also slowed the development of the PSTN from a voice-only network to a broadband network. Among others, the dot-com bubble and the shattering of the telco industry created a negative environment for investment in this sector. However, a slowdown in core networks in terms of capital spending and the failure of the basic telco business model have once again rejuvenated interest and investment in the first mile access area.

Technological Aspirations Forecast for First Mile Access Networks

The dearth of revenue in the core of worldwide networks, and the absence of a business model that alleviated the debt traps of the telco industry, created a plethora of reasons for wise investments in the first mile access area. The most logical implementation of the first mile access area networks was the direct adaptation of MANs. However, MANs are typically characterized by their superlative performance and stringent behavior patterns expected (and are thus costly). Leading companies and groups the world over conducted research in this area and defined a direct need for technology that could solve the first mile access issue (that is, provide connectivity to the end user). The absolute identification of the first mile issue hastened the process of developing newer access area technologies.

Networks continued to adapt MANs to deal with first mile access issues, and a business case could be made in their favor. Despite this, however, a positive response met the efforts of the research community to streamline access network technology.

Broadly, two well-defined strata of access network technologies can be identified: wired technologies and wireless technologies. Conventional networking fomented and supported the belief in wired networks (over different media and using different protocols). Wired networks were conventional, low cost, and functioned on an infrastructure that generally already partially existed (or in whole in some cases). The main objective was to provide somewhat basic connectivity to the end user. Wire technologies included, among others, DSL and its variants, cable technology, PONs, and basic dial-up services. On the other hand, the past two decades justified a strong investment and deployment scenario in wireless networking. From basic wireless telephony evolved a requirement to transfer wireless data to end users. The idea of having connectivity to end users, through a wireless medium, also justified the revolution in portable devices such as laptops and handheld computers. This created a parallel access technology in the wireless medium.

In the following sections we discuss each of these technologies to show the basic capability of each incumbent to provide the necessary bandwidth to end users and consumers.

Technologies Deployed in Access Networks

As mentioned before, two main technological classifications can be seen in access telco networks: wired and wireless technologies.

Wired networks generally mean a physical connection to the end user, whereas *wireless* networks mean an absence of such a physical connection. At the time of this writing, the technological classifications of wired networks are as follows:

- DSL
- Coaxial cable
- Fiber (PON and coarse WDM [CWDM])
- Copper (for example, Ethernet)
- Power-line communication

In the wireless domain, the following three technological innovations were seen as key attributes to access networking:

- WiFi
- Basic cellular technology
- Infrared and free space optic communication

DSL Technology

In its most primitive form, DSL technology can be considered as the coalescing of data and voice on the same PSTN. By using the medium (STP) as a band-pass filter and by multiplexing (in this case, sharing) the voice and data formats on individual slots (bands), we can optimize the use of the STP in the access. Further variants of DSL technology are asymmetric DSL (ADSL), very-high-data-rate DSL (VDSL), and high-data-rate DSL (HDSL), each of which is explained in Chapter 7, "DSL Technologies." The high-bandwidth core networks can feed data to the access networks. The end users segregate the voice from the data by using a low-pass (voice) and a high-pass (data) filter. By virtue of its physical property, copper does not allow for high-bandwidth communication. Moreover, copper wire also has a high attenuation. Therefore, the signal at high frequencies (data) cannot travel large distances, a severe drawback for DSL technologies. A DSL modem, typically a filter, can segregate the voice and data from the band-pass channel. Further coding of the data waveform by a keying or coding technique, such as Quadrature Phase Shift Keying (QPSK), can create and allow transmission of higher-bit-rate signals.

Because of its dependence on copper as a medium for communication, DSL is restricted in data rate and distance. In addition, it is not easy to increase the distance of DSL transmission by installing repeaters. A DSL repeater needs to extract the signal and then process it, and hence this cannot be accomplished technologically and does not justify the costs involved. This limits DSL transmission to the last mile, and further limits it to small distances on account of the high attenuation of the signal traversing a copper channel.

Another issue with DSL and its family of transmission technologies is the directional power of communication. Typically it is desired that a technology be able to provide a seamless communication flow, from the core network to the end user (and vice versa). This kind of transmission is called *bidirectional* or, in telco language, *full duplex*. The bit rate achieved from the end user to the core should be the same as the bit rate achieved from the core to the end user. However, Internet traffic analysis generally reveals that the amount of data in upstream communication (from the end user to the core) is much less than the amount of data that flows from the core to the end user, by a typical ratio of 2:7 to 2:9. Hence DSL, to surmount some of its technical disadvantages, offers a symmetric and an asymmetric version.

In the symmetric version, the data rate from the end user to the core is the same as the data rate achieved from the core network to the end user. In the asymmetric version, the data rate available from the core to the end user is different (typically more) than the data rate available from the end user to the core. This creates a system that is optimized for normal operation of IP traffic. However, this creates severe buffering issues when there is reverse transfer of data (typically uploading or sending large files). In short, the asymmetric system works well for normal operations such as web surfing or even downloading multimedia, but there can be severe delay penalties in reverse operations such as uploading files or sending multimedia contents (and, especially, when using bandwidth-intensive applications such as videoconferencing).

Fiber

Optical fiber can be considered the biggest breakthrough (along with wireless) in communication technology of the past century. Possessing a bandwidth of 25,000 GHz, a single strand of optical fiber is as thin as a human hair. Glass as a medium for communication was proposed in the early 1960s, but it was not until recent innovations in laser, detector, and optical amplifier technology that fibers were used for commercial communication.

Different methods of communication have evolved for fiber as a medium of transport. Initially, a time-division multiplexed (TDM) electronic signal (at lower speeds on to a high-speed TDM signal) was modulated by a laser diode and fed (coupled) to a fiber. These optical signals were detected at the far end, across a large distance, by a photo detector. The converted electronic signal was further demultiplexed in the time domain to get individual slower signals at lower bit rates. Figure 1-3 shows a schematic.

Figure 1-3 *Optical Transmission System*

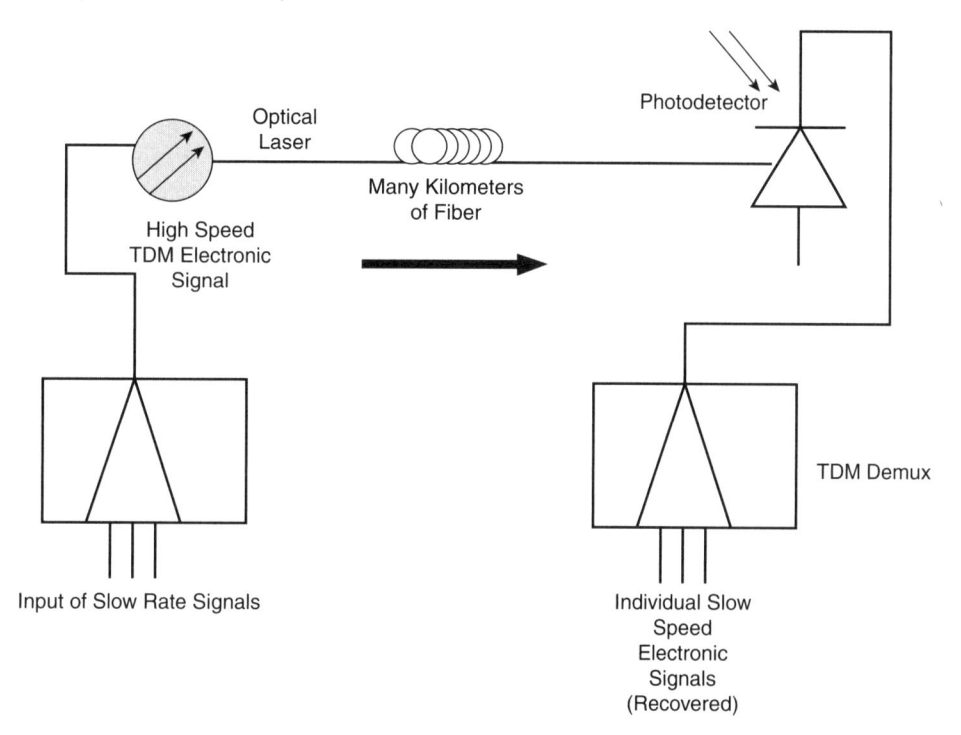

Fiber communication can be classified further based on the physical characteristic of the optics involved. Typically, the laser, which injects the optical signal, emits light at a closely knit group of wavelengths. A laser that emits its optical power in a narrow stream such that most of the optical power is concentrated around one wavelength is called a *narrow-aperture laser*. Such a laser is also said to possess negligible chirp, and according to classical physics the emitted light travels in a fiber that supports this wavelength. Although negligible, the attenuation of an optical signal in a fiber is still finite, and this signal depends on the wavelength it is traveling on. In other words, the attenuation in a fiber is wavelength sensitive.

Finally, in the early 1990s a scheme developed to maximize the use of the near-infinite capacity of fiber. This scheme, called *wavelength-division multiplexing* (WDM), is a method to pack multiple signals (each on different wavelengths) in the same fiber by the spatial diversity of the wavelengths. The main motivation for the development of this scheme was the optoelectronic bottleneck—that is, the inability of electronic systems to modulate more than 10 Gbps of data even though fiber could allow up to 25,000 Gbps of data. This discrepancy was in some way alleviated by WDM solutions. Although WDM is a typical metro and core solution, some aspects are applied even to the access area. This low-cost, easy adaptation of the WDM solution in access is called *coarse WDM* (CWDM) (in contrast to dense WDM [DWDM] in the core).

Attenuation Windows in a Fiber

Authors have always argued about the number of operating windows of wavelengths or bands that can exist in an optical communication network. To the designer or systems engineer, this is not much of an issue for argument, because practical optical networks currently function in three discrete bands: the conventional (C), long (L), and short (S) bands.

The C band is approximated from 1525 to about 1565 nanometer (nm). It has a low loss of about 0.2 decibels per kilometer (dB/km). C stands for *conventional* and most metropolitan as well as long-haul networks use this band. The band is about 40 nm and can accommodate 80 different wavelengths, each 100 GHz (or 0.8 nm) apart. The spacing between the wavelengths is a standardized value. Currently, for dense-division multiplexing, the spacing is standardized at 0.8 nm or 0.4 nm.

The L or long band starts from about 1570 nm and extends to 1620 nm. It has slightly higher loss than the C band but has similar characteristics to the C band. Much research has been carried out in this band and signs of early commercial deployment are evident. The future could see a lot of vendors positioning their DWDM products and technologies in this band.

The S or short band is spread around the 1310-nm window. It is of strategic importance due to its close proximity to the zero-dispersion wavelength (a wavelength around 1300 nm that has minimal dispersive effects due to the cancellation of material and wave guide dispersions by each other). It has a higher loss than the C band at about 0.5 dB/km and hence is not the best solution to long-haul communications. The evolution of wider technologies for the C band, such as doped amplifiers, switch matrices, and filters, makes the S band rather underused.

Apart from these three standard bands, there is also the traditional 850-nm band, which was first used for optical communication systems. The 850 to 980-nm band is mostly used for multimode systems and for short LANs. It has a high loss characteristic of almost 2 to 3 db/km. Experimental research is focusing on the 1400-nm segment for new methods to eradicate the OH^- (hydroxyl) molecule. The best a design engineer can hope for is to have a continuous band from 1300 nm to 1650 nm yielding about 400 wavelengths 0.8 nm apart and 800 wavelengths 0.4 nm apart.

Figure 1-4 shows the attenuation curve in a fiber.

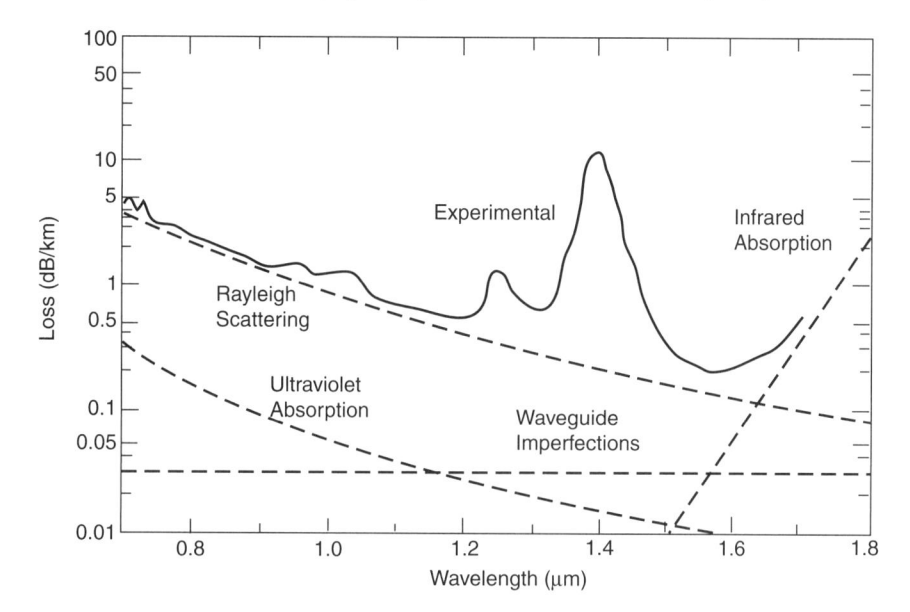

Figure 1-4 *Attenuation Curve in a Fiber (Reprinted from* IEEE Electronics Letters, *1979)*

Passive Optical Networks

Passive optical networks are a technological innovation that can alleviate the first mile bottleneck issue in access networks. As the name implies, passive optical networks are typically *passive*, in the sense that they do not have active components for data transport. They may be spread across different physical topologies. PON development, although propelled by the surge in bandwidth requirements, also answers a definite need for low-cost optical networks for end-user applications. During the initial phase of PON development, some of the primary desirable features of a PON were as follows:

- **Low-cost network, low-cost components**—Because the revenue was in the number of consumers (quantity) rather than the pure service delivery to each consumer (quality), the amount of investment each end user would have to make had to be kept to a bare minimum. In metro and core networks, at each network element a composite WDM signal could drop an entire wavelength or a group of wavelengths. In contrast, in the access first mile area, each network element at the consumer site had bit rates typically of the order of 100 Mbps or even less.

- **Ease of management**—The first feature of low cost also initiates the second point of management complexity. Management complexity creates undue surges in network equipment cost. What is desired is a simple, efficient, and scalable management system that can manage the network and guarantee the network users of some network parameters such as QoS, delay, fairness, throughput (service level agreement [SLA]), and resiliency.

- **Upgradability, in-service upgrade, and interoperability**—The rapid development of newer technologies creates a need for ease of upgradability in PONs. In soft upgrades, the basic fabric remains the same, but a software upgrade enhances the features of the network. Because most of the end users are residential customers or enterprise users, in-service upgrades are important so as not to disrupt real-time services in PONs. Finally, due to the high-volume nature of PON users, there is a strong probability of PON networks having multivendor equipment in them. To facilitate ease of communication and create fair-competition interoperability, standards must exist for PON network elements to talk to one another.

- **Guarantee of basic network features**—The PON must be able to guarantee the end users some degree of network parameters, which are promised at inception (SLA). Although low-cost networks and simple protocols are generally designed for best-effort service, the quantum leap from traditional broadband to PON represents a sea-change shift in the end-user paradigm, and end users must get the desired services through the PON to justify the cost of deployment.

- **Security**—A PON topology is typically that of a star; hence, all the nodes in a star have access to the broadcast information sent by the hub. It's imperative to secure the information by secure services and methodologies such as 802.1x and MAC layer encryption along with virtual private network (VPN) services.

Under current implementations, the following three kinds of PONs exist:

- **ATM PON (APON) and Broadband PON (BPON)**—These types of PONs support extra overlay capabilities for high-speed delivery. APON is the PON transmission technique based on the ATM signaling layer; it was developed by five service providers as part of the Full Service Access Network (FSAN). BPON is approved as International Telecommunication Union (ITU) spec G.983x. It supports data rates to 622 Mbps out to an endpoint (upstream) and back from the customer to the service provider's remote aggregation point (downstream). (Courtesy Lightreading.com.)

- **Gigabit PON (GPON)**—This PON type can support up to 2.5 Gbps of traffic and uses Generic Framing Protocol (GFP), which is SONET compatible, so it allows the creation of an end-to-end unified system based on TDM hierarchy. At this time, GPON is not standardized but is under the umbrella labeled G.984.x.

- **Ethernet over PON (EPON)**—This type is the simplest type of PON, consisting of broadcasting Ethernet frames in the star. It requires a multipoint-to-single-point protocol to guarantee collision-free communication in the upstream because of its shared nature.

In the simplest architecture, a PON network is a star connection. At the core network side, there is an optical line terminal unit (OLTU, or OLT for short), whereas the end-user side has an optical network unit (ONU). Multiple ONUs are connected to a single OLT in a star configuration. This creates a broadcast medium by virtue of the fact that a single optical splitter splits the optical power from the OLT to each of the ONUs. Different mechanisms, such as ATM, Ethernet, and so on, guarantee a good throughput to each ONU (that is, a guarantee of a good share of the total bandwidth to each ONU). In Chapter 2, "Passive Optical Networks in the First Mile," we cover each of these mechanisms in detail and examine the solutions from a technology point of view. Figure 1-5 shows a basic PON configuration where the central office is the OLTU and access nodes are ONUs.

Figure 1-5   *Basic PON Configuration*

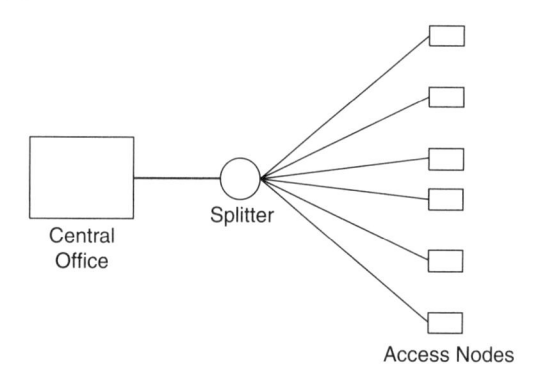

Cable Networks

Community Antenna Television (CATV), commonly known as cable TV, was invented to solve the issue of poor TV reception in rural areas. Ever since, CATV rose above its challenge and today almost 95 percent of U.S. communities have access to CATV networks. A typical CATV distribution network is comprised of coax cable, headend systems, and customer end devices. Frequency-division multiplexing (FDM) is used to transport programs or data over the hybrid coax cable. CATV usually makes use of the 50 to 550 frequency spectrum, with each channel spreading over 6 MHz of the frequency band.

At the headend (collection point), programs received from satellite or microwave systems are converted to one of the preset channels, Then the channels are scrambled (coded) according to the desired quality (paid or free) and multiplexed onto a single broadband analog spectrum or band (by FDM) before being broadcast to the CATV subscribers. The headend along with coaxial cables and subscribers (set boxes) comprise a typical cable system. Subscribers are connected to the feeders or trunk with a drop cable and tap (connector). Figure 1-6 shows the architecture of a one-way cable system. The signal strength attenuates by the square of frequency as the signal propagates through the cable. To maintain signal strength, amplifiers are installed every 1 kilometer or so.

Figure 1-6 *A Simple Coax CATV Network*

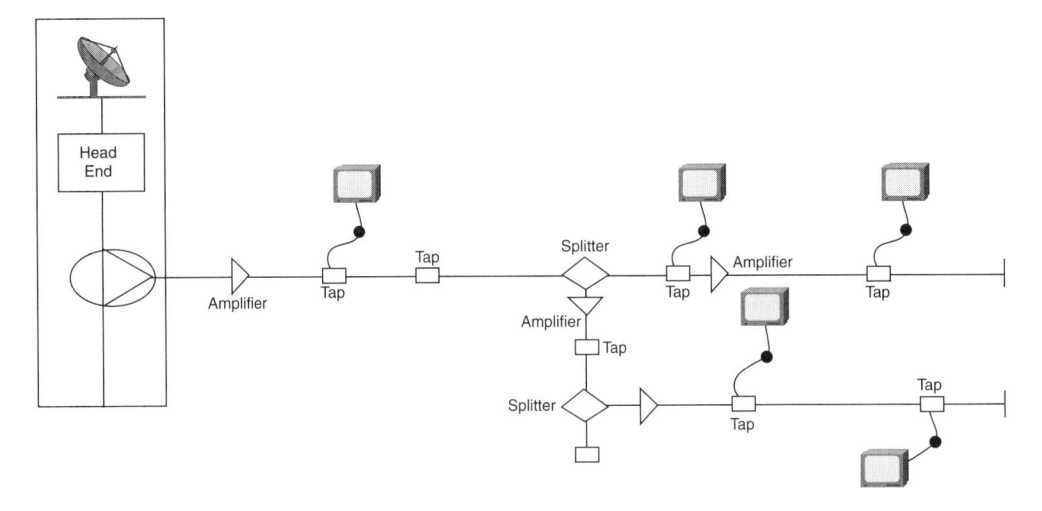

Pure coax systems are incapable of driving high-speed residential broadband services. Coax systems can only provide up to 40 channels, which puts the CATV system in direct disadvantage when competing against direct-broadcast satellite systems. Coax cable systems lack robustness and are very difficult to design and maintain. The limitations of amplifiers (capacity) and of the maximum distance the signal can travel without degradation are also restraining factors to providing scalable broadband services. To overcome these issues that relate to the inherited coax systems, cable operators proposed to use fiber as a trunk medium. The total system is comprised of both fiber and coaxial cables, hence the term *hybrid fiber coax (HFC) networks*. Figure 1-7 shows a typical HFC plant. Between the subscriber and the fiber node (junction), the assembly of coaxial cables, amplifiers, splitters, and taps is same as the pure coaxial cable networks.

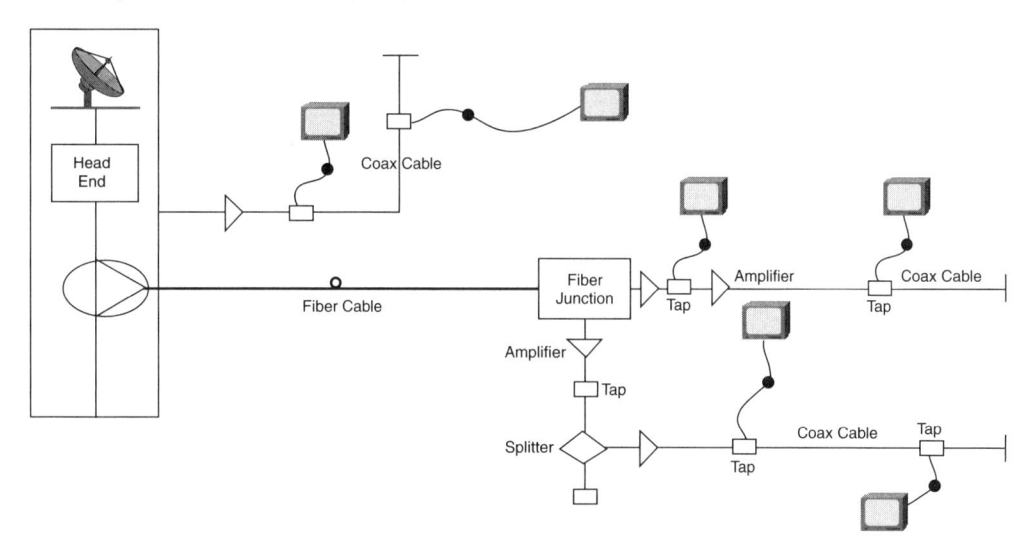

Figure 1-7 *Hybrid Fiber Coax Cable (HFC) Network*

Today, most cable revenue comes from content delivery. The extended service offering using IP-enabled signals delivered over standards-based infrastructure (Data Over Cable Service Interface Specifications [DOCSIS]) helps service providers explore new revenue opportunities.

The PSTN

The Public Switched Telephone Network (PSTN) forms the backbone of most of today's wired voice communication. Figure 1-8 shows a generic PSTN.

Figure 1-8 *PSTN: From Phones to the Voice Routers*

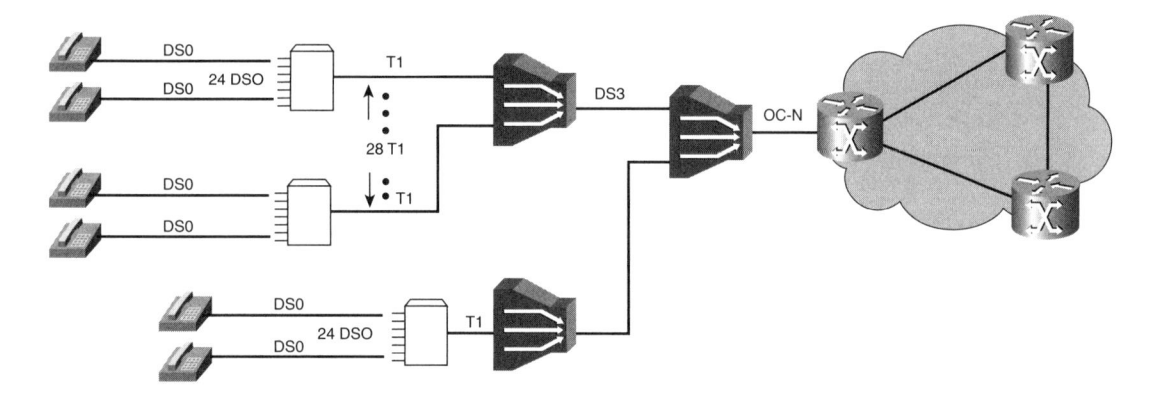

The PSTN started as human-operated analog circuit-switching systems (typical form of plugboards) and progressed through electromechanical switches. By now this has almost completely been made digital, except for the final connection to the subscriber (the "last mile"): The signal coming out of the phone set is analog. It is usually transmitted over STP as an analog signal. At the telco office, this analog signal is usually digitized, using 8000 samples per second and 8 bits per sample, yielding a 64-kbps data stream (DS0). Several such data streams are usually combined into a fatter stream: In the U.S., 24 channels are combined into a T1; in Europe, 31 DS0 channels are combined into an E1 line. This can later be further combined into larger pipes for transmission over high-bandwidth core trunks. At the receiving end, the channels are separated, and then the digital signals are converted back to analog and delivered to the receiver phone.

Although all these conversions are inaudible when voice is transmitted over the phone lines, they can make digital communication difficult.

In other words, the PSTN offers the end user a fair amount of connectivity, both in terms of voice and data. Before the advent of fiber, as a communication medium, the STP along with coaxial cable were the sole wired media that could be used in modern networks.

Wireless Networks

The idea of being able to access information while not compromising on mobility gave tremendous impetus to wireless networking. Users were and are fascinated by the idea of being able to communicate without being connected to the backbone network through a wire. This led to the development and successive deployment of wireless and cellular networks. Initially, wireless networking was relegated and confined to voice communication. The initial idea was to provide the public with wireless telephony, and its use was restricted to emergencies and short calls. Later, wireless telephony became a strong topic of research, and the opening of new frequency spectrums created two parallel spheres of business. Broadband wireless offered in 2.4-, 5.5-, and 11-GHz bands for data and voice coalesced. Cellular technology developed wideband code-division multiple access (WCDMA), global system for mobile communication (GSM), and so on for voice communication. Wireless technology to the last mile was a new approach floated recently by multiple vendors and part of the recently standardized IEEE docket 802.11b. At high frequencies such as 2.4 GHz and 5.5 GHz, large spectrums of frequencies could provide end users the necessary data rates for first mile applications. Three main areas of technological development sustain wireless broadband networking:

- Fixed wireless
- WiFi
- Infrared and free space optics

Fixed Wireless

Where a wire could not be placed, or where there was a need for a high QoS data rate, fixed wireless was deployed. Fixed wireless technology typically is based on microwave communication. Two microwave antennas (both geographically fixed) are located at some distance from each other. The microwave signal is created by a cavity method (for example, klystron cavity) and fed through the antenna, which broadcasts the same. The receiving antenna in the line of sight receives the signal and may act as a repeater for increasing the distance. Microwave communication is restricted to communication of signals whose wavelengths (λ) are in the micrometer range. Therefore, although large volumes of data can be transmitted on such waves, they are severely limited in transmission distance on account of the attenuation they experience. The typical application of microwave communication is to link two base stations or cell sites to each other. From the first mile access point of view, however, microwave communication can considered an excellent choice for reaching end users without the need to lay high-bandwidth wire, which in this case is fiber. Of course, the penalties are the distance the signal can travel, the maximum bandwidth available, and the line-of-sight limitations in microwave communication.

Figure 1-9 shows the wireless spectrum.

Figure 1-9 *Figure 1-9 Wireless Spectrum*

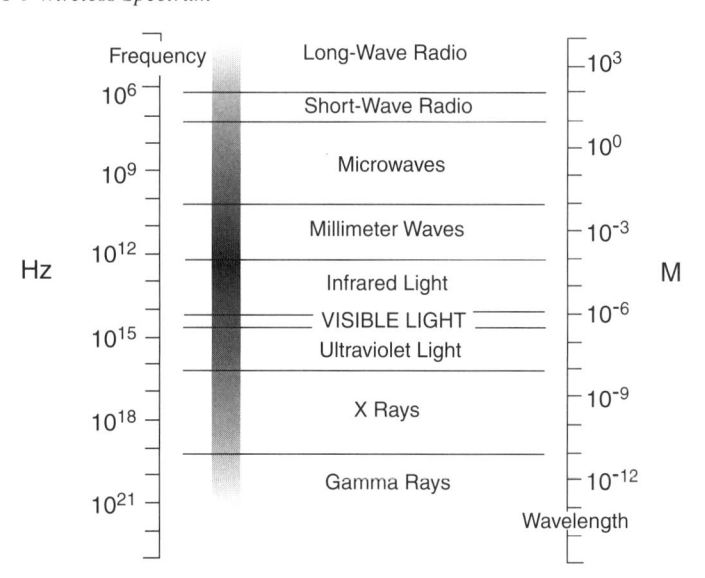

WiFi: Wireless Fidelity

The standardization of the 802.11b format for wireless networking has paved the way for tremendous research and deployment efforts for WiFi networks. A WiFi, in short, is a high-performance high-bandwidth network built on the wireless LAN (WLAN) concept. A wireless transmitter at one of the Industrial, Scientific, and Medical (ISM) frequencies (2.4, 5.5, and 11 GHz) emits a signal that can be received by a small wireless receiver. The high bandwidth and ease of communication renders WiFi a good choice for last mile wireless networks. Multiple vendors have demonstrated new products that are 802.11b compliant and are able to provide networks with high QoS to the end user. Further details are provided in Chapter 4, "Data Wireless Communication."

Infrared and Free Space Optics

Because the wireless spectrum seemed to be too small for data traffic, and there seemed to be a bottleneck between the data that emerged from a fiber and that which could be fed into a wireless channel, there was a new development effort to use other parts of the electromagnetic spectrum. Two such efforts were infrared and free space optics. Principally similar to each other, both techniques include the modulation of data waves onto an electromagnetic wave (light or infrared) and achieving phenomenal bit rates.

In free space optics, the motivation is to use the several thousand gigahertz of available bandwidth at those high (THz) frequencies. Free space optics means point-to-point communication and is further limited by line of site. Typically, a laser diode sends an optical signal through free space. This signal travels through air (undergoing attenuation, fading, and so on) and reaches the receiver. A portion of the transmitted power is received by the receiver. The receiver then detects the signal and communication is achieved. There is high loss in such a system, but the bandwidths achieved are quite high.

When distances are small, instead of free space optics it is more convenient to use infrared waves. Infrared frequencies are much easier and cheaper to generate and are generally able to give a good throughput (bandwidth). Typical applications include portable digital assistants (PDAs) and handhelds within a room.

Comparison of First Mile Access Technologies

Table 1-1 compares some of the technologies deployed in access networks that are covered in this chapter. In this chapter we have covered multiple technologies and have discussed each technology in some detail. The forthcoming chapters highlight each technology and showcase circumstances when the appropriate technology can be used.

Table 1-1 *Comparison of First Mile Access Technologies*

Service	Medium	Intrinsic Bandwidth	Per-User Bandwidth	Standard	Issued By
DSL	24-gauge twisted wire	10 kHz	8 Mbps	G.992	ITU
Cable modems	Coax (HFC)	1 GHz	10 Mbps	DOCSIS 1.1	Cabelabs
APON BPON (ATM)	Fiber	25,000 GHz	10-1000 Mbps	G.983 FSAN	ITU
EPON (Ethernet)	Fiber	25,000 GHz	10-1000 Mbps	EFM	IEEE
WiFi	Wireless	2-4 GHz	1-5 Mbps at distance	802.11b	IEEE
GPON	Fiber	25,000 GHz	Up to 100 Mbps	ITU G. 984	ITU

Summary

This chapter showcased multiple technologies and solutions that can be used for first mile access. We have in this chapter defined the basic problem faced in first mile networks and explained the need for multiple solutions in this area. Based on this diversity of needs and applications, there is a generic evolution rather than revolution of new technologies from core and LANs that solve the first mile access area problem. First mile access networks are thus an eloquent solution to solve the future business needs of telco networks.

We have, in this chapter, briefly discussed the multiple technologies that underline first mile advanced access networks. Among others, we covered PON, fiber (CWDM), DSL, wireless, and PSTN as discrete contenders for first mile solutions. In the following chapters, we deal with each of these technologies and design first mile access networks, to fill the void created by the emergence of high-speed core and high-speed PCs.

Review Questions

1. Define first mile networks.
2. What are the key enabling technologies in first mile networks?
3. What are the business justifications for first mile networks?
4. What are the applications in first mile networks?
5. How would you differentiate technologically in a first mile network?

6 What are the two primary reasons for the massive growth of first mile networks?

7 Define broadband and explain how broadband can solve the digital divide.

8 Compare GPON, EPON, and BPON from technological and standardization perspectives. What are the key reasons for each development?

9 How do multiple technologies work over a PSTN?

10 From Table 1-1, differentiate each technology solution in terms of services. For example, for pure voice lines a PSTN is sufficient, but for videoconferencing we need higher speeds (and hence ADSL is required). List all the possible applications that a technology can provide.

11 Based on Question 10, list all the salient applications that a particular technology can provide that no other technology can provide.

12 Define bandwidth management. How is it related to bandwidth on demand? Considering that there n users in a network (where n is sufficiently large) and there are k service types that a user can subscribe to, prove that allotting bandwidth in real time is not possible in this network without a heuristic well-defined algorithmic solution. Hint: Show that the solution that does not have any standard algorithm is NP complete.

References

Betsekas, Gallaghaer. "Data Networks."

Chlamtac, Imrich, and Ashwin Gumaste. "Bandwidth Management in Community Networks." Keynote address, International Workshop on Distributed Computing (IWDC). Calcutta, India, December 2003.

Dutta, Achyut K., Niloy K. Dutta, and Masahiko Fujiwara, (Editors). *WDM Technologies: Active Optical Components*. New York: Academic Press, 2002.

Gibbons, Alan. *Algorithmic Graph Theory*. Cambridge: Cambridge University Press, 1985.

Gumaste, Ashwin, and Tony Antony. *DWDM Network Designs and Engineering Solutions*. Indianapolis, Indiana: Cisco Press, 2002.

IEEE/ACM "Trans on Networking." 3:5, October 1995.

IEEE. "EFM Ethernet in First Mile." Web draft www.ieee.org.

IEEE. "Special Issue on Powerline Communication," *Communications Magazine*, 17, April 2003.

ITU. Document G.709

ITU. Document G.983, ITU. Working document G.984

CHAPTER 2

Passive Optical Networks in the First Mile

As Internet traffic doubles every six months or so, there is a tremendous surge in end-user bandwidth requirements. 14.4-kbps modems were replaced by higher-speed modems, which in turn were replaced by digital subscriber line (DSL) and cable modems. However, even these advances could not alleviate the basic bottleneck in access networks. The bottleneck stems from two causes. First, at this time, there is ample capacity in the core and metro area due to the emergence of wavelength-division multiplexing (WDM) as a premier technology for high-speed transport; WMD maximizes the use of the near-infinite bandwidth offered by the optical fiber by sending multiple data streams on multiple wavelengths. Second, the ratio of sink users to sources in access is quite high. Moreover, home and work PCs are able to operate at gigabit-level speeds, often creating a void between the core networks and the PCs for seamless data flow. In other words, multiple end users are connected to a single terminating line, each trying to extract and squeeze every possible bit of data from the line.

These two effects create a bottleneck in the access area, which, of course, we refer to as the *first mile issue*. This term can be replaced with the *last mile issue* without change of meaning. Why the access area attributes such importance to revenue can be understood by the fact that it is the end users who generate the revenue on which the business chain of service providers, enterprises, and system and components vendors functions. To create excellent transport methodologies in the access area, newer technologies are tried and deployed. The business proposition of the access area can be understood by the fact that there are multiple end users, each generating small amounts of revenue. The amounts may be small, but the volume of end users is truly enormous, creating a very solid value proposition.

The first implementations of commercial enterprise solutions led to the deployment of broadband access, namely DSL, asymmetric DSL (ADSL), very-high-data-rate DSL (VDSL), ATM, and other solutions. However, each was limited by bandwidth and scalability issues among others. The advent of optical fiber as a means for transport of data at a low cost and high speed (bandwidth) led to showcasing fiber to end-user applications as a possible and pragmatic solution. Fiber to the enterprise or user deployment, although slow (due to the high initial deployment cost of fiber) is the best and possibly only way to circumvent the bandwidth bottleneck between the end user and the metro access network.

If we consider the excellence in characteristics provided by an optical fiber in terms of its longevity and protocol transparency, we realize the long depreciation cycle that actually justifies and in most cases lowers the cost of fiber as compared to legacy copper solutions. The solution to providing bandwidth to end users has to be low cost, efficient, easy to manage, and scalable (among other intricacies such as resiliency and interoperability and the solution must provide degrees of quality of service [QoS]).

Passive optical networks are a technological innovation that can alleviate the first mile bottleneck issue in access networks. As the name implies, passive optical networks are typically *passive*, in the sense that they do not have active components for data transport. They may be spread across different physical topologies. PON development, although propelled by the surge in bandwidth requirements, also answers a definite need for low-cost optical networks for end-user applications. During the initial phase of PON development, some of the primary desirable features of a PON were as follows:

- **Low-cost network, low-cost components**—Because the revenue was in the number of consumers (quantity) rather than the pure service delivery to each consumer (quality), the amount of investment each end user would have to make had to be kept to a bare minimum. In metro and core networks, at each network element a composite WDM signal could drop an entire wavelength or a group of wavelengths. In contrast, in the access first mile area, each network element at the consumer site had bit rates typically of the order of 100 Mbps or even less.

- **Ease of management**—The first feature of low cost also initiates the second point of management complexity. Management complexity creates undue surges in network equipment cost. What is desired is a simple, efficient, and scalable management system that can manage the network and guarantee the network users of some network parameters such as QoS, delay, fairness, throughput (service level agreement [SLA]), and resiliency.

- **Upgradability, in-service upgrade, and interoperability**—The rapid development of newer technologies creates a need for ease of upgradability in PONs. In soft upgrades, the basic fabric remains the same, but a software upgrade enhances the features of the network. Because most of the end users are residential customers or enterprise users, in-service upgrades are important so as not to disrupt real-time services in PONs. Finally, due to the high-volume nature of PON users, there is a strong probability of PON networks having multivendor equipment in them. To facilitate ease of communication and create fair-competition interoperability, standards must exist for PON network elements to talk to one another.

- **Guarantee of basic network features**—The PON must be able to guarantee the end users some degree of network parameters, which are promised at inception (SLA). Although low-cost networks and simple protocols are generally designed for best-effort service, the quantum leap from traditional broadband to PON represents a sea-change shift in the end-user paradigm, and end users must get the desired services through the PON to justify the cost of deployment.

Network Profiling for Passive Optical Networks

In the preceding section we identified passive optical networking as the key to high-bandwidth delivery to the end user. The sudden growth (and later stagnation) of WDM networking led to a void in the access area. The access area needed high-speed bandwidth, at a low cost, and needed to be able to provide reliable connection through a robust management platform. Fiber seemed to be the choice of many for providing the high-speed links to the end users. However, the cost of optical-to-electronic interfaces at end-user premises was the inhibiting factor for the industry, slowing deployment of fiber-based networks in the access area. In the mid-1990s, universities in North America proposed a low-cost and flexible scheme for broadband access networking called *passive optical networks (PONs)*. The term PON generically means optical networks that are pervasive and passive in nature. The passivity arises from the design contrast with metro networks, in which switching, amplification, and regeneration of the optical signal is carried out en-route from a source to a destination. Therefore, PONs typically do not require active optical components, namely switches, amplifiers, and so on, which are the key to the upward-driven prices seen in generic optical networking.

Characteristically, a PON consists of low-cost components that are engineered to provide reliable high-speed communication. One of the main motivating factors for achieving low cost through PON is the network topology itself. Unlike in a metro core and a long-haul network, where rings and meshes are the dominant topology often leading to expensive network gear for routing of connections, in the access area the use of fiber is limited to tree formation only, thus creating a natural broadcast medium that does not need routing protocols per se (only for the tree).

The Network Topology

Physically, a PON is based on a tree topology. The main reason for the tree topology is the formation of the network whereby a single central office is communicating with multiple end users geographically dispersed yet clustered around in the same area. This means that the relationship (graphical) between the single central office and the multiple end users can be described as one–is-to-many or simply one-to-many. When we relate this kind of one-to-many semblance to optical networking, the topology apposite to our requirement is that of a tree, where there is a single root and multiple leaves connected through the branches. Translating this to access area networks, this means that there are primarily two kinds of network elements in PONs in tree formation: the first at the central office of the service provider and the second at each of the end users.

Figure 2-1 shows a generic diagram of a PON network. Multiple end users are connected to a single network element (NE) at the central office through a tree of optical fibers—hence the passive network is formed. The NE at each end user is called the optical network unit (ONU), whereas the NE at the central office is called the optical line terminal unit (OLTU, or sometimes just OLT). The OLT is connected on one side to these ONUs and likewise is connected on the other side to the metro core network, invariably a Synchronous Optical Network/Synchronous Digital Hierarchy (SONET/SDH) or Gigabit Ethernet (GigE) network that drops bandwidth of granularity as required in the passive network. The OLT has the strategic importance of interfacing between the first mile access components (namely the ONUs) and the metro core network. Note that the OLT is a rather complicated device having to sometimes do protocol transfer and management functions. The ONUs, on the other hand, are connected to end users through multiple methods. In one embodiment, ONUs may be interfaced to the end user through a user network interface (UNI). In another scenario, ONUs may be connected to homes using a wireless solution. Yet another implementation has ONUs connected to homes and small businesses using copper solutions and taking advantage of long-reach Ethernet (LRE) concepts.

Figure 2-1 *PON Topology*

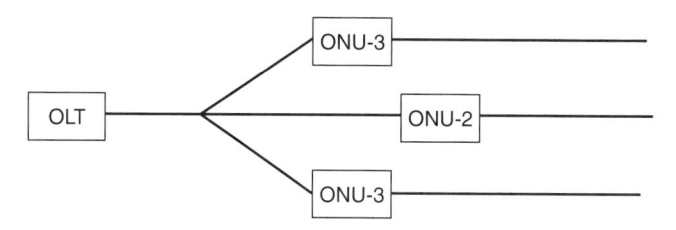

As mentioned previously, the structure of the network is that of a tree. The OLT is connected to the multiple ONUs through the fiber tree such that from the OLT a single fiber emanates toward the ONUs. The signal from the emanating fiber is split (in power) by an optical splitter. Multiple fibers emanate from the optical splitter, each carrying a portion of the signal from the ONU. Each of these multiple fibers is connected to an ONU. In some cases, these fibers may be further split, sharing two or more ONUs between them. Figure 2-2 explains this case of the tree topology. The manner of communication from the single OLT to the multiple ONUs is considered to be in the downstream manner, quite in conformance with the top-down approach cited in most networking environments. Also, the way communication is done, by using the single-to-multipoint splitter, which inherently is a passive device, sets about the motivation for the nomenclature of this network as a passive optical network. The reverse communication, that from the multiple ONUs to the single OLT, is considered to be in the upstream. Here again the power from multiple ONUs is added up in the splitter (which now acts as a combiner). Of course the story of this chapter is how to ensure upstream communication, the real problem in PON solutions, and ways to deal with it, because it can now be understood that successful upstream communication is the result of providing a collision-free environment when multiple ONUs transmit.

Figure 2-2 *General Tree Topologies*

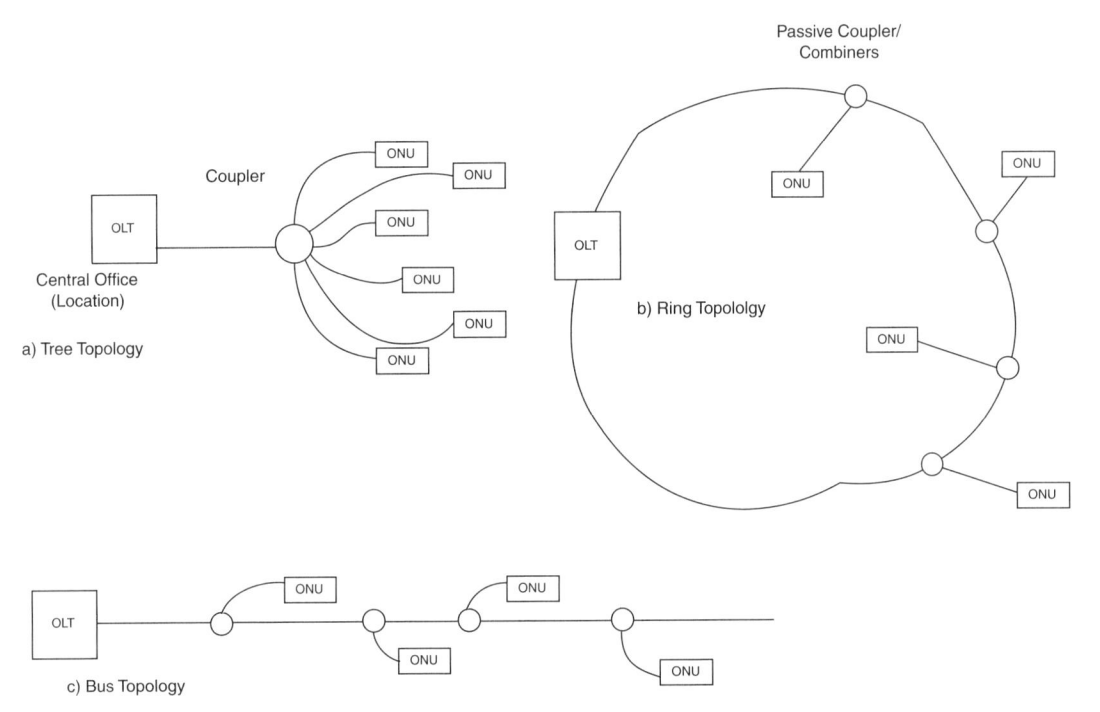

a) Tree Topology

b) Ring Topololgy

c) Bus Topology

Passive Optical Networking: Problem Formulation

In the preceding section we covered the working of a PON and the main reasons for its nomenclature, the passivity of the medium. We now can digress to outline the main issue in PONs. We have noted and understood the importance of the star topology in passive optical networking and generally also for access networks: the need for connecting a large number of users to a single access point, namely the central office. This leads us to the obvious intuitive solution that a star topology is the best way to create passive optical networking. The passivity in the star-shaped network creates a bus kind of architecture, such that this bus is point to multipoint. This means if the central office (say connected to the center of the star) sends a signal into the star, all the end users get a copy of the signal. This is because the splitter splits the optical signal in such a way that each end user gets a replica of the signal.

Optically speaking, this is done as follows: Assume the splitter has a ratio of 1:N such that there is an input fiber and N output fibers connected to N users, and N > 1. The splitter is formed by the hot-fusing of the N output fibers with the input fiber such that when an optical pulse with some power P arrives, the power is almost evenly split into the N output ports, so that each of the N output ports gets P/N amount of power. Of course the power level drops down, but each port still gets a replica of the same signal. We use the following equation to calculate the optical power at any of the output ports of an optical splitter. If P is the input optical power, and there are N output ports (hence N users), further if Δ is loss of signal in the coupler, the power at any of the N output ports is as follows:

$$P_{port} = (P - \Delta) - 10\log_{10} N$$

The authors developed the preceding equation, and the reader is encouraged to prove its correctness through analytical methods.

Earlier in this chapter, the behavior of a PON network was described as in a downstream direction—that is, from the central office to the end user. We also make special mention of the fact that the direction is downstream and, most importantly, the passivity creates a broadcast medium, one that is specially adapted for random traffic having broadcast nature, such as Ethernet.

In contrast, when we consider the other direction of communication—namely, from the end user to the central office (ONU to OLT in the previously defined nomenclature)—we see that the system is slightly more complicated: We observe that there are now N users (N ONUs) connected through our passive splitter (now acting as a combiner) to the central office (OLT). We see that the medium is to be shared in a ubiquitous way between the N users.

In contrast to the downstream case where the broadcast nature made communication very simple, the upstream case has the biggest problem, that of avoiding possibility for collision. Assume that any two of the N users want to send some data to the OLT. Now these two users do not know whether the other wants to send data in the time period they individually want to send. In other words, if users A and B want to send data to the OLT in a time period starting from t, both A and B do not know whether the other node has data to send commencing from t. This creates a possibility of collision between the transmissions from A and B in the same time slot starting from t. This means that in a PON network, there is an absolute uncertainty of transmission in the upstream direction; one uncertainty stems from the passive nature, and one needs a degree of intelligence to be avoided. A point to note here is that, unlike in wireless communication, detection of collisions in fiber is not feasible. Experiments have shown that although it may be possible to detect collisions by comparing the power of the collided signal with the initial signal power, it still is not a foolproof method for collision detection. This showcases the need for an efficient method to alleviate the problem of upstream communication, that of being able to transport data from ONUs to the OLT in a collision-free environment.

The previous discussion raises an important point about PONs: the need for a protocol for upstream communication—a protocol that guarantees collision-free upstream communication yet also ensures that each node gets a fair share of the bandwidth. There are numerous methods of providing upstream communication in the PON. WDM, time-division multiplexing (TDM), statistical TDM (STDM), and hybrid solutions are some of the approaches. The most logical (but not cost-effective) solution to be considered for upstream communication is to use separate noninterfering channels for communication from the ONUs to the OLT. The obvious way to implement such a system is to have each ONU communicate to the OLT on a separate optical wavelength channel. This means each ONU would have its own operating wavelength, in which a laser would be tuned—one that emits modulated data in optical format. This also means that at the OLT side, for receiving data from N ONUs we would need N receivers (photodiodes) coupled with filters that filter just the desired frequency. Although this whole system is very effective, there is minimal deployment as of today and as projected for the near future. The reason for its minimal deployment is simple: cost. Using WDM (the method of multiplexing multiple wavelengths in the fiber), the system requires an arbitrarily large number of optical components that are generally very expensive. For example, an N-node network having N ONUs connected in star formation to N OLTs, needs N optical receivers and filters at the OLT, in addition to N optical transmitters, each at a different wavelength, one each at every ONU. This makes the system cost exorbitantly high, and there is particularly no justification in access networks, especially the first mile area, for such high-cost solutions.

The other two generic solutions proposed are using TDM and STDM. First, let's consider TDM. To multiplex more and more voice signals in a trunk, TDM was invented, whereby using high-speed sampling, one could multiplex multiple signals in the time domain by spatially separating the sampled signals. In other words, N number of signals could be delivered through just one channel by slotting the channel in N recurring slots and writing the data of the first signal in the first slot and so on, thus creating a time-division system. One of the benefits of such a system is the requirement of comparatively fewer components.

Let us now analyze the requirements of a TDM system for Ethernet over PON (EPON) upstream communication. Assume N end users (ONUs) are connected to an OLT through a passive star. The upstream channel is a single optical wavelength able to support the cumulative bit rates of the N nodes. The upstream channel is slotted such that one slot is destined for one node, meaning that if there are N nodes, node 1 gets slot 1 and the next slot it receives is slot N + 1 and so on. In other words, the periodicity of slots is N. When the system is functional, the OLT transmits data on these slots, which are predefined. A node detects a slot only if the slot is meant for the node. Otherwise the node just discards the slot. Therefore, it's imperative that the node knows when a slot starts and when it ends. This means the NEs (the N ONUs and the one OLT) need to be synchronized. Synchronization in PON networks for TDM is essential for successful communication. Synchronicity can be provided by using a standard clock reference (at the OLT, for instance); all the ONUs align their clocks accordingly. Note that while aligning to the OLT, the ONUs may be at different distances, and hence there may be variable delay from the OLT to the ONU, creating a slight difference in synchronization. This can be rectified by using mathematical tractable solutions to the far-near problem—a case when two users communicate to an OLT, but both are at different distances. Hence, the closer user gets a packet ahead in time to the farther user, which creates a discrepancy.

Moreover, the very reason for deployment of PONs worldwide was to facilitate the huge surge in data networking, propelled by the absolute exponential growth of IP-centric traffic. It has been seen that IP traffic is bursty in nature, meaning that the interarrival times of IP packets is not uniform, and further that when IP packets arrive they come in large numbers followed by large time gaps, in which fewer or no packets at all arrive. This kind of traffic behavior means that bandwidth requirements of end-user applications are dynamically variable. This leads to the requirement of packet-oriented communication that supports IP traffic and one that can be made to be flexible to support variable traffic needs. Therefore, if we were to have a TDM system with fixed-sized time slots, one each for every node, this would not be efficient, because of the bursty behavior of IP traffic creating many voids. In other words, a large number of slots are empty because there is no data to send, and for nodes that have a high volume of data (burst), the periodic nature of the slots does not allow the nodes to cater to IP-centric traffic. This means that a TDM formation—that is, allotting slots of a TDM system—is not effective for a PON solution. Another issue is the cost involved in synchronizing the nodes of the PON. This is the motivation for a slight variant of TDM called STDM.

Consider a case in which six ONUs are connected to an OLT and are named A, B, C, D, E, and F. Now assume that a burst of Ethernet frames at the OLT of time duration t_1 seconds arrives and is to be routed to node A. In a TDM scheme, the burst would be broken into slots proportional to the TDM slot size and scheduled periodically. Of course, we note that in a passive broadcast architecture as that of the PON, when the OLT sends any information all the nodes receive the information (optical signal) and only the node(s), for whom it is meant creates an address matching and decodes the information. Coming back to the TDM system, it would take $t_1/T + Kt_1/T$ seconds to transmit the burst of information t_1 (burst size), where T is the TDM slot size, and K is the periodicity (after how many seconds does the slot for the node come back). This means that the system is inefficient on account of the long wait it would have and would typically need large buffers to fill in the waiting time at the OLT. Consider if t_1 is 10 ms, slot size T is 1 ms, and the periodicity is 3 ms, the total time needed to transfer the data out of the OLT is 40 ms. Now further assume that the system is such that there is no data for other ONUs at the OLT; this means that most of the downstream bandwidth is not occupied. To alleviate this efficiency issue in PON communication, two logical proposals led to communication protocols: Transmission Upon Reception (TUR) and Interleaved Polling with Adaptive Cycle Time (IPACT) (Gumaste and Chlamtac). The protocols are discussed in the concluding sections of this chapter, but before that the generic solution—the STDM approach—needs some attention. The bandwidth in the PON—assume for downstream communication—was slotted in periodic fixed time slots in the TDM approach. Unlike this scheme, in the STDM approach the downstream bandwidth is also slotted but the slots are not of fixed duration and, hence, are not periodic. The bandwidth is slotted into asynchronous variable time slots, such that these slots can be made to fit the size of a burst of frames/packets to cater to the IP-centric nature of data traffic. The immediate performance gain by using STDM is the ability to cater to bursty or any other traffic distribution and to be able to provide the much-needed dynamic flexibility for bandwidth-killer applications such as bandwidth on demand, video distribution, and so on. It is easy to understand both logically and intuitively the superlative performance benefits of using an STDM kind of scheme rather than a generic TDM scheme, but we have to also note the level of difficulty in implementing the scheme. Before we talk about the negative aspect of an STDM scheme, if any, however, we should look at its positive aspect and why it is a most natural contender for PON application.

STDM represents a seamless approach for first mile networks because of its flexible nature in being able to allocate bandwidth on demand, especially considering the randomness of the requirements for a large fraternity of consumers.

By removing the constraints of fixed-size time slots, in an STDM system we also remove the constraints associated with the need for node synchronization. This can prove particularly important from a cost perspective, especially as data rates increase and there is a need to use synchronization (leading to high-speed and high-performance clocks, at each and every node in the network).

Finally, an STDM solution is particularly necessary for honoring SLAs between users and service providers; in an STDM solution, the bandwidth allocation can all be either dynamic or pre-allocated, depending on the SLA between the user and the provider.

As this discussion makes obvious, an STDM solution is far better than a TDM solution in terms of cost, performance, and appropriateness for IP-centric traffic. However, the biggest issue is the deployment technique of an STDM solution.

Consider Figure 2-3, in which upstream communication and variable-length bursts of data are being sent to different nodes from the OLT. The issue here is how the ONUs (nodes) can know the start and end of the bursts, and, second, how do we allocate this bandwidth dynamically. Finally, we also need to outline a method for upstream communication, the initial problem seen in basic PON deployment.

Figure 2-3 *Exemplifying the Need for Synchronization*

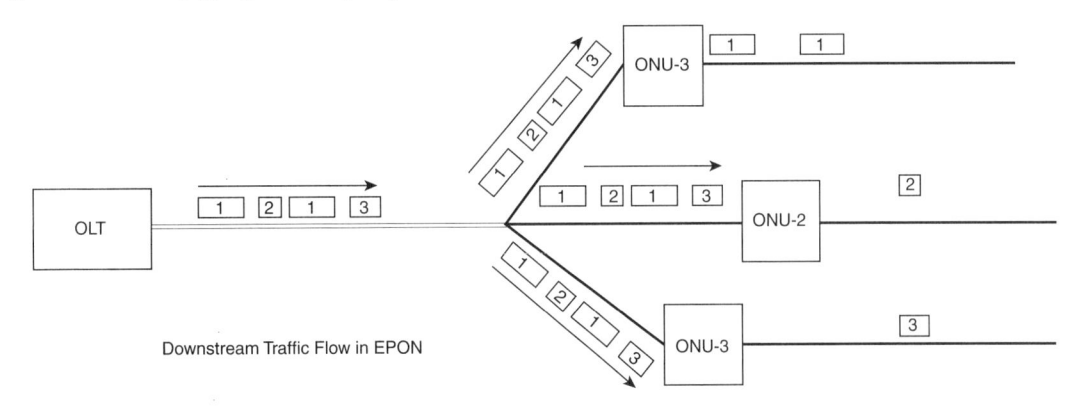

These issues are resolved later in this chapter when we cover the implementation of TUR (a protocol for EPON communication). Before digressing further, however, it's important to identify, define, classify, and compare the various types of PONs.

PON Classification—APON and EPON

As of today, there are three documented types of PONs: APON/BPON, GPON, and EPON. APON means ATM over PON, BPON stands for broadband PON, GPON means gigabit PON, and EPON is called Ethernet over PON. APON and BPON are variants of each other, whereas GPON is a vendor-proposed implementation not standardized as of this writing (but it is expected to be so soon). EPON, on the other hand, is the most significant type of PON that we discuss in detail as we progress through this chapter.

APON, the voice-centric PON prominently generating revenue in networks worldwide, evolved into a network that facilitated voice-type communication. In this regard, ATM over PON was a promising approach due to its circuit-centric nature. Through a consortium of carriers and vendors, the International Telecommunication Union (ITU) issued a series of recommendations—namely G.983.1, G.983.2, and G. 983.3—for possible deployment of APON (BPON) solutions. The recommendation of the BPON network was fully discussed in 1996 with regard to the formation of Full Services Access Network (FSAN). (You can find more information about this at http://www.fsanet.net.)

NTT and BellSouth (two carriers) formed the first collaborative effort for the deployment of BPON in 1998. This effort was called the *Common Technical Specification* (CTS). The number of partners finally joining hands grew to five when NTT and BellSouth were joined by British Telecom, France Telecom, and SouthWestern Bell Company (SBC) (each of these three being service providers).

BPON essentially means the same thing as APON. ATM is used because it can guarantee a good degree of QoS to the end user. The ATM layer resides on SONET interfaces, which are fed to an OLT that is connected through a tree to multiple ONUs. In other words, it is ATM over SONET in the last mile. ATM guarantees a committed bit rate, and SONET gives the necessary reliability and resilience for high QoS applications. The data rates seen in BPON are SONET/SDH equivalents—namely, 155 Mbps and 622 Mbps (OC-3 and OC-12, where OC means optical circuit).

In the upstream direction, the basic frame contains 53 cell slots of ATM, each 56 bytes long. Each cell slot consists of a 53-byte ATM cell and a 3-byte overhead. The overhead consists of a guard time, a preamble, and a delimiter. In the downstream direction, the basic frame consists of 56 cell slots, and each is 53 bytes long (same ATM format). In the downstream direction, each ONU gets the full complement of ATM cells but discards those cells (after making an address match) that are not destined for the incumbent ONU. In the upstream direction, the OLT gives ONU permission to send data by using a "grant" via the downstream cell. This cell is the physical layer operation, administration, and maintenance (PLOAM) cell. Despite the variance in the distances between different ONUs from the OLT, we have to ensure a mechanism that creates a collision-free communication. We outlined this as the far-near problem. This is solved in the following way: The OLT measures the distance to each ONU and then instructs every ONU to insert an appropriate delay so that all OLT-ONU distances are now virtually the same. This act of balancing the distances and solving the far-near problem is called *ranging*.

Advantages of BPON

The BPON architecture and paraphernalia represented the first way to implement passive optical networking for end users. What should be noted here are the motivating factors for BPON: First, there was a need to lower costs; second, the system needed to exhibit absolute protocol friendliness. On one hand, we see the absolute abundance of SONET/SDH networking and the capability to guarantee high-performance data (and voice and video) transport over a SONET network using its packet adaptation—namely, Packet over SONET (POS) (creating an amalgamation of ATM and SONET) in the core of networks. At the same time, and encouraging the initial deployment of ATM, there was a surge in the use of DSL and similar technology in the first mile, requiring a common denominator (ATM) that could cater to voice traffic as well as to bursty data traffic (of course, to a certain extent, because of its inherent TDM nature). These two factors prompted the need for APON and hence gave rise to the concept of BPON.

Note that despite the backing of multiple vendors and carriers in the initial deployment of BPON, several drawbacks have forced BPON to be deployed only partially; therefore, BPON cannot be seen as the best possible PON solution.

Of the many BPON drawbacks, the most relevant is the cost. BPON deploys primarily electronic-based technology—namely, SONET and ATM. Although effective and well proven, the problems with ATM and SONET technology are the associated cost and the heavy reliance on being a total electronic solution. The former creates a situation in which the economics do not support or validate BPON. The latter—the use of high-cost, low-performance electronic systems—contrasts negatively with the advantages of passive optics. The overdependence on precise high-speed electronics means that the cost of deploying a BPON network is high. As indicated in Chapter 1, " Introduction to First Mile Access Technologies," the solution proposed for first/last mile access must have a low cost, one that suits the many and varied consumers who make up the revenue pool. The attempt to keep costs low is to a great extent nullified by the use of an electronic solution such as ATM and SONET. The higher costs significantly shrink the market for BPON from all end users (residential and small enterprises) to just small enterprises and homes with very high-capacity requirements (which are minimal). Therefore, despite its capability to technically solve the first/last mile problem, BPON is limited by cost. The other issue, that of not being able to use the passive optics, is another serious drawback for the deployment of BPON. PONs in a star topology represent a very viable solution for first mile networking. High costs associated with optical networks per se severely limit their use in the core and backbones. However, PONs represent one way to bring this cost down, because they use smarter technology and take advantage of the optical characteristics of coupler-like components that are technologically mature and extremely inexpensive. This direct advantage is nullified by the electronic bias of a BPON solution, which results in an expensive and inflexible system.

Gigabit PON

In 2001, the FSAN group (http://www.fsanet.net) initiated a new effort to standardize a high-bit-rate PON, now known as GPON. Apart from the need to support higher bit rates, the overall protocol was opened for reconsideration. The solution needed to be the most optimal and efficient in terms of support for multiple QoS-associated services such as voice and video. In other words, the GPON standardization process is basically a carry forward from the circuit-based processes seen in BPON, with some degrees of correction as compared to the earlier BPON version. Note that GPON uses Generic Framing Procedure (GFP) for framing and, hence, transport over the optical medium. The advantage of GFP is that multiple services can be mapped and adapted into a single (and generic) frame. This leads to less processing than other formats (such as ATM) and also enhancement of bit rates through forward error correction (a coding method to reduce transmission losses).

As a result of this FSAN activity, a new solution (that is, GPON) was showcased in the optical access area offering higher bandwidth (in the gigabit range) and enabling the same transport services (namely, circuit-based services such as voice).

As part of the GPON effort, a gigabit service requirement (GSR) document is implemented based upon the collective requirements of all member service providers (representing the leading Regional Bell Operating Companies [RBOCs] and Incumbent Local Exchange Carriers [ILECs]). The document is currently under submission to the ITU under the title "G.GPON.GSR."

The salient features of GPON, summarized by the groups proposing GPON, are as follows:

- Full-service support, including voice and variants (TDM SONET and SDH), for data services through 10/100BASE-T
- Physical reach of at least 20 km, with a logical reach able to support 60 km (with regeneration)
- Support for various bit-rate options using the same protocol, including symmetrical and asymmetrical implementations using combinations of 622 Mbps, 1.25 Gbps, 2.5 Gbps, and so on
- Strong operation, administration maintenance, and provisioning (OAM and P) capabilities
- Security at the protocol level for downstream traffic (because of the multicast nature of PON)

When the vendors supporting such a hierarchy originally proposed GPON, multiple promises motivated its development:

- Higher data rates than previous versions of PON
- Better efficiency (typically for protocol considerations)
- GFP encapsulation of any type of service on standard SONET frames (125 ms)

- High efficiency with no overhead transport to native TDM traffic
- Dynamic allocation of upstream bandwidth via bandwidth maps (pointers) for each ONU

Let's digress from the GPON issue to consider GFP, specifically the importance attributed to GFP for optical transport networks. GFP provides a generic mechanism to adapt traffic from higher-layer signals over a transport network. The transport network could be of any type (such as SONET). Client signals may be packet based (IP) or circuit based. GFP has been officially standardized in the ITU document "G.7041." Because GFP provides a generic mechanism to facilitate the transport of multiple services in an efficient manner, over SONET, it ideally suits the basic GPON system, which is also an optical network with nodes synchronized with each other.

Briefly, GPON design characteristics are as follows:

- GPON is a frame-based, multiservice transport over PON.
- Upstream bandwidth allocation is accomplished via slot assignment.
- Database reports, security, and other issues are integrated into the physical (PHY) optical layer.
- Line coding is of a nonreturn to zero (NRZ) type, and therefore there is no need to return to zero after every pulse (1/0, for instance).
- GPON supports asymmetric line rate (namely, up to 1.25 Gbps).
- The upstream burst mode preamble includes clock and data recovery (CDR).
- The PHY layer also supports QoS.

Emergence of EPON—An Effective PON Solution, Particularly for IP-Centric Communication

The advent of data networking led to the emergence of a protocol that allowed computers to be linked and to "speak" using a universal and simple method. This method, Ethernet, was gradually incorporated into 95 percent of networks worldwide (initially by LANs but now by core networks, too).

SONET defines a standard for the transmission and multiplexing of data in the time domain through a network layer. SONET/SDH was created to cater to large volumes of voice traffic over circuit-switched networks and was standardized in the 1980s. SONET/SDH uses TDM and provides users a guaranteed level of service by assigning virtual channels within each frame. This guaranteed service ensures users that they can meet the critical requirements of voice traffic (namely, QoS), creating a platform for excellent timed delivery. Over the years, SONET has evolved to carry both voice and data traffic, and SONET data rates have increased from the original EC-1 (51 Mbps) to OC-192 (10 Gbps) today. Lately, some variations of

SONET/SDH for data networking have been proposed (such as Packet over SONET [POS]), typically to match the growth from Ethernet.

It was very natural for vendors to consider SONET and SONET-like circuit-switched technologies (ATM, for example) as possible candidates for PON. This solution of circuits over PONs is good but is not the best. Why?

EPON has a relationship analogous to fish and water. First, the passivity in PON is the most conducive technical environment for Ethernet. Second, the cost involved in SONET networking was extremely high. The expensive electronics at each node and the level of precision required made SONET networking a very difficult proposition for first mile problems. In contrast, Ethernet was a ready solution; it was low cost and had good performance, especially for IP-centric communication (for bursty traffic arrivals). Ethernet also is a ubiquitous and well-understood topology, and it exhibits a plug-and-play nature that makes management and installation easier.

Despite these direct advantages, some industry pundits regard Ethernet in the access (and in the core) as a disruptive technology, viewing it with suspicion and deploying it only when alternatives dry up. We discuss in the following section the disruptive aspect of Ethernet and why it can withstand as a superior technology despite claims by legacy networking companies and pundits.

Ethernet as a Disruptive Technology

Legacy service providers that have generated income through SONET/SDH circuit-switched technologies (primarily), which entail synchronization and expensive management—often criticize the deployment and development of packet technology, specifically because it is not revenue yielding.

Casson, Christensen, and others published a seminal paper titled "Ethernet in the MAN-A Case Study in Disruptive Technology" explaining the idealistic nature of SONET/SDH and ATM and how Ethernet proves to be a technology that is an able replacement for circuit-switched networks, especially in the first mile.

Performance oversupply is a principle that holds that mainstream technologies in certain industries improve at rates faster than the market can absorb. In Casson's paper, this is illustrated by the fact that this oversupply of new technology causes a paradigm shift in the market, creating a situation in which customers are given a wide choice of products. Christensen states that performance oversupply is the key prelude to signaling a future change in the basis of competition. Recent circuit-switched network performances have focused on increasing the distance and speed of WANs. As a result, these advances have overshot the needs of the access area. Figure 2-4 shows the pricing advantage of Ethernet as compared to a SONET (TDM) hierarchy.

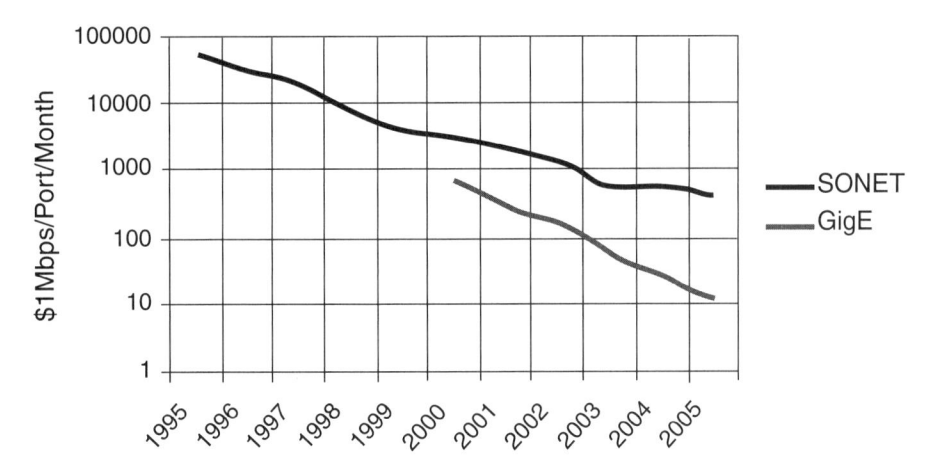

Figure 2-4   *Pricing of Ethernet and SONET*

Most incumbent firms fail to realize the benefits of an emerging, disruptive technology. The ones that do are generally smaller startup kind of companies that do not focus on mainstream markets but create their own area of dominance (away from Christensen's theory of "resource dependence"). They instead find niche markets in which the attributes of the new technology find business value. Invariably, these attributes are the same ones that make it less attractive to mainstream customers. Consider, for instance, the LAN market. Ethernet was first introduced into the LAN in the early 1980s by a number of small startups that initially focused on the DEC minicomputer market, and later the PC market. However, the large incumbent firm, IBM, delayed introducing its Token Ring technology to the LAN market until 1986. By that time, Ethernet already had an installed base of 30,000 networks containing at least 417,000 nodes. In the case of MANs and access area, Ethernet is also emerging in niche market segments where its attributes are valued. (Consider the metro Ethernet effort in Japan, for instance, an area technologically very superior to most parts of North America.) One MAN niche targeted by these companies has been small and medium-sized businesses and access area. Unlike large enterprises, access area generally does not require circuit-switched high-quality performance and reliability and values cost and flexibility in provisioning over other attributes. Another market niche targeted has been the largest metropolitan areas in the U.S. These large cities have a high density of potential customers and a large amount of unused, installed fiber that can be leased at a relatively low cost. The third and probably most important attribute is the management aspect and ability to provide service on demand. We all know the nuances of DSL and how difficult it is to provision a new service over an existing one without several software and hardware plug-ins. Circuit-switched technology inherently causes a lot of management problems, especially in provisioning networks. Packet-oriented networks, such as Ethernet, based over good topologies, such as PON, facilitate high-speed provisioning in the network domain.

Finally, disruptive technologies usually have lower profit margins than mainstream technologies. This factor often makes them unattractive to incumbents, whose cost structure and market valuation are based on higher-margin products. The low margins and profits, and associated lack of incumbents in the market, are the key factors that make the ensuing disorder and dislocation so pronounced.

All this leads to an important conclusion: Technologically, Ethernet is a superior candidate as compared to its circuit-switched counterparts. Financially, Ethernet costs much less to deploy than circuit solutions. This means that the end consumer (the end user) benefit immensely from the deployment and promulgation of Ethernet as a method for providing broadband access in terms of both cost and performance. It is the carriers and providers who need to refocus and relay cost-effective and performance-oriented strategies in the last mile. At times service providers may be dissatisfied with the costing structure (perhaps because of a temporary loss of revenue). However, it's in the best interest of the incumbent to incorporate the change and be able to sustain business for a long time than not to incorporate the change and face being wiped out of business altogether. It is in this context that EPON can be seen as the premier enabling technology, creating a very market-conducive technologically superior solution to solve the first mile access problem.

Standardization Efforts in EPON—The EFM Push

Because EPON appeared to be a natural and performance-oriented solution for the first mile access, a need arose to standardize the networking environment, especially the multiple subsystems that play a part in the deployment of EPON. In 2002, a task force called Ethernet in the First Mile (EFM) attempted to standardize EPON. The efforts of EFM culminated in a new standard for point-to-point as well as point-to-multipoint communications in the first mile using Ethernet as the central mechanism. Supported by the IEEE and to be nomenclated as IEEE 802.3ah, the EFM group's standardization has three main objectives:

- Point-to-point (P2P) Ethernet over fiber
- P2P Ethernet over copper for the last mile
- Point-to-multipoint (P2MP) Ethernet over fiber (as in PON)

The preceding three objectives are very important in the last mile and are discussed in detail in the following sections. The second objective—providing Ethernet over copper—is sometimes also considered the long-reach Ethernet (LRE) issue, which attempts to address the following considerations:

- Provide a family of physical layer specifications for the following:
  - 1000BASE–X (for transmission up to 10 km over single-mode fiber [using GigE])
  - 1000BASE–X (providing temperature extension in optics)
- Support far-end OAM in subscriber access networks (making sure there is excellent delivery of data to the end user)
  - Remote failure indication (ONU failing)
  - Remote loopback (MAC acts like a switch, looping back on the client side)
  - Link monitoring

The basic P2P architecture can be considered for seamless delivery of high-speed data over fiber. The application can be enterprise-level connectivity or transporting from an OLT to a central splitter.

Among others, some of the strategically important issues that have been upheld by EFM are as follows:

- **Finalization of upstream and downstream wavelengths**—In the upstream, a lower wavelength (for instance, 1310 nm or 1490 nm) is generally chosen. In the downstream, however, a 1550-nm wavelength is chosen. The reason being that, first, it is better to have lower-cost equipment at consumer premises, which means 1310-nm lasers rather than the 1550-nm ones (and therefore a cost savings). Second, by choosing different wavelengths for upstream and downstream communication, we best utilize the same fiber without optical signal collision.
- **Temperature aspect**—Most ONUs, and for that matter any first mile customer apparatus, are kept at sites without indoor temperature controls and such. Therefore, the ONUs (and other apparatus, especially for optical components) must be able to operate in a wide range of temperatures without significant performance change. Laser instability because of temperature variation is a chief drawback to optics. An associated chirp (wavelength variation) of a laser's emitted light is dependent on temperature stability. Newer systems in the first mile have to take into account this chirp and develop a way to avoid rapid and large changes in frequency that may result from temperature changes.

Multipoint Application

Figure 2-5 shows the extension to P2MP networks.

EFM also introduced the concept of EPONs, in which a P2MP network topology is implemented with passive optical splitters, along with optical fiber physical media dependent (PMD) layers that support this topology. In addition, a mechanism for network OAM is included to facilitate network operation and troubleshooting.

Figure 2-5 *Multipoint Networks, Hierarchical Classification*

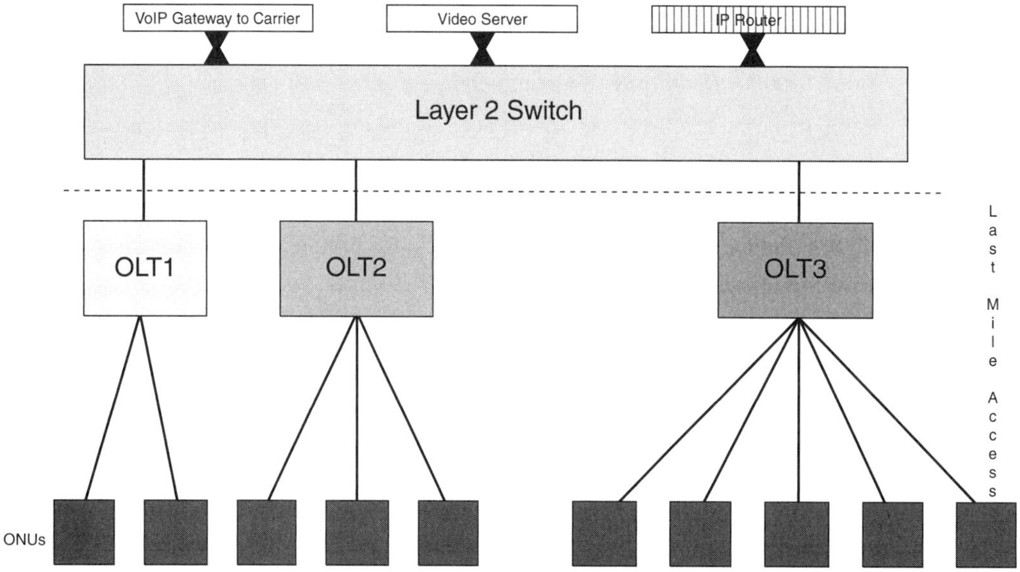

EPON is based upon a mechanism named *Multi-Point Control Protocol* (MPCP), defined as a function within the MAC control sublayer. MPCP uses messages, state machines, and timers to control access to a P2MP topology. Each ONU in the P2MP topology contains an instance of the MPCP protocol, which communicates with an instance of MPCP in the OLT.

At the foundation of the EPON/MPCP protocol lies the P2P emulation sublayer, which makes an underlying P2MP network appear as a collection of P2P links to the higher-protocol layers (at and above the MAC client). It achieves this by prepending a logical-link identification (LLID) to the beginning of each packet, replacing two octets of the preamble.

In this book, due to the nonstandardization of the MPCP protocol, we cover two very similar protocols: TUR and IPACT.

Transmission upon Reception

The key to successful communication in access PONs is to provide a solution that alleviates issues of throughput and fairness for upstream communication. The approach has to be low cost and easy to deploy. To alleviate issues faced by TDM and WDM in PON, we propose a protocol called Transmission Upon Reception (TUR). This protocol ensures both downstream and upstream communication and is able to accommodate the variations in traffic between nodes and guarantee fairness to the end-user nodes. TUR also ensures a collision-free system. TUR is a simple protocol that guarantees each end-user node some time to transmit data; the amount of time it gets is proportional to the amount of traffic it gets (from the central office [CO]), with some variations for ensuring fairness.

The idea is as follows: An end user, upon receiving a frame at time t_1 of size S_1, gets the transmitting time from time $t_1 + S_1$ to time $t_1 + S_1 + S_1 - d$, where d is some constant of the system. The constant d neutralizes the differences in propagation delays among different nodes in the system (analogous to the far-near problem). The protocol thus is simple and intuitive but can create undue fairness issues for nodes that have frames to transmit but none to receive. As can easily be observed, there is a "lag" between downstream and upstream communication. (You can see details of such in Figure 2-6.) Consider, for example, a 10-node system: If a node N_x receives a burst of duration T_x at system initialization, it would be able to transmit for duration $T_x - d$ starting from time T_x.

Figure 2-6 *System Block Diagram for PON*

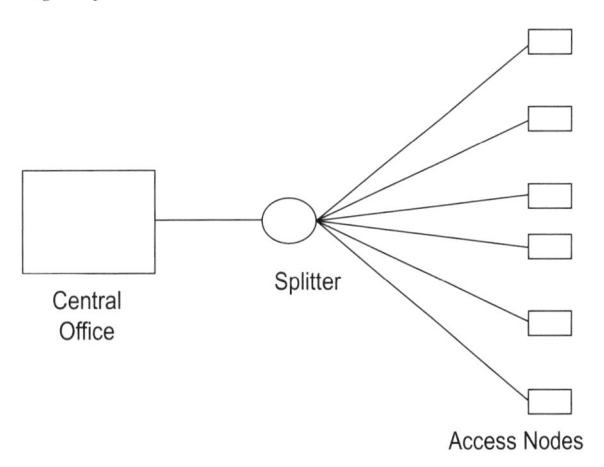

If the next burst of duration T_y is such that $T_{y/} < T_x$, and is destined for node N_y, node N_y can begin transmission from time $T_x + T_y$ to time $T_x + T_y + T_y - d$. However, node T_x is transmitting data from T_x till $2T_x - d$. If $2T_x - d > T_x + T_y$, there would be a collision of the two frames, although we are still adhering to the basic principle of TUR. To avoid this issue, we propose the following corollary to TUR: At initialization, the CO sends a counter called the *start point indicator* to each end user, whose value is zero. As downstream transmission begins, each node starts incrementing the start point indicator by the length of the latest received frame, whether or not this frame is destined to itself. This way every end user knows the status of the upstream channel in terms of the time when there is no transmission. If node N_x receives a burst of frames of duration T_x, all the nodes (inclusive of N_x) increment the counter value to $T_x + T_x$. This means that the earliest value at which the upstream channel is free is $T_x + T_x$. Subsequently if node N_y receives the next burst of frames of duration T_y, all the nodes increment the counter value to $2T_x + T_y$. Also, because the frames were destined for node N_y, the node gets the right of transmission from $2T_x$ to $2T_x + T_y$, although it ends its transmission at time $2T_x + T_y - d$ for reasons explained later. TUR basically allots time slots (of variable length) to nodes on the upstream channel. This corollary to the basic TUR protocol ensures collision-free access of the channel but creates some voids in the upstream channel. The voids are created when the present burst of frames is greater in duration than the previous burst of frames (to a particular user) and this difference exceeds the lag time between upstream and downstream channels. Because we assumed a burst protocol as opposed to a frame-based protocol, the bursts are quite large compared to the voids, and from simulation we observed that voids represented just 3 percent of the channel capacity even after transmission of 10 million bursts (of Ethernet frames). Note here that TUR is not a TDM protocol in the real sense but is a statistical TDM scheme, which does not require synchronization of the end-user nodes with the CO or each other.

Concerning multicasting, if a burst of frames is destined for two or more end users in the downstream direction, then, as per TUR, all the receiving nodes will attempt to transmit from the earliest time the upstream channel is free. To avoid collision due to multicasting, the CO sends a multicasting frame and scheduling information for each member of the multicast group at the beginning of the burst. The end users that are part of the multicast group transmit at their turn. For applications such as video on demand, the members of the multicast group may not even be required to send any information (except small acknowledgements) in the upstream, and hence upstream capacity during this interval can be devoted to some other node (procedures for such are discussed later).

Figure 2-7 shows a receiver block diagram. Note PD has a built-in 1310 filter with a band separator.

Figure 2-7 *Receiver Block Diagram*

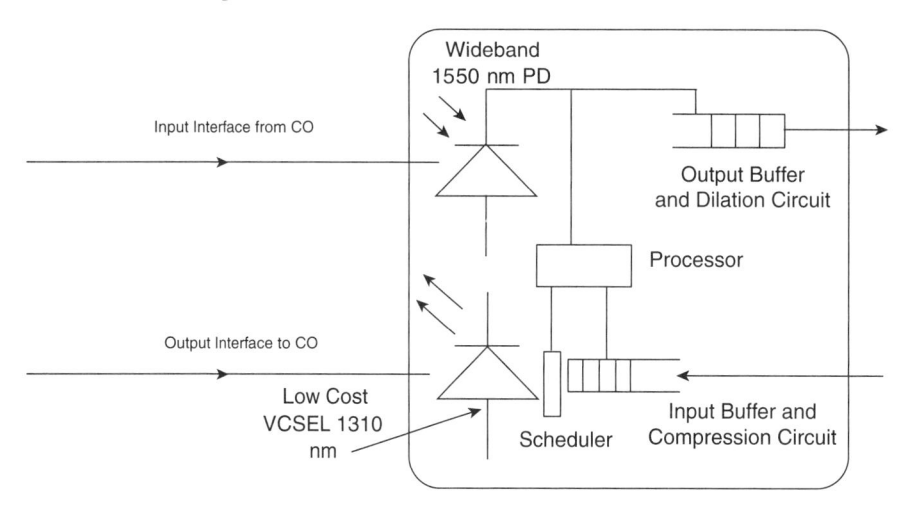

System Operation

Consider the network shown in Figure 2-6. In the downstream direction, the CO sends bursts of Ethernet frames to the end-user nodes. Whereas in the upstream, each end user sends frames in a way that frames sent from different nodes should not collide with each other. To avoid collision the burst size is constrained by two issues: the upstream burst size from a receiver R_x is equal to the length of the burst that R_x last received; secondly the size is also curtailed by the difference in propagation delays from each receiver (end user) to the CO. At time t_1 if end user R_x receives a burst of frames of duration t_b, it can begin transmitting from $t_{stpt} + t_b$ and continue till $t_{stpt} + 2t_b$, where t_{stpt} is the start point indicator mentioned in the preceding section. To avoid contention, however, we have to note that the burst would reach other nodes at different times depending on the propagation delays from the CO to the different nodes. Let us assume that the node we are looking at is not the closest node from the CO. Let the difference in propagation delay between the incumbent node and the node closest to the CO be _. Hence node R_y (closest to CO) would get the frame from $t_1 - $_, and the frame would last till $t_1 + t_b - $_. Under the tenets of the TUR protocol, node R_y would assume node R_x sends a burst from time $t_{stpt} - $_ to time $t_{stpt} - $_ $+ t_b$, although R_x actually sends a burst from t_{stpt} to $t_{stpt} + t_b$. Hence there is the likelihood of collision. To avoid collision, we reduce the upstream transmission time by a factor of 2_, which essentially takes into consideration the difference in propagation delays from the CO to other nodes.

Let t_i be the amount of time node i gets a burst. Let $\_j$ be the difference in propagation delay between node i and the CO and node j and the CO, where j is the node closest to the CO. Then the total amount of time node i can transmit is $TX_i = t_i - 2 \_j$. If F is the average frame length (= 1500 bytes in the case of Ethernet), $N_i = t_i/F$ is the number of frames that were transmitted to node i from the CO. Then the total transmission of node N_i is as follows:

$$D = Ct_i \left[1 - \frac{2\delta}{F} \right]$$

(C is the line rate.) Because _ is negligible as compared to frame duration:

$$D \cong Ct_i$$

Here even if we assume the closest end user to be 1 km from the CO and the farthest end user to be 20 km away, _ would still be a fraction of the transmission of an Ethernet frame, and hence the protocol is very tight in terms of bounds on the ratio of upstream to downstream communication.

TUR allots upstream channel access to a node that has received a burst of frames. This may seem to result in long queues for nodes that do not receive sufficient frames from the CO. If so, this could lead to a fairness issue for upstream access among the end users. We solve this issue as follows: The CO has an array of counters with a counter for each end user. The counter notes the time since the last transmitted frame to this particular node. The CO, upon realizing that a node has not received any frames for some time t_{idle}, sends a "status" frame of length L_{status} to this node. The node sends back to the CO the state of its output buffer. With this exercise, the CO is periodically able to know the status of buffers and is able to intelligently multiplex frames, both real and dummy, to ensure some degree of fairness. The drawback, of course, is when a node receives frames but has no frame to send and hence wastes upstream bandwidth. We solve this issue by allotting "empty-buffer" frames to the end user nodes. An end user upon getting bursts of frames can inform the CO that its buffer is empty. For the next few frames (less than the QoS requirement), the CO tags a "do not send" label behind the upstream frame to this node while sending an "out of turn send" label to an end-user node whose output buffer condition is comparatively severe. In this way, the protocol guarantees degrees of fairness and throughput to the network. The protocol can also be provisioned for SLAs, guaranteeing some basic QoS requirements to an end user.

Interleaved Polling with Adaptive Cycle Time

Another academic protocol proposed for creating high-efficiency collision-free communication in PONs is called *Interleaved Polling with Adaptive Cycle Time* (IPACT). The main principle of IPACT is as follows: An OLT is the central point of intelligence. An OLT polls a grant message to an ONU allowing it to transmit data. Simultaneously, the OLT in the downstream transmission sends data to the ONU. At the end of transmission to a particular ONU, the OLT so plans the upstream transmission that this incumbent receiving ONU informs the OLT of the size of data it has in its buffer. This way the OLT knows how much data is there in each buffer of every ONU.

This protocol is superbly designed for centralized management. However, the protocol lacks in being able to cater to the dynamic variation of IP traffic. Note that IPACT also cannot serve QoS-sensitive traffic very well, which in contrast is taken care of by TUR. Second, in the downstream the OLT has to synchronize the data size being sent to an ONU while granting a request. At the end of the transmission, the ONU responds with its buffer size information. This means that the other ONU, which currently had access to the grant from the OLT, is now transmitting data and, hence, a collision could occur between the transmitting data of the granted ONU and the transmitted control frame of the incumbent ONU. Although IPACT is essentially a STDM protocol, it works in these cycles of data transmission interleaved with grants and other control information. The good aspect of such communication as seen even in TUR is the absolute nonrequirement of a control layer, which is submerged within the data transmission.

Numeric Evaluation of EPONs

We conducted a simulation program to demonstrate the operation of the TUR protocol with bursts of Gigabit Ethernet frames. We assumed bursts rather than single-frame transmission because IP traffic is bursty and, second, because the protocol is intended for burst transport rather than frame transport to maximize the performance. Although, of course, Ethernet frames are transported and the performance is not degraded even if we consider just pure Ethernet frames (assuming fast vertical cavity surface emitting laser [VCSEL] technology at the end users).

Figure 2-8 shows the working of the TUR protocol for four end-user nodes.

Figure 2-8 *Working of the TUR Protocol for Four End-User Nodes (A, B, C, and D)*

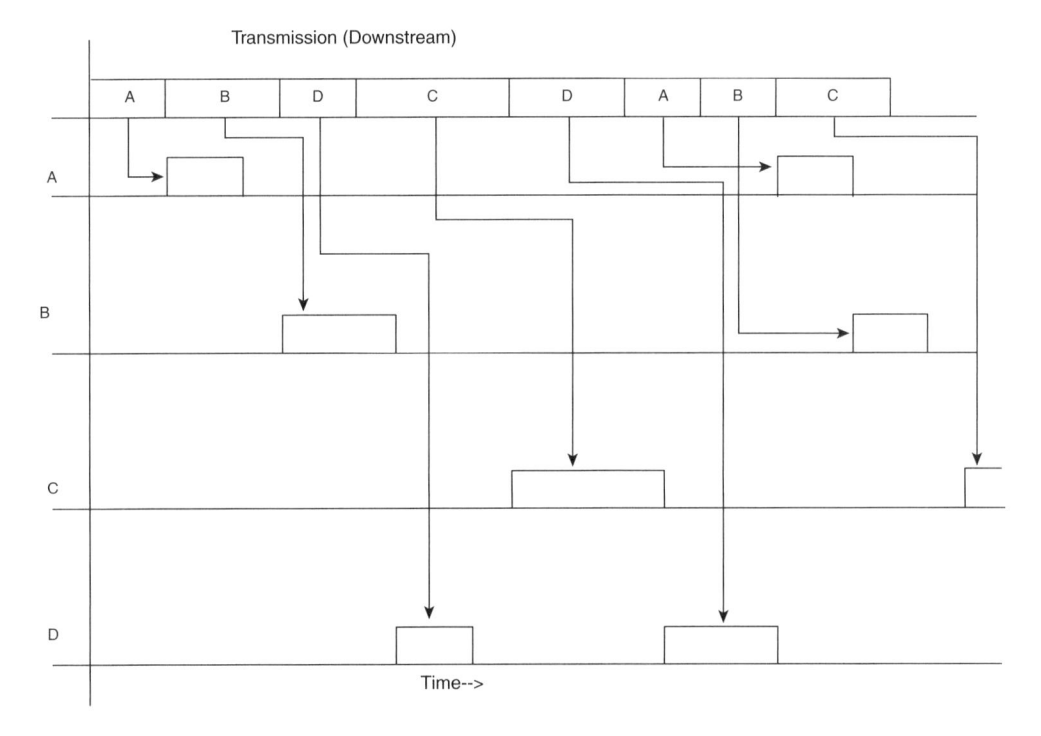

Figure 2-9 shows the efficiency of TUR for different nodes. We assumed a 1-GB Ethernet line rate in both upstream and downstream directions. Frames are generated in a way that bursts of several frames are destined for a randomly selected end user. The frame generation phenomenon for both the CO and the end users follows a Pareto distribution. The mean number of frames that make up a burst in our simulation was 20, and the variance of burst size was 16. We measured throughput of the system. Throughput is defined as the ratio of frames accepted by the end users for transmission to the CO, to the total number of frames that come to the end users. The buffer status of the end users was also measured for various configurations of number of nodes. We observed that the system performs particularly well for 10 to 16 nodes. The buffer requirement was also quite low for that number. As the number of nodes exceeded 16, we observed severe penalty for throughput as well as fairness at the end users. By increasing the line rate proportionally, this issue can be alleviated. The percentage of voids created was less than 3 percent for 10 million frames.

Figure 2-9 *Efficiency of the Protocol with the Number of Nodes and the Corresponding Receiver Buffer Status for a Measurement of H = 0.9*

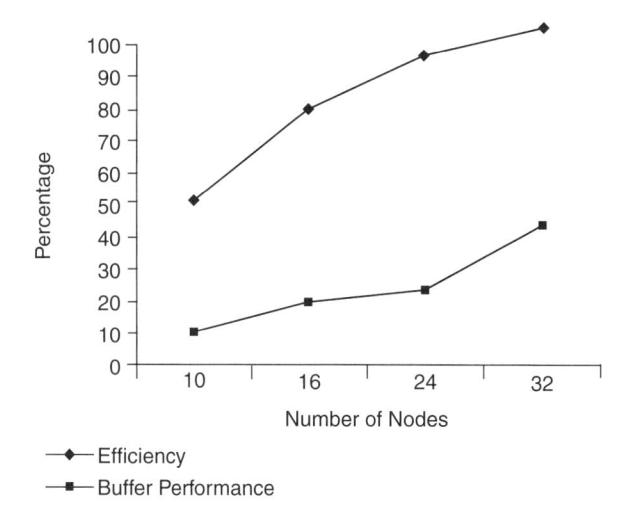

Because we were considering bursty traffic (predominantly IP), we measured the receiver buffer performance for several degrees of burstiness, given by the Hurst parameter in Figure 2-10. As the burstiness of the input source increases, the buffer requirement increases nonlinearly. Note that the buffer requirement does not increase exponentially as noted by an initial study of self-similar traffic (see Kramer, Mukherjee, and Pesavento).

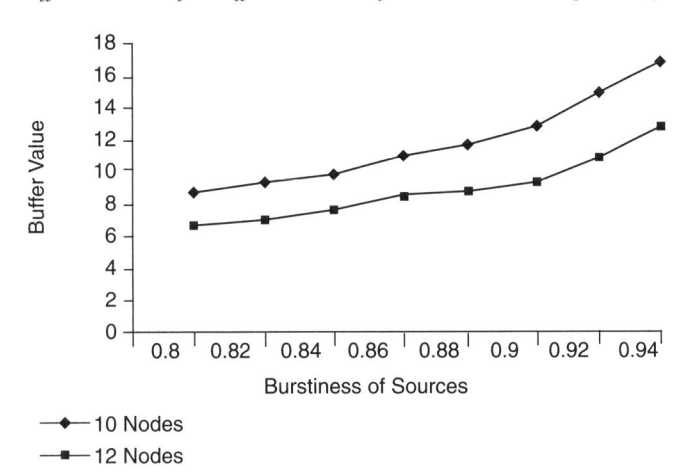

Figure 2-10 *Receiver Buffer Condition for Different Values of Burstiness at the Input Interface*

This means that TUR is an efficient protocol for bursty communication. In Figure 2-11, we have assumed the correction for fairness to be present during simulation. In Figure 2-11 we consider the effect of having correction in TUR for cases when a receiver gets a burst of frames but has no frames to transmit. For normalized loads in the range of 0.3 to 0.75, we observed maximum benefit by using correction in TUR.

Figure 2-11 *Blocking Probability in Terms of Dropping of Frames for Schemes With and Without Correction for Normalized Load*

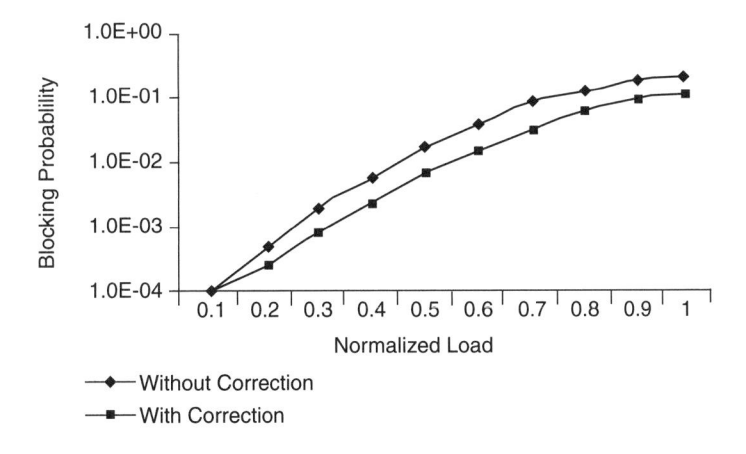

The blocking (percentage of packets dropped at the end user) is reduced by 20 percent in these cases. The number of end users in the measurement of Figure 2-11 was 12, and we observed a similar result for 10 and 16 nodes.

Figure 2-12 shows the fairness issue of TUR in PONs. For different numbers of nodes, we measure fairness for very bursty (H = 0.92) and comparatively less-bursty (H = 0.8) traffic. Fairness here is measured as follows: It is the ratio of times a node gets the upstream bandwidth to the times it has bursts to send scaled and normalized traffic with respect to a TDM scheme. In other words, if for a TDM scheme with 10 nodes a node gets every tenth slot for upstream communication, correspondingly we measure the performance for our scheme with more realism by considering pure bursty traffic.

Figure 2-12 *Delay Performance of TUR and IPACT*

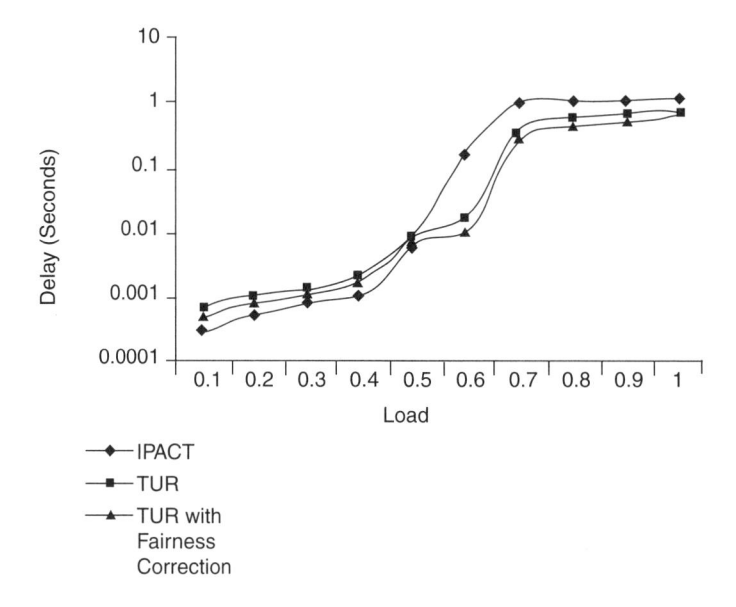

Figure 2-13 shows the performance of TUR for QoS traffic. Some fraction of the traffic is assumed to be delay sensitive. We keep a round-trip delay under 50 ms. TUR performs explicitly well for QoS schemes, too. If we compare TUR versus a TDM protocol for QoS issues, we would expect to find that TUR is initially less proactive to delay requirement, but as loads increase TUR performs better than TDM schemes. This is because through simulation we observed that for a 16-node system, the percentage of empty slots for TDM was about 14 percent. On the other hand, the percentage of empty slots in TUR was merely 2.9 percent. TUR makes good use of the available bandwidth and maximizes network use.

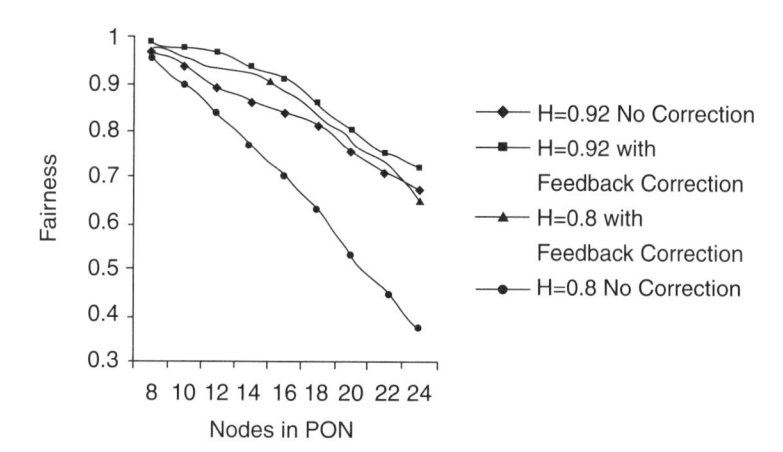

Figure 2-13 *Fairness with Correction for Various Bursty Schemes*

- H=0.92 No Correction
- H=0.92 with Feedback Correction
- H=0.8 with Feedback Correction
- H=0.8 No Correction

Comparison with Contemporary Schemes

The proposed TUR protocol for access PON is efficient and low cost and is built from very mature technologies. A number of solutions have been proposed that consider various protocols and technologies for implementation. Among them, three generic solutions that prompt interest for comparison are pure WDM, pure TDM, and IPACT.

Table 2-1 shows comparative performance with other PON schemes.

Table 2-1 *Comparative Performance with Other PON Schemes*

Feature	TUR	IPACT	Pure TDM	Pure WDM
Synchronization for end-user nodes	No	No	Yes	No
Cost	Low	Medium	High	Very high
End-user delay	Medium	Medium	Low	Very low
Throughput	High	Medium	Variable	Underutilized
Scalability	Good	Average	Bounded scalable	Very good (depends on the number of wavelengths)
Efficiency to burstiness in IP traffic	Yes	No	No	No
Enabling technology	Mature	Quite futuristic	Mature	Mature but expensive

In Gumaste and Chlamtac's "A Protocol to Implement IP Centric Communication in EPON," we see a very similar PON implementation with the CO polling different users and creating a pseudo-TDM scheme based on end-user feedback. This scheme, called IPACT, is an efficient and scalable solution. Figure 2-14 compares the delay requirements of frames at the end users using IPACT and TUR schemes respectively. As shown, the delay performance of TUR and IPACT is quite similar, but TUR has a lower delay requirement than IPACT for medium to high loads. Load is calculated based on the burstiness of the source and the ratio of generated load to total capacity of the system. Table 2-1 also compares the performance of TUR with pure WDM, pure TDM, and IPACT. To implement dynamic variations in traffic, we see TUR represents a good solution for PON. The technology is very mature, and implementation costs are relatively low. From the table, you can see that the proposed solution is quite in conformity with contemporary protocols.

Figure 2-14 *Performance of TUR for QoS Requirements*

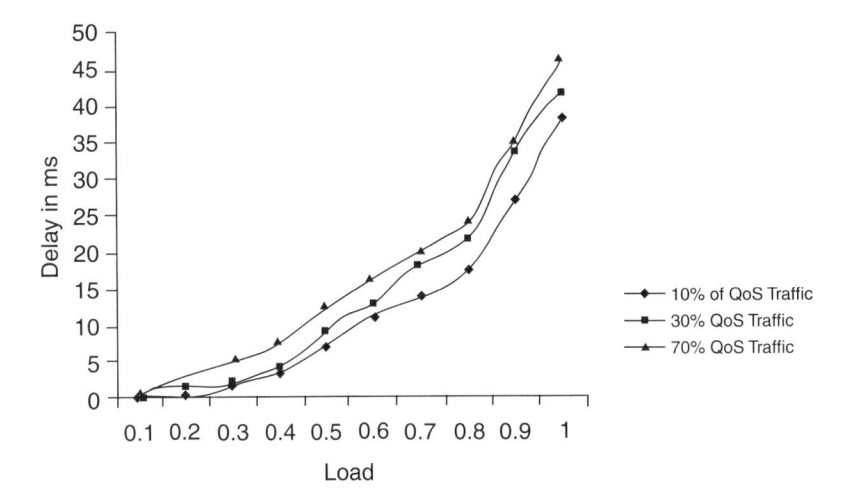

Summary

In this chapter we discussed the multiple PON technologies (including APON, GPON, and EPON) and elements that affect first mile access networks. This chapter also covered the benefits of EPON as compared to APON and GPON. This chapter focused on the aspects of cost, performance, and protocol issues and, therefore, outlined EPON as the key technology of the future in the first mile. In this chapter, we demonstrated the disruptive aspects and protocol behavior of Ethernet. The chapter concluded with an in-depth look at an academically proposed protocol called TUR.

Review Questions

1. Differentiate between the three types of PONs and explain the advantages of each.
2. Explain the advantage of EPON over GPON and BPON.
3. Compare BPON and GPON for a network in which there is a wide differentiation of services.
4. Explain how you would map voice and video in the same physical transmission layer using GFP.
5. What are the key management issues in EPON?
6. How is Ethernet a disruptive technology, and how can you increase its advantage as an incremental service?
7. With regards to cost, compare EPON with GPON and BPON for the same line rates and identical network sizes.
8. Calculate the efficiency of a PON protocol using TDM for bursty traffic. Develop an analytical formula for evaluating the efficiency of the protocol.
9. Based on Question 8, develop a correction protocol that enhances the efficiency of the system by using some sort of adaptation for bursty traffic.
10. Discuss the differences between TDM and STDM and show, with a numeric example, how STDM is more suited for IP-centric traffic.
11. For a 1:32 coupler, if the input power is 0 dB, calculate the power at each ONU that is 10 km away from the coupler. Assume 0.2 db fiber loss.
12. Discuss the far-near problem in PON.
13. Two ONUs are 10 km and 14 km away from an OLT. How much is the differential delay that the further ONU must cater to so that the far-near problem is solved and both ONUs appear equally far from the OLT?
14. Design a last mile network with 12 end users and 1 central office. The bandwidth requirement for each end user is 3, 7, 9, 11, 1, 2, 4, 16, 32, 10, 8, and 16 Mbps respectively. Label these users A to L. Users A, C, F, G are mobile and have to be connected through a wireless LAN. The remaining users are fixed. Optimize the network such that each ONU can cater to 32-Mbps traffic. How many ONUs are needed? From each ONU there is a distribution network to the end users. Optimize the placement of the ONU. Now assume arbitrary distances (min = 8 km, and max = 20 km) for each end user from the OLT. Develop a linear program to maximize performance of the system. Discuss the problems of BPON if used in this system. Now solve those problems faced when using BPON by using TUR in EPON. Compare numeric results.

References

Casson, et al., "Ethernet in the MAN-A Case Study in Disruptive Technology."

Chlamtac, Imrich and Ashwin Gumaste. "Bandwidth Management in Community Networks." Keynote address, International Workshop on Distributed Computing (IWDC). Calcutta, India: December 2003.

Green, Paul, "Fiber to the Home" Keynote address, Opticomm 2002 in association with ITCOM, Boston: 2002.

Gumaste, Ashwin and Imrich Chlamtac, "A Protocol to Implement IP Centric Communication in EPON," Proc of Optical Network Design and Modelling, ONDM, Budapest, Hungary: 2003.

Gumaste, Ashwin and Imrich Chlamtac, "A Protocol for EPON Implementation," Proc of ICC, Anchorage, Alaska: 2003.

Kramer, Glen, et al., "IPACT: A Dynamic Protocol for an Ethernet PON (EPON)," *IEEE Communications Magazine*, Volume 40: No. 2: February 2002.

Pesavento, Gary, "Ethernet Passive Optical Networks," *IEEE Communications Magazine*, Volume 40: No 2: February 2002.

Tolley, Bob, "Point-to-Point GBE and OAM for PON," Networld + Interop, Chicago: 2002.

CHAPTER 3

Enterprise Fiber Solution

Commercial optical fiber, a strand of which is as thin as a human hair, has an allowable bandwidth of 30,000 GHz and is an excellent medium for communication on account of its low loss and high throughput. The optical characteristics of fiber-optic transmission are discussed in Chapter 2, "Passive Optical Networks in the First Mile," and can also be found in the books *Optical Networks: A Practical Perspective*, Second Edition, and *DWDM Network Designs and Engineering Solutions*.

In the past decade or so, optical networking and fiber-optic technology for communication have experienced rapid growth. Most backbone networks are built around fiber as the basic medium to transfer data. Fiber is the transport medium choice on account of its excellent characteristics such as low loss, cost, reliability (life of a fiber is 40 to 50 years), availability (silica is easily available), and so on. The massive bandwidth offered by fiber cannot be matched by electronic technology, which actually is used at the two ends of a fiber link. In fact, to send data through an optical fiber, we need to modulate a laser with the data, and this technique is achieved by high-speed electronics. Considering the state of electronics today and the developments to come, there is an absolute mismatch of electronic technologies compared to the optical bandwidth available through an optical fiber. Even the fastest of commercially deployed systems do not function faster than at a 10-GHz rate, where the clock pulse is of the order of 100 pulses per second (ps). At such stringent timing levels, it is very difficult to accurately create an electronic system. Common technology, such as Combined Metal Oxide Semiconductor (CMOS) and N-type Metal Oxide Semiconductor (NMOS) gates, are replaced by microstrip technology beyond 1-GHz speed. Because there is an absolute dearth of low-cost interfacing technologies that can be used as possible interfaces between the fiber and electronic ports, there is a need for solutions that are low cost, yet able to inject the maximum amount of data into the fiber.

One way to maximize the use of bandwidth offered by the fiber is through wavelength-division multiplexing (WDM). By definition, WDM is a technology that offers the use of optical fiber as a multilane highway, such that each lane can be used as a platform/carrier for data transport. Each lane of this express highway is a wavelength, and data is modulated onto this wavelength. In other words, using multiple wavelengths and modulating different data streams on each of these wavelengths can maximize the fiber use to a great extent. The spacing between two wavelengths, often termed the *interwavelength distance* or just *spacing*, calculated in nanometers differentiates multiple WDM systems (such as coarse and dense WDM [CWDM and DWDM, respectively]). Theoretically, to maximize bandwidth the channels (wavelengths) should be as close to each other as possible, to maximize the number of channels that can be put in the fiber. The idea is to have each channel at a slow bit speed. However, because there are a large number of channels, the aggregate speed achieved through the fiber is high. To pack more channels in a fiber, there is a requirement to process the electrical data onto the optical medium. We need sharp lasers that modulate data streams such that their profiles are well defined. Such lasers need to have characteristics such as low chirp, high power, and sometimes even tunability. Moreover, optical components for dispersion compensation are very expensive for signals that are close to each other. A much easier and more inexpensive way to multiplex multiple signals in a fiber in the wavelength domain is by deploying CWDM. As a note, it's important to observe that closely packed channels have an effect on each other like cross-phase modulation, which distorts signals because of the proximity of the surrounding channels. For more detail, refer to Chapter 5 of *DWDM Network Designs and Engineering Solutions*.

Coarse Wavelength-Division Multiplexing

CWDM per se is multiplexing multiple optical signals in the same fiber by using the property of wavelength domain diversity. We have multiple signals each at some distance (in terms of wavelength) from each other, and these signals are multiplexed together in the same fiber. Unlike in DWDM where multiple signals are packed with each signal at a relatively short spacing from its adjacent neighbors, in CWDM the signals have a larger spacing (relative to DWDM), hence relaxing the optical requirements.

In a commercial DWDM system, the channel spacing as defined by the International Telecommunications Union (ITU) is 0.8 nm, or 1.6 nm typical, with some implementations of 0.4 nm also seen. Because so many channels are not needed in access networking and to reduce equipment cost, however, we deploy CWDM networking where the channel spacing is large—for instance, from 2 nm to 20 nm between channels (see Figure 3-1). The ITU standard for CWDM is 20-nm spacing between two consecutive channels.

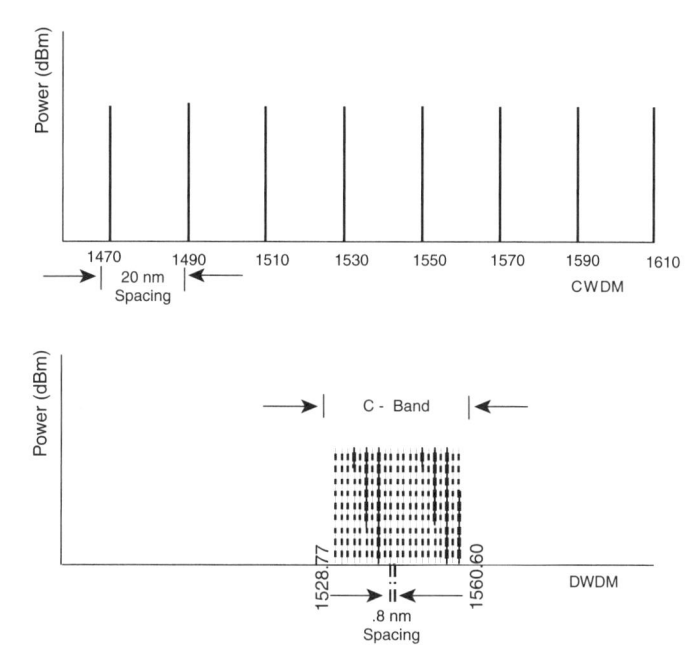

Figure 3-1 *CWDM and DWDM: Notice the Spacing Between Channels*

Business Case for CWDM and Access Networks

CWDM implementation, although low cost, does not yet justify deployment to the end user on account of its magnitude in terms of bandwidth offered. CWDM, on the other hand, is a pure metro access solution. This means that the beneficiaries of such a solution are enterprises and customers who may subsequently pass on the data after de-aggregation to other consumers. Typically, CWDM networks are installed in ring topologies with wavelengths dropping to individual nodes. The main business case validation for CWDM networks is that multiple wavelengths can be used in the same fiber at low costs; hence, these networks can reach a higher number of users at longer distances but are still low cost. Typically for enterprises, a CWDM network makes a good value proposition. A service provider can provision services to the enterprise by keeping a network element at the premises of the enterprise. CWDM solutions are typically one-fifth or less the cost of DWDM solutions and have multiple times capacity as compared to regular solutions such as Synchronous Optical Network/Synchronous Digital Hierarchy (SONET/SDH). The capability to use wavelengths also provides protocol flexibility, and the user can run anything (X) over WDM. This ensures that a plethora of protocols can be deployed, such as SONET, ATM, IP, Fibre Channel (storage protocols), and so on.

CWDM Network Configuration

Typical CWDM networks are configured in two main topologies: rings and point to point. However, some implementations of mesh can also be obtained if required. Most typical CWDM applications are ring topologies due to the benefits such as protection and configuration that are obtained in ring networks. Moreover, ring networks, besides their resilience to failures, are also standard topologies in metro cores. In addition, ring topologies are cost efficient (on account of the minimum number of fibers needed to create a fully connected logical topology).

Figure 3-2 shows a typical metro access CWDM network.

Figure 3-2 *Typical CWDM Ring with 8 Mux/Demux and Single-Channel Optical Add/Drop Multiplexer (OADM)*

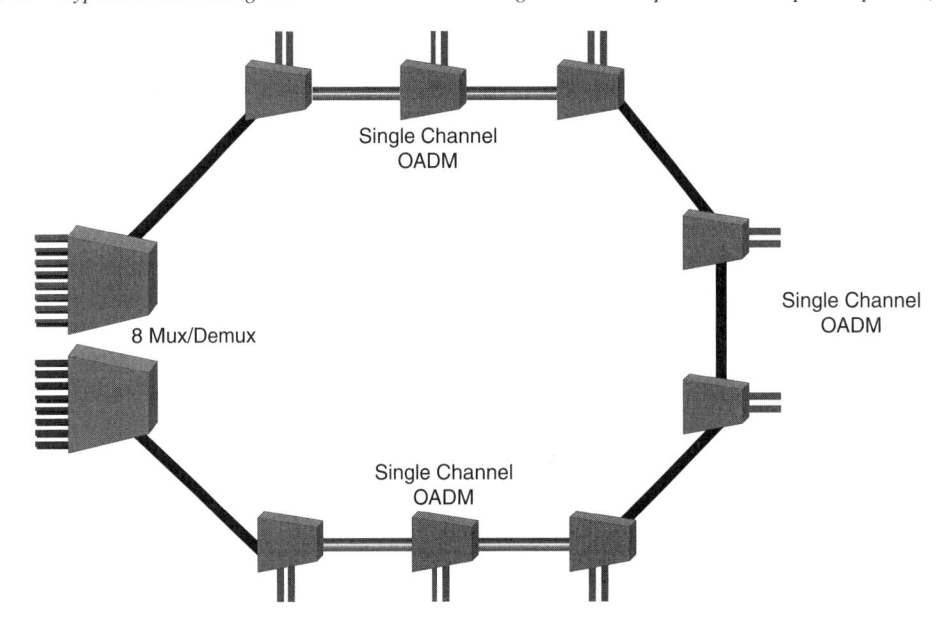

One or two of the nodes of the network are connected to the core network through central office (CO) equipment. The other nodes in the ring are typical drop filters, which drop one wavelength from the CWDM band. These filters, along with the corresponding added ports, are called OADMs. Generally, there is no need for optical amplification here because the distances are around 20 to 60 km. Besides, the bit rates here per wavelength are typically 1 Gbps or sometimes even less, but can go up to 2.5 GHz if needed. The other nodes (filters) are at enterprise locations or at hubs, which then have an optical filter and a compatible CWDM transponder. A transponder is a device that detects the optical network signal and transforms it to a client signal (often, in this case, on the same wavelength). More often than not, a transponder is a regenerative device that detects, amplifies, and then regenerates the optical signal. A figure of merit when designing CWDM networks is the bit error rate (BER) of a signal. By definition, this means the number of corrupted bits during transmission. Mathematically it means the number of uncorrupted bits that can be transmitted successfully before the arrival of a corrupted bit (during transmission) at the receiver. BER gives an idea of network performance. It is difficult to calculate BER, so the optical signal to noise ratio (OSNR) of the signal is calculated and a formula is available to translate OSNR to the Q factor and then the Q factor to the BER. The Q factor is a qualitative merit of an optical signal with respect to noise (*DWDM Network Designs and Engineering Solutions*).

Technicalities of CWDM

Service providers are currently evaluating architectures that enable them to offer efficient metro services and last mile transport services, typically to their end customers. The plug-and-play feature of Ethernet, along with its low cost, enables service providers to offer new services rapidly and economically. Ethernet's capability to provide seamless integration across LANs, MANs, and WANs is motivating enterprises and service providers to migrate from traditional circuit-based networks to an Ethernet or packet-based network. Due to low costs and commercial availability, CWDM systems are becoming widely accepted as an important transport architecture.

Bandwidth on an optical fiber can be increased by transmitting data faster or multiplexing wavelengths into a single fiber, what we call WDM. WDM systems are realized by using a multiplexer (mux) to combine signals at the transmitting end and using a demultiplexer (demux) at the receiving end to separate the signals. The channel spacing between the signals plays an important role in determining how many signals can be multiplexed, which in turn determines the total capacity of the WDM system.

The main distinction between DWDM and CWDM is the channel spacing between the individual wavelengths. DWDM systems use channel separation of 200 GHz (1.6 nm), 100 GHz (.8 nm), or 50 GHz (.4 nm) currently, and this is defined by the ITU (ITU G.694.2). Channel spacing between CWDM systems is typically 20 nm. The distributed feedback (DFB) lasers used as sources in DWDM systems are cooled to stabilize the wavelength from drifting outside the pass-band of the mux and demux filters as the temperature fluctuates. CWDM systems use DFB lasers that are "not cooled" due to its wide channel spacing with operating temperatures of 0 to 80 degrees centigrade and wavelength drifts of about 6 nm over this range. Because the spacing is 20 nm apart, the tolerance to manufacture CWDM components is *relaxed*, and hence CWDM is cheaper to manufacture than DWDM systems. The cost of the CWDM laser is also reduced due to the lack of a cooling requirement for thermal stability.

CWDM systems offer advantages over DWDM systems for applications that require a channel count on the order of fewer than eight channels in the band. The benefits of CWDM (as compared to DWDM systems) include low cost, small size, and low power requirements.

CWDM Network Elements

To understand how to implement CWDM in access networks, we must cover a certain set of technologies that enable us to describe the functionality of CWDM systems. With this in mind, this section covers certain component technologies that basically differentiate CWDM systems from their DWDM counterparts. Among these are low-cost vertical cavity surface emitting lasers (VCSELs); gain-flattened, thin film filters; and so on.

Vertical Cavity Surface Emitting Laser

A VCSEL is a semiconductor laser diode that emits light perpendicular to the p-n junction (unlike semiconductor lasers, which emit light in the plane of the p-n junction). VCSELs can be integrated as onboard components with other electronic components without prepackaging. For very short reach (VSR) applications at 850-nm as well as 1310-nm band wavelengths, VCSELs have become the preferred choice at even gigabit speeds.

A VCSEL is made of many specialized layers (like edge-emitting lasers). The main parts of a VCSEL are the active region and the mirrors. A sandwich of active region (between the mirrors) is created by stacking the subcomponents vertically on top of each other as shown in Figure 3-3. For fiber-optic communication, VCSEL uses indium gallium arsenide (InGaAs) for wavelengths of 850 nm and 1310 nm.

Figure 3-3 *Vertical Cavity Surface Emitting Laser*

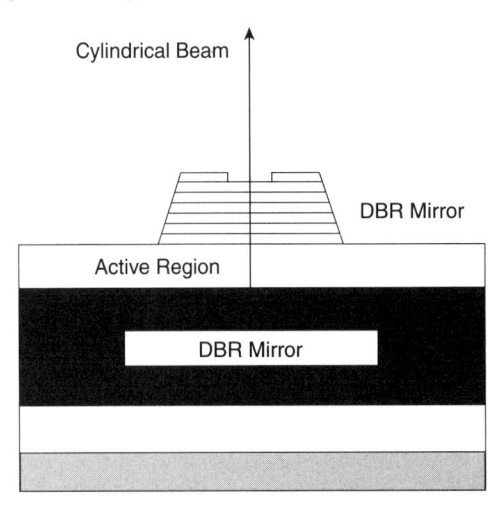

When a small current is applied across the device, light is created in the active region of the laser (same operation as semiconductor lasers). This light is reflected back and forth between the mirrors, while a small percentage of the light "leaks" through the mirror to form the laser beam. These lasers are very efficient and have low voltage requirements due to their high gain and the small size of the VCSEL cavity structures. VCSELs are capable of fast direct modulation speeds (up to 10 Gbps), which means no additional external components, such as modulators, are required. Currently VCSELs are used in wavelengths of 1310 and 1550 nm (occasionally).

VCSEL has a number of advantages over conventional lasers:

- High performance, low cost, and smaller size and power.
- The structure can be integrated into a two–dimensional array.
- Easier manufacturing and packaging.
- Efficient fiber coupling.

Line Width of a Laser

The line width of a laser is the power spread of the light that is emitted by a laser with respect to wavelength. The narrower the line width, the more fine the emitted light. In other words, light from a laser with a narrow line width is confined to a smaller spectral spread (initially) than the light from a laser with a larger line width.

Ideally, the line width of a laser should be as narrow as possible to prevent the data spectrum from overlapping the spectrums of adjoining channels. Generally a laser spectrum has a dominant central frequency with multiple smaller sideband frequencies.

Figure 3-4 shows the line width of a laser.

Figure 3-4 *Line Width of a Laser*

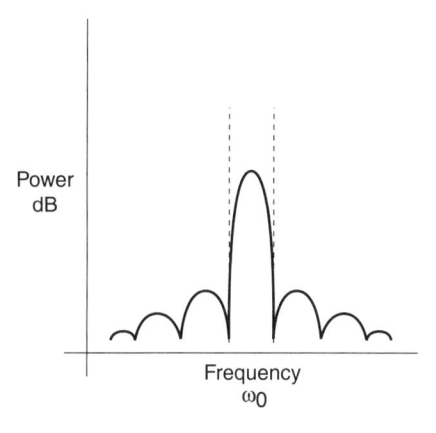

The typical line width for CWDM systems is in the range of hundreds of megahertz. Laser line width is a major issue in CWDM systems because the greater the line width, the greater the cross talk with adjacent channels and the greater the nonlinear effects.

Thin Film Filters

To recover the CWDM signal or to differentiate from a composite CWDM signal, we need optical filters that filter out multiple wavelengths and allow a single wavelength (a single waveband) to pass through from the CWDM band. Of course for basic demultiplexing of an 8-channel or 16-channel CWDM signal, we need other technology—for example, arrayed waveguide gratings. Thin film filters are similar to cavity filters in the sense that a resonant cavity selects the wavelengths that are allowed to traverse through. The cavity is formed by the thin films with interfaces that act as reflectors. The wavelength or group of wavelengths that are selected depends upon the length of this cavity. Thin film filters are commercially available and perform functions of optical band-pass filtering as well as single wavelength filtering. Multi-cavity cascaded thin film filters have excellent response in the sense that due to cascade, the filtered band approximates to more and more rectangular characteristics (ideal filter). Thin film filters are typically made of quarter-wave-thick layers ($\lambda/4n$) of alternating high and low refractive indices. The principle is that of coalescing multiple layers forming multiple resonant cavities one on top of the other. There are three main regions in a thin film filter:

- The spacer
- The transition layer
- The reflective stack

The spacer consists of multiple quarter-wave layers of either a high-refractive or low-refractive index but not both. It is between the reflective ends of the formed cavity. The transition layer is composed of a single quarter-wave layer and its function is to produce flat-top filtering. The reflective stack is made of alternating high and low indices and forms a dielectric mirror.

Standards for CWDM Deployment

The ITU G.694.2 recommendation provides a grid of wavelengths (20-nm channel spacing) for target distances of up to 50 km on single-mode fibers. The CWDM grid explicitly defines only 18 wavelengths, from 1270 nm to 1610 nm. Regular fibers with water attenuation peak will lose 3 or 4 channels due to attenuation caused by hydroxyl ions (–OH). New fibers such as All Wave and SMF-28e, which do not have the water attenuation peak, can use the entire band of fiber (see Figure 3-5).

Figure 3-5 *Fiber Spectrum Attenuation Profile (Source:* IEEE Spectrum, *August 2002)*

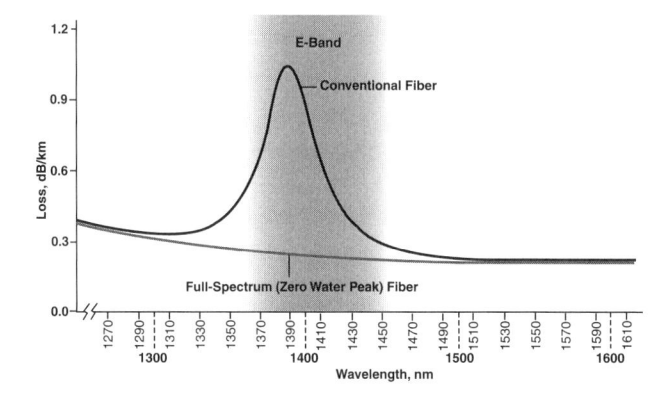

CWDM specifications are based on the fact that there is relaxed tolerance on spectral width and thermal drift of CWDM lasers compared with DWDM lasers. Wider channel spacing specified by the standard allows the use of cheaper optical components, which in turn reduces optical system costs for metro applications.

A CWDM grid under consideration is as follows:

1270 nm, 1290 nm, 1310 nm, 1330 nm, 1350 nm, 1370 nm, 1390 nm, 1410 nm, 1430 nm, 1450 nm, 1470 nm, 1490 nm, 1510 nm, 1530 nm, 1550 nm, 1570 nm, 1590 nm, 1610 nm

The initial implementation of CWDM systems is in the S + C + L band (short + conventional + long) and defines eight wavelengths starting from 1470 nm. In the future, CWDM will be able to support up to 18 wavelengths.

CWDM Benefits

Due to its low cost, there is a tremendous interest in CWDM technology. We briefly summarize some of the benefits achieved by CWDM networking here:

- **Comparatively cheaper hardware**—Manufacturing tolerances required to build CWDM lasers and transceivers are not as tight (as compared to DWDM components). The CWDM filters are also less expensive to manufacture due to the wideband (20 nm) nature of CWDM filters. This leads to filters with approximate bands working as well.
- **Low power requirements**—Typical CWDM lasers are not temperature stabilized and do not require feedback. The temperature stabilization circuits and the use of cooled lasers (cooled DFB lasers) increase the power requirements for DWDM systems. Another CWDM benefit is its smaller size due to the use of uncooled lasers, and CWDM has very good reliability compared to DWDM.

The biggest disadvantage is scalability and lack of ability to amplify all the CWDM channels. Usually CWDM systems are not used in scenarios that require amplification. Currently, erbium-doped fiber amplifiers (EDFAs) work only in the C band. Hence, EDFAs cannot handle the entire CWDM band. This is one of the primary reasons why there are no amplifiers for CWDM.

CWDM Services

The typical CWDM solution includes eight-wavelength mux/demux, single-wavelength OADM, and CWDM gigabit interface converters (GBICs), allowing easy scalable Gigabit Ethernet services over a CWDM architecture. The main applications addressed by CWDM systems include the following:

- Campus intra- and interconnections via point-to-point CWDM links
- Serving business customers via CWDM access rings (small to medium metro/customer premises equipment rings)

- Transport service between service providers
- Points of presence (POPs) in a small to medium metro ring

Ethernet services delivered using the CWDM transport platform enable customers to alleviate the bandwidth bottleneck between their sites where fiber is not available. Services offered by this CWDM architecture include the following:

- Full-wire-speed Ethernet/SONET point to point over CWDM (protected)
- Full-wire-speed Ethernet/SONET point to point over CWDM (unprotected)
- Full-wire-speed CWDM ring, with add/drops at multiple locations (unprotected)
- Full-wire-speed CWDM ring, with add/drops at multiple locations (protected)

CWDM Network Protection

At the optical layer, a lightpath (an all-optical, end-to-end path on a given wavelength) is set up using a control mechanism that involves the setting up of switches and transponders at the ingress, the egress, and the intermediate nodes and equipment in addition to the fibers to set up the lightpath. Failure of equipment, fiber, or nodes can cause the lightpath to be disrupted, causing huge revenue losses. Protection of lightpaths is a means whereby failure of fiber or equipment can be overcome by other means. Protection essentially means adding some degree of redundancy or diversity to the network. The excess redundant portion of the network provides network capacity in the event of a failure or fault. This excess capacity is called *protection bandwidth* and may be in terms of multiple wavelengths or even redundant fibers.

In principle, protection is a very fast phenomenon, such that the failed lightpath is switched onto the excess allocated capacity in the shortest possible interval of time. The interval of time required to protect a failed network part is often critical and for carrier-class networks needs to be less than 50 ms. In an optical network, there may be various different kinds of failures. The most common failure is that of equipment. Equipment failure accounts for almost 70 percent of all failures and results from various factors, such as aging, malfunction, and human error.

Protecting networks from equipment or subsystem failures is a hard task, if not an impossible one. The only way to protect from equipment failure is to deploy redundant equipment, which then serves as protection equipment, and, in the even to failure, switch automatically from the normal equipment to the redundant gear. There are algorithms that actually describe and facilitate the way in which the signal is transferred from the normal equipment to the redundant equipment as well as when this change is to occur. Equipment failure can also occur when the entire node fails. This may happen because of power outages or even human error. Nodal failure is quite difficult to protect against, and unless the failure is partial the node is usually down until rectification exercises are carried out.

The next common failure after equipment failure is fiber failure, more commonly known as *fiber cut*. Fiber cuts are failures caused by the damage rendered to operational fibers, either due to physical cuts or severe bending (thus increasing their losses to unbearable values). Fiber cuts can be taken care of by using redundant fibers along with "work" or normal-use fibers and switching signals from the work to the redundant fibers. Protection in CWDM networks can be classified inherently into two types: line- and path-level protection:

- Line protection means protecting the entire fiber or the entire band of CWDM channels in the event of a fiber cut or failure (see Figure 3-6).

Figure 3-6 *Line (1:1) Protected Link (Note Both Transmitter and Receiver Have to Switch)*

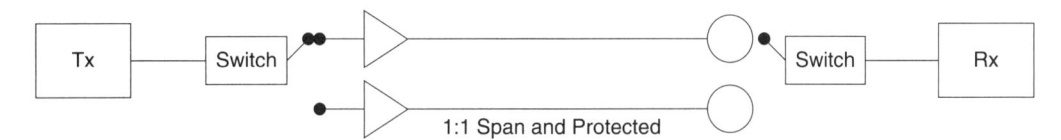

- Path protection means protecting just the single lightpath (on some arbitrary failed wavelength) that has failed (see Figure 3-7). Path protection is thus more specific and, hence, more difficult to implement, whereas line protection is more generic and easier to implement. For most of the discussion on protection and its mechanisms, we consider only fiber cuts as the predominant failure in optical WDM networks because of the simple redundant techniques present to solve the protection issue for equipment failures.

Figure 3-7 *Path (1+1) Protected Link (Note That Only the Receiver Has to Switch to the Protection Signal)*

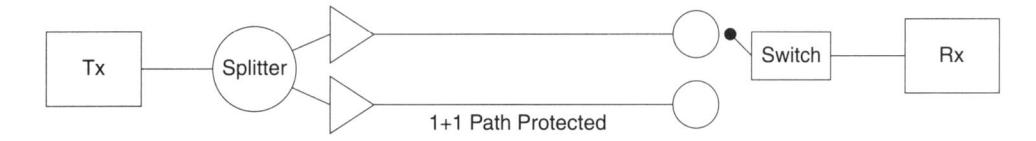

In conventional SONET networks line and path protection has been incorporated into the ring topology by two distinct schemes called *unidirectional path-switched ring (UPSR)* and *bidirectional line-switched ring (BLSR)*. The same scheme can be extended to WDM networks. For path-switched protection in point-to-point or mesh topologies or even WDM rings, the protection is known as *1+1 protection*, whereas for line-switched protection the scheme is called *1:1 protection*.

For path-switched 1+1 protection, essentially the transmitter or ingress node transmits the signal (CWDM lightpath on a specific wavelength [λ]) into the work and the redundant path simultaneously. At the receiver or egress node, the receiver chooses the signal from either the work or the redundant path depending on the signal quality. So essentially this kind of scheme is very easy to manage as the changeover decision (for choosing either work or redundant fiber) is taken only by the receiver section and not by multiple sections. Note here that it is best to have the work and redundant paths on different fibers to create physical diversity.

On the other hand, in the 1:1, or line-switched, scheme as in Figure 3-6, the signal is sent only in the work section or work channel, while one redundant channel serves as a backup to multiple geographically diverse work channels. In the event of a failure, both sender and receiver need to coordinate with one another and switch the signal in the failed section to the redundant channel. This involves dual switching at both the transmitter and receiver sections as shown in Figure 3-7.

Naturally the 1+1 protection format is a much easier way to guarantee fast protection (restoration time is shorter), but the cost involved is often more because more bandwidth (resource) is needed to facilitate it. 1:1 protection, on the other hand, is not that fast in terms of restoration time (time needed to restore a failed link or lightpath), but the cost involved is often quite less, although the signaling procedure is often cumbersome on account of the fact that full-duplex signaling is involved.

However, 1:1 line protection has the added advantage of optimizing the protection bandwidth among many work paths. This kind of scheme is also referred to as *1:N protection*, whereby one redundant path can provide for protection of a single fault in any of the N fibers (of course, one fault at a time only). See Figure 3-8. When the protection algorithm is 1:1—that is, for every channel there is a given resource to ensure protection—the scheme is also called *dedicated protection* or, more technically, *dedicated 1:1 protection*. If the protection algorithm shares many work channels for a single redundant channel, this kind of scheme is called *shared protection* (see Figure 3-8). Shared protection envisions one redundant channel for N work channels, assuming that only one of the N work channels will be cut at any given time. Shared path protection is also an interesting concept for according individual lightpaths in CWDM metro access rings. Ongoing research for optimizing the amount of bandwidth allotted for protection compared to that allotted for work in a 1:N protection network is being conducted. This ratio is called the *share to work (S/W) ratio*.

Figure 3-8 *1:N Protection (Shared Protection)*

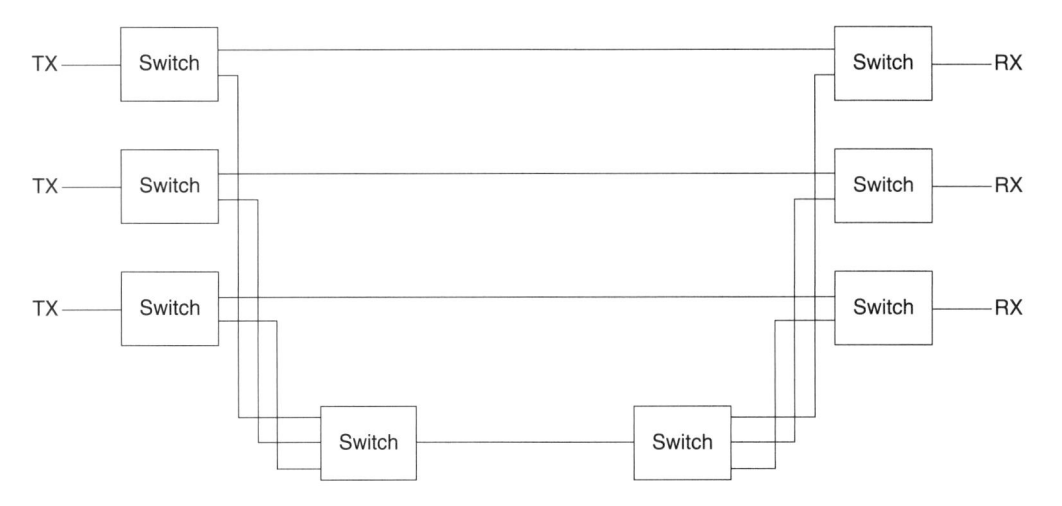

Protection in Ethernet over CWDM Networks

CWDM passive components do not provide any forms of optical protection. Protection is achieved by using Spanning Tree Protocol (STP), EtherChannel, and unidirectional link detection (UDLD) for Ethernet devices. Layer 3 protocols can also provide Layer 3-based protection (equal-cost path).

Protection Using STP and UDLD

The IEEE 802.1w Rapid STP (RSTP) provides rapid convergence of STP by assigning port roles and by determining the active topology. RSTP natively includes most of the Cisco proprietary enhancements to the 802.1D spanning tree, such as BackboneFast, UplinkFast, and PortFast. As result of link failure, spanning-tree reconfiguration takes around 1 second in a well-designed network, depending on load, network diameter, and configuration, which still is a major improvement over the 50 seconds for its predecessor IEEE 802.1D.

Multiple STP (MSTP) uses the 802.1w engine to provide rapid convergence and enables virtual LANs (VLANs) to be grouped into a spanning-tree instance, with each instance having a spanning-tree topology independent of other spanning-tree instances. This architecture provides multiple forwarding paths for data traffic, enables load balancing, and reduces the number of spanning-tree instances required to support a large number of VLANs.

UDLD is a Layer 2 protocol that works with Layer 1 mechanisms such as autonegotiation to determine the physical status of the link. At Layer 1, autonegotiation handles physical signaling and fault detection. When both autonegotiation and UDLD are enabled, Layer 1 and Layer 2 detection features can work together to prevent physical and logical unidirectional connections and malfunctioning of other protocols.

Protection Using EtherChannel and UDLD

EtherChannel is a feature that allows up to eight physical links to be bundled into a single logical link that provides the aggregated bandwidth of up to eight physical links. The IEEE 802.3ad protocol facilitates EtherChannel bundling of ports with similar physical interface, speed, and duplex settings. Moreover, 802.3ad provides the logic for load balancing across logical links based on source and destination MAC addresses.

The Cisco proprietary Port Aggregation Protocol (PAgP) facilitates the ports of similar physical interfaces, speed, and duplex to be bundled/unbundled from the EtherChannel group. Moreover, PAgP provides the logic for load balancing across logical links based on source and destination MAC addresses.

CWDM Network Architecture

The CWDM GBIC solution supports any point-to-point, ring, and mesh architecture with up to an 18 maximum number of wavelengths. A typical eight-wavelength protected CWDM architecture (Gigabit Ethernet) is shown in Figure 3-9. This topology uses two Cisco 7600s (Layer 2/3 switch with optical interface) at the hub end and Cisco 3500s (Layer 2/3 switch with optical interface—as smaller version) as satellite nodes. Each Cisco 3550 is connected by a CWDM GBIC to the east and the west side to achieve protection. Protection is achieved by either Ether-Channel or Layer 3 (Open Shortest Path First [OSPF]).

Figure 3-9 *Protected 8-Channel CWDM Network*

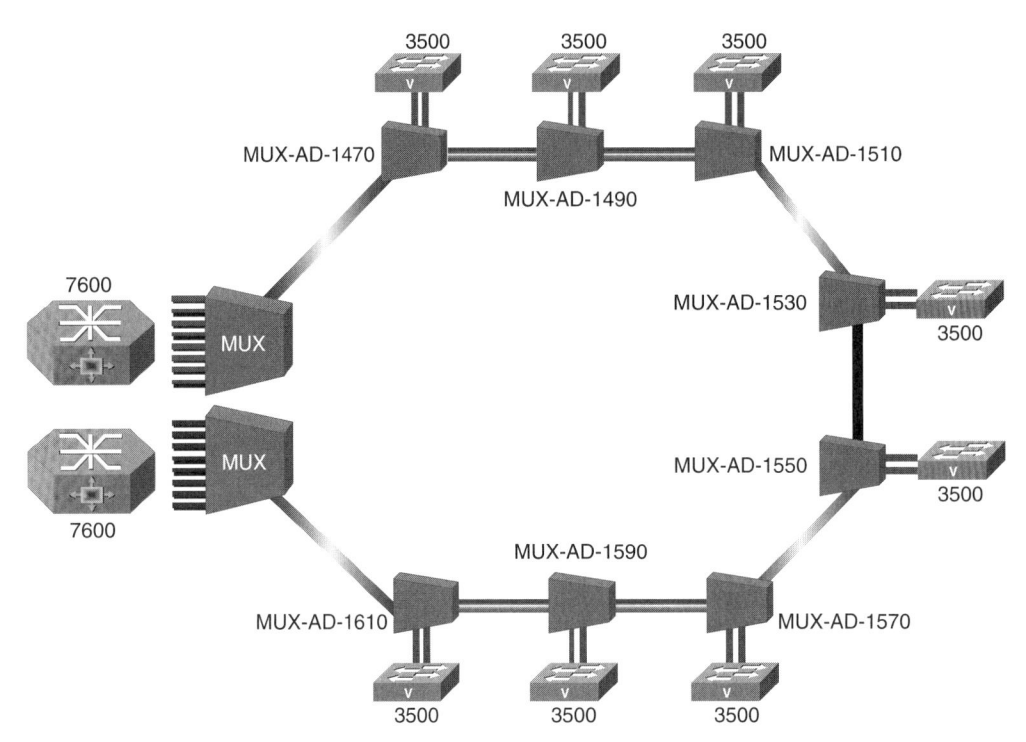

Mux/Demux and OADM Characterization

The mux/demux and OADMs are characterized mainly by the following parameters:

- **Nominal wavelength**—The wavelength for which the system is designed to operate. With CWDM, there are multiple operating wavelengths. (CWDM wavelengths: 1470 nm, 1490 nm, 1510 nm, 1530 nm, 1550 nm, 1570 nm, 1590 nm, 1610 nm.)

- **Peak wavelength**—The wavelength measured at the output of an optical device with the minimum loss between the input and output (see Figure 3-10). The x-axis shows the wavelength in nm, whereas the y-axis shows the power.

Figure 3-10 *Peak Wavelength in CWDM*

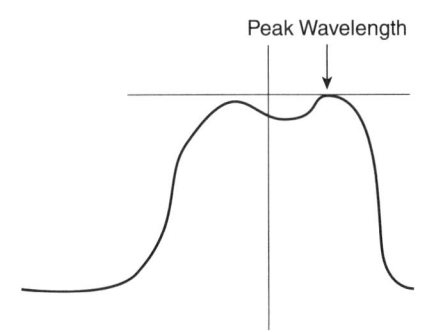

- **Bandwidth**—The width (pass-band) of a filter at a specified level below the maximum peak.
- **Mean center wavelength**—The mean value of the wavelength measured at the edges of the band. (Band edges are the wavelengths below the 1-dB peak). In ideal cases, the nominal wavelength and mean center wavelength should be the same.
- **Pass-band**—The reference band on which measurements are taken and good communication occurs. It refers to the useable operating bandwidth and is always in reference to ITU-defined wavelengths. If a pass-band is expressed as P (in nm), the pass-band equals mean central wavelength + P and mean central wavelength − P. (See Figure 3-11.)

Figure 3-11 *Pass-Band*

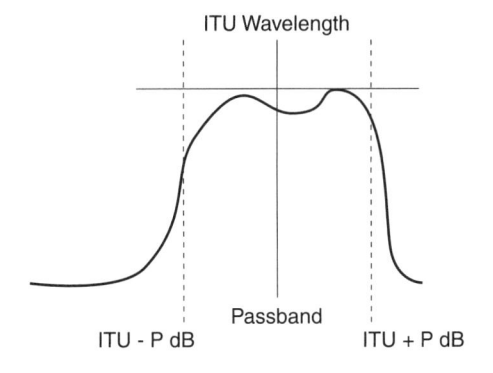

In Table 3-1, we list the typical characteristics of a transponder used in CWDM.

Table 3-1 *Typical Transponder Characteristics*

Characteristic	Minimum	Typical	Maximum
Transmit power	+1 dBm	+3 dBm	+5 dBm
Receiver sensitivity	–31 dBm	–28 dBm	–7 dBm

The transponder characteristics are as follows:

- **Minimum transmit power**—The minimum power level in dBm you can expect during the lifetime of the laser
- **Typical power**—The expected (average) power during the operational phase of the laser
- **Maximum power**—The best-case power that is available from the laser (never used in power-budget calculations)
- **Receiver sensitivity**—The range of input power required at the receiver end to support the BER specified
- **Insertion loss**—The worst-case optical power loss over the pass-band
- **Isolation**—The difference in power (dB) between the insertion loss over the pass-band and the point where the band intersects the adjacent band at λ_{ITU} +/– bandwidth / 2, where λ_{ITU} means a wavelength that is ITU specified

Design Rules for CWDM Networks in Metro Access

As a general guideline, when implementing a CWDM network, as in any optical network, we measure the optical power at the receiver and the transmitter end. We also ensure that the transmitter power is sufficient to support the required transmission distance. We also ensure that the power at the receiver is within the specification of the receiver—that is, conforming to receiver sensitivity (greater than –31 dB and less than –7 dB). If the power is above the specified range (greater than –7 dB), it needs to be attenuated or the receiver will be damaged due to excess photons hitting the p-n junction at the receiver. If the power is under the specified minimum level (less than –31 dB), the receiver cannot distinguish between 1 and 0 and bit errors will increase. Therefore, to meet the specified BER of the transmission system, keep the power levels within the specified range at the receiver end. CWDM networks do not support amplifiers; therefore, adequate power levels at the transmitter end are required to drive the distances. (Currently, there is no pre- or postamplifier support.)

Power-Budget Calculations

As the signal passes through the transmission medium and other passive components, the signal undergoes attenuation. The attenuation due to the fiber can be calculated using the attenuation coefficient factor (in dB/km) published by the fiber manufacturer. (The typical attenuation coefficient factor [$\propto$] for single-mode fiber [SMF] is .25 dB/km to .3 dB/km depending on the wavelength. Loss due to the add/drop units can be obtained from the data sheets of each component.) In an actual network, we also need to consider losses due to connectors (typically less than .3 dB per connector) and splices.

Raw power budget = minimum transmitter power − minimum receiver sensitivity

For example, from Table 3-1:

Minimum transmitter power = +1 dB
Minimum receiver sensitivity = −31 dB
Raw power budget = +1 − (−31) = 32 dB

Maximum Distance Transmitted in a Point-To-Point System

To determine the maximum distance we can transmit using a given fiber in a point-to-point system, we use the following formula:

Maximum Distance (theoretical) (km) =

$$\frac{raw\_power\_budget(db)}{attenuation\_factor\_of\_the\_given\_fiber(dB/km)}$$

In our case, we obtain the following:

32 / .3 = 106.66 km
(Assume the attenuation coefficient for SMF fiber = .3 dB/km.)

In this case, we did not consider other losses, such as splices and connectors. Usually 2 to 3 dB is deducted for those factors, including aging. This 2 to 3 dB is typically called *design margin*. It is important to have a tolerance margin in a given design because we might encounter unanticipated issues.

Hence the new formula for maximum distance is this

$$\frac{raw\_power\_budget(dB) - Margin(dB)}{attenuation\_factor\_of\_the\_given\_fiber(dB/km)}$$

So in our case

$$\frac{(32dB - 3dB)}{0.3} = 99.66 km$$

This can be used as a general guideline; usually the margin varies with the type of fiber and the entire optical layer physical setup (in its entirety, called the *fiber plant*). For older fiber plants, the margin and attenuation factors may differ. In the field, one may see attenuation factors ranging from .23 dB/km to .35 dB/km for SMF fibers. (As a best practice, the measurements are taken in the field rather than estimated.)

Maximum Distance Transmitted in a Ring Using OADMs

The calculations for determining the maximum distance in a ring design using OADMs are similar to the calculations for point-to-point systems. Here additional losses due to passthrough, add/loss, and drop/loss attributed to OADM devices must be considered. Each wavelength should be treated separately, and we should verify that it supports the transmission distance. A quick approximation is to verify the worst-case transmission scenario.

Transmission distance (km) =

$$\frac{raw\_power\_budget(db) - add\_loss(dB) - pass\_loss(dB) - drop\_loss(dB)}{attenuation\_factor\_of\_the\_given\_fiber(dB/km)}$$

In an 8-node ring with add loss (max) = 4 dB (8-λ mux), pass loss (max) = 2.0 dB, drop loss (max) = 2.3 dB (assume single-wavelength OADMs and one 8-wavelength OADM at the hub end):

$$\frac{32 - 4 - (7 \times 2) - 2.3}{0.3} = 39 km$$

OSNR and dispersion calculations are not necessary in CWDM designs. Because no optical amplifiers are used in CWDM networks, OSNR does not play a major role. Dispersion is not an issue for 1-Gigabit Ethernet speeds for the short distances for which CWDM networks are proposed.

CWDM Design Examples

When designing CWDM networks, we need to know the number of channels required in the network, the distance (node-to-node span distance and total circumference), and the fiber type. From an optical parametric design point, we need to know the power budget (that is, the transmitter power and the receiver sensitivity). In these examples, we do not consider the effects of noise or fiber nonlinearities and they are not necessary to include in CWDM system designs.

Example 1

Corporation ABC wants to connect two sites, 65 km apart, using CWDM. They want to use 1-Gigabit Ethernet connections but want to grow to four or more Gigabit Ethernet connections later. (Use the data sheet for CWDM characteristics that an equipment vendor provides, or perhaps available on company websites.) Assume that the customer uses SMF with .3 dB/km loss. (Use Table 3-1 for transmitter characteristics.)

Example 1 Solution

Initially, consider the single-channel case; here we do not require mux and demux. When the customer grows to four or more Gigabit Ethernet connections, we need to add mux and demux, which will add other losses. Assume the customer adds an eight-channel OADM later.

Case 1

We need to first estimate the power budget and see whether this supports the distance.

Figure 3-12 shows the design for a point-to-point link for a single channel.

Figure 3-12 *Point-To-Point Link for a Single Channel*

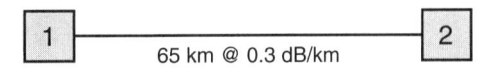

```
[1]---------------------------[2]
      65 km @ 0.3 dB/km
```

There is no add/drop or passthrough loss because it is a point-to-point link.
Minimum transmit power = 1 dB
Receiver sensitivity = -31 dB
Raw power budget = minimum transmitter power − minimum receiver sensitivity =
  1 − (−31) = 32 dB
Distance between the two sites = 65 km
Attenuation coefficient for the fiber is given as .3 dB/km
Hence total loss in the link = 65 km * .3 dB/km = 19.5 dB
Power budget > all the losses (add, drop, passthrough, and link losses)
Or power budget − all the losses is greater than or equal to 0:
32 − 19.5 = 12.5 dB

In this example, the power budget (32 dB) is greater than the total link loss (19.5 dB); therefore, the system is within the specifications.

We also need to ensure that the power received is well within the receiver sensitivity range. If the power is greater than the receiver sensitivity, we need to attenuate the signal. The required attenuation can be calculated as follows. (This is a best estimate. Due to changes in environment, the attenuation required may differ. *Always* use a power meter to check that the power at the receiver end is well within the receiver's dynamic range.)

The power at the receiver end can be calculated as follows:

Maximum transmit power – all the losses =
5 dB – link loss = 5 dB – 19.5 dB = -14.5 dB

This is well within the receiver sensitivity range (–7 dB to –31 dB); therefore, no attenuation is needed.

Case 2

Now the customer wants to add more Gigabit Ethernet connections to the network. In this example, we need to add a mux and a demux and consider the losses associated with those passive devices. (Assume add loss = –4 dB and drop loss = –5 dB for an eight-channel mux/demux.) See Figure 3-13.

Figure 3-13 *Point-To-Point Link: 8 Channels*

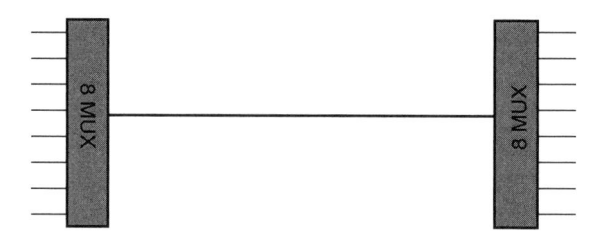

Minimum transmit power = 1 dB
Receiver sensitivity = –31 dB
Raw power budget = minimum transmitter power – minimum receiver sensitivity:
    1 – (–31) = 32 dB
Loss calculation:
Distance between the two sites = 65 km
The attenuation coefficient for the fiber is given as .3 dB/km.
Hence, total loss in the link = 65 km * .3 dB/km = 19.5 dB
Add and drop losses:
Add loss of 8-λ mux is = –4 dB and drop loss is –5 dB.
Hence the total loss = (4 + 5) = 9 dB
The total loss = link loss + add/drop loss = 19.5 dB + 9 dB = 28.5 dB
Power budget – all the link losses should be greater than or equal to 0:
32 – 19.5 – 9 = 3.5 dB, which is greater than 0; therefore, the system is viable.

You can determine how much attenuation is needed by calculating the power received at the receiver end. However, it is advisable to use a power meter to determine the attenuation needed.

Example 2

XYZ is planning to deploy a hub-and-spoke CWDM network with eight λs (protected) to connect their eight satellite locations to their CO. Design the network and determine whether the CWDM network supports the distance (shown in Figure 3-14).

Figure 3-14 *Hub-And-Spoke Protected CWDM Network*

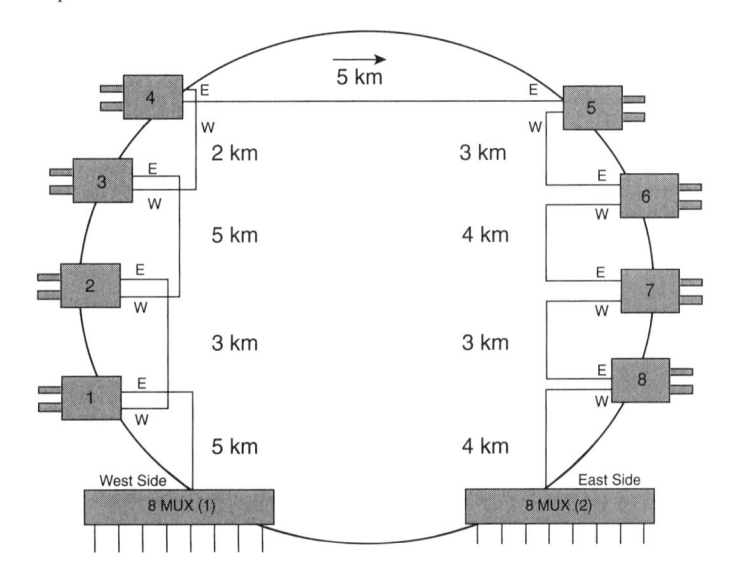

Table 3-2 lists the characteristics of a typical OADM and mux/demux.

Table 3-2 *Characteristics of a Typical OADM and Mux/Demux*

Description	Max Insertion Loss			Typical Insertion Loss		
	Add (dB)	Drop (dB)	Pass (dB)	Add (dB)	Drop (dB)	Pass (dB)
1-λ OADM	1.90	2.30	2.00	1.20	1.60	1.50
8-λ mux/demux	4	5	N/A	3	4	N/A

Example 2 Solution

We need to calculate the power budget for both the east and west side for the longest path. The calculation is usually done on the longest path. If the system supports the distance for the longest path (wavelength), it can be safely assumed that all the intermediate nodes will also function within the specifications. However, it is necessary to determine the power budget in both directions (transmit and in the reverse direction).

Figure 3-14 depicts the path from 8 Mux (1) to the node 8. First we calculate the power requirements in this direction and then from node 8 back to the mux.

East side power budget (for the longest path; see Figure 3-14):

> 8 Mux (2) – 8 (node) – 7 (node) – 6 (node) – 5 (node) – 4 (node) – 3 (node) – 2 (node) – 1 (node) (clockwise)
> 1 (node) – 2 (node) – 3 (node) – 4 (node) – 5 (node) – 6 (node) – 7 (node) – 8 (node) – 8 Mux (2) (counterclockwise)

West side (for the longest path; see Figure 3-14):

> 8 Mux (1) – 1 (node) – 2 (node) – 3 (node) – 4 (node) – 5 (node) – 6 (node) – 7 (node) – 8 (node) (clockwise)
> 8 (node) – 7 node) – 6 (node) – 5 (node) – 4 (node) – 3 (node) – 2 (node) – 1 (node) – 8 Mux (1) (counterclockwise)

Figure 3-14 depicts the path for a CWDM network from Mux 1 to node 8. The term *mux* here means the eight-channel CWDM multiplexer. Nodes are just filters that can add and drop a smaller (fewer than eight) number of wavelengths. First we calculate the power requirements in this direction and then from node 8 back.

Raw power budget:

> Minimum transmit power = 1 dB
> Receiver sensitivity = -31 dB
> Raw power budget = minimum transmitter power – minimum receiver sensitivity:
> 1 – (–31) = 32 dB

Calculations (Clockwise)

Fiber loss:

> Assume attenuation coefficient of .3 dB/km
> 8 Mux (1) to Node 1 = 5 km * .3 dB/km = 1.5 dB
> Node 1 to Node 2 = 3 km * .3 dB/km = .9 dB
> Node 2 to Node 3 = 5 km * .3 dB/km = 1.5 dB
> Node 3 to Node 4 = 2 km * .3 dB = .6 dB
> Node 4 to Node 5 = 5 km * .3 dB = 1.5 dB
> Node 5 to Node 6 = 3 km * .3 dB = .9 dB
> Node 6 to node 7 = 4 km * .3 dB = 1.2 dB
> Node 7 to Node 8 = 3 km * .3 dB = .9 dB
> Total fiber loss = 1.5 + .9 + 1.5 + .6 + 1.5 + .9 + 1.2 + .9 = 9 dB

CWDM Design Examples

Loss due to nodes:

Add Loss 8 Mux (1) + passthrough loss (1, 2, 3, 4, 5, 6, 7 Nodes) + drop loss (Node 8) =
4 + 7 * 2 + 2.30 =
4 + 14 + 2.30 = 20.30 dB

System check:

Power budget – all the losses (link + node) must be greater than or equal to 0:
32 dB – 9 dB (link Loss) – 20.30 dB (nodal loss) =
32 – 29.30 is greater than 0; therefore, the system is viable.

Calculations (Counterclockwise)

Here the distance is the same, and hence the fiber loss is 9 dB.

The nodal loss is as follows:

Add (Node 1) + passthrough (1, 2, 3, 4, 5, 6, 7 Nodes) + drop 8 Mux (1) =
1.90 + 14 + 5 = 20.90 dB

System check:

Power budget – all the losses (link + node) must be greater than or equal to 0:
32 dB – 9 dB – 20.90 dB = 32 dB – 29.90 dB is greater than 0; therefore, the system is viable.

Calculating the Attenuation

You can calculate the attenuation required by calculating the available power at the receiver.

This can be done by subtracting all the losses from the maximum transmit power. The power at the receiver should be between the receiver sensitivity range (greater than –32 dB and less than –7 dB). Even though this calculation gives a ball-park value of the attenuation required and can be used for determining how much attenuation is needed (from Bill of Materials perspective), it is always advisable to measure power at the receiver side to qualify.

Case 1 attenuation requirements:

8 Mux (1) to the node (8)
Total power at the receiver should be within the dynamic range of the receiver (receiver sensitivity).
Maximum transmit power – typical loss of nodes (add + passthrough + drop) – link loss =
5 – (3 + 1.5 * 7 + 1.6) – 9 = –19.1 dB
This is well with it dynamic range; therefore, no attenuation is need.
You can use the same approach to calculate the attenuation requirements for all nodes.

Case 2 attenuation requirements:

8 Mux to Node 1
Maximum transmit power – typical loss of nodes (add + passthrough + drop) – link loss
5 – (3 (add) + 1.6 (drop) – 1.5 = 1.1 dB
This is not within the range (between –32 dB and – 7 dB).
We need to add a 10-dB attenuator at the receiver end.
So after attenuation, we get 1.1 – 10 = –8.9 dB, which is well within the range.

Exercise

Design and validate the given CWDM network. Use 8 Mux at the hub node and single-channel OADMs at drop locations. (Use Figure 3-15 as a reference.)

Figure 3-15 *Linear CWDM Network*

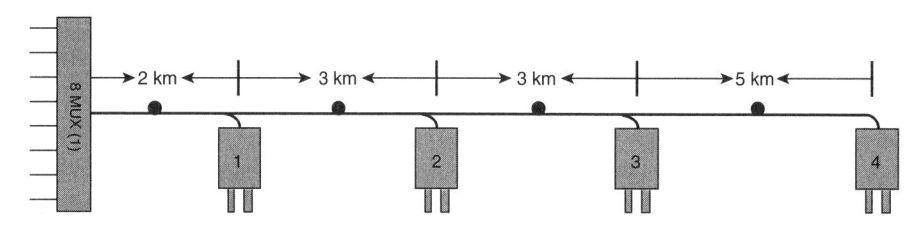

Use Table 3-3 for CWDM characteristics.

Table 3-3 *Nodal Characteristics Table*

Description	Max Insertion Loss			Typical Insertion Loss (dBm)		
	Add	Drop	Pass	Add	Drop	Pass
1-λ OADM	1.90	2.30	2.00	1.20	1.60	1.50
8-λ Mux/Demux	4	5	N/A	3	4	N/A

Summary

In this chapter, we have shown CWDM as a technology for enterprise access networks. The advantage of CWDM is its capability to exploit the massive bandwidth offered by fiber and yet keep network equipment costs low. This chapter covered CWDM from a deployment perspective rather than from a technology perspective. (Most of the technical details are covered in *DWDM Network Designs and Engineering Solutions*.) In this chapter, we discussed how CWDM networks are beneficial to access area networking, especially considering the fact that these networks are built from low-cost components. With the help of design guidelines, we introduced the reader to CWDM network deployment and demonstrated the intricacies of network designs.

Review Questions

1. Differentiate between CWDM and DWDM.
2. Explain the design procedure of a CWDM network.
3. Why don't we use amplifiers in a CWDM network?
4. Discuss the implications of CWDM to the first mile.
5. How is CWDM a first mile solution for enterprises?
6. Discuss CWDM filter and transponder characteristics.
7. Explain the term *restoration* in CWDM networks.
8. Differentiate between 1:1 and 1+1 protection in CWDM networks.
9. Numeric: For an 8-channel CWDM ring network of radius 7 km, calculate the minimum and maximum input signal power if the ring is unidirectional; there are 8 nodes and each node is equidistant. The nodes have a passthrough loss of 2 dB and an add/drop loss of 4 dB. Receiver sensitivity is 28 dB, and fiber has a 0.2 dB/km loss.

   Make the same calculation if the ring is bidirectional; all other parameters are the same.

10. Design an interconnected-ring CWDM network with the following characteristics:

    There are two rings of size 10 km diameter and 8 km diameter named A and B, respectively.

    Ring A has 6 nodes and B has 8 nodes, all equidistant.

    The network uses SMF-28 fiber type.

    The interconnection point in the ring has an optical cross-connect, and there is no regeneration of signal at the only interconnection point.

    Calculate the power levels required to set up the lightpath in the longest segment in the network.

    Calculate the dynamic range of the receiver (the max/min power a receiver can take).

References

Cisco 7600 product manual, San Jose, California: Cisco Systems.

Cisco 3550 product manual, San Jose, California: Cisco Systems.

Cisco CWDM product manual, San Jose, California: Cisco Systems.

Cisco GBIC product manual, San Jose, California: Cisco Systems.

Gumaste, Ashwin and Tony Antony. *DWDM Network Designs and Engineering Solutions*. Indianapolis, Indiana: Cisco Press, 2002.

IEEE 802.1w/s Security DocumentITU standard 694.

Ramaswami, Rajiv and Kumar N. Sivarajan. Op*tical Networks: A Practical Perspective*, Second Edition. San Diego, California: Morgan Kaufmann, 2001.

CHAPTER 4

Data Wireless Communication

Mobile communication, and the unsurpassed flexibility it provides, is a luxury that has been long sought. The ability to communicate while being portable gives a degree of flexibility that cannot be surpassed. Marconi's invention of radio as a one-way broadcast technology using electromagnetic waves has led to massive research in wireless net-working. The one-way broadcast methodology gave way to two-way wireless communication, initially over a simplex channel (where source and destination could not talk at the same time) and finally to full-duplex voice communication. Electromagnetic wave propagation is subject to a dispersive effect (attenuation), such that the power of a signal in free space attenuates as it travels because of multiple causes (among which, fading and scattering are the primary causes).

Since then, wireless communication has evolved to produce a level of maturity to support the entire telecommunications industry. Wireless communication initially included voice communication, but since the advent of mobile computing has now spread itself to data communication also. The ability to access network traffic yet be mobile without the overhead of a wired line is a strong enough motivating factor for the successful deployment of such a communication methodology. The technological maturity of mobile computing devices and the need for access at different locations were strong motivating factors behind the rapid spread of wireless technology into data networking.

In this chapter we introduce the basic concepts of wireless technology, including a look at the physical layer aspects of wireless technology. This chapter also discusses the evolution of both kinds of code-division multiple access (CDMA) technology: direct-sequence spread spectrum (DSSS) and frequency-hopping spread spectrum (FHSS). The focus of the chapter then turns to the seamless migration of this technology to first mile wireless networks; the IEEE 802.11 series of standards form the basis of this discussion. This chapter concludes by covering 802.11b wireless access scenarios.

The Wireless Medium and the Three Wireless Technologies for Communication

Before delving into the three technologies deployed for communication in wireless networks, it is important for you to understand the basis of the wireless medium—air—as a dielectric and as a medium for transport of electromagnetic disturbances. Air is a lossy conductor of electromagnetic waves, and there is heavy attenuation of an electric signal as it propagates through free space. This loss can be attributed to the following reasons:

- Air is a bad conductor of electromagnetic disturbances, and air is not a good waveguide. This means that there is no confinement of the signal from source to destination during propagation. Note that a waveguide is a physically confined medium that allows waves of some form to go through it (transport) with predictable properties such as loss. In our case, air as a waveguide essentially means the capability of air to transport electromagnetic waves. While doing so, there is loss, scattering, and dispersion, which basically can be predicted by the calculation of waveguide equations in the simplest forms. These parameters limit the transmission distance.

- Another issue, highlighted later, is the relationship of the frequency of an electromagnetic radiation through air and the length of the antenna (dipole) responsible for the production of this electromagnetic disturbance.

To summarize, the frequency spectrums that allow seamless wireless communication between users is quite limited. Previous chapters, some of which focused on fiber as a medium for data transport, discussed the near-infinite bandwidth offered by an optical fiber. In contrast, wireless offers a fraction of the bandwidth. Needless to say, wireless communication can be deployed where fiber cannot (for example, a remote island that needs immediate connectivity). In addition, wireless networks provide the best solution for places where connectivity has to be established and laying a fiber is not possible or feasible. Because of the limited band available for wireless communications, the number of users that can be admitted in a network is limited. Moreover, unlike optical fiber—where two parallel fibers can carry twice the data without interference—interference is a distinct possibility in wireless communication if two transmitters that are too close geographically use the same frequency for transmission. This creates a frequency-sharing problem.

However, the following three multiplexing techniques allow a sufficiently large number of users to intelligently and efficiently share the limited spectrum:

- Time-division multiplexing (TDM)
- Frequency-division multiplexing (FDM)
- Code-division multiple access (CDMA)

In the following sections we discuss each of these because of the prominent role they play in wireless communications per se, let alone first mile communications.

Time-Division Multiplexing

TDM is a legendary time-sharing scheme. Multiple users get access to the entire bandwidth, but each user can access the entire bandwidth only at certain times. In other words, N users get access to the entire bandwidth (B) for 1/Nth of the time. This time-sharing procedure ensures that all users have access to the entire bandwidth, but each only for a specified period of time. The advantage of TDM is its simplicity and its capacity to handle a greater number of users. On the flip side, TDM has multiple disadvantages: TDM requires synchronization among all the users; also, the slots (TDM slots, one each for every user) may not be efficiently used. In some cases, a slot for a user arrives but there is no data to send. This is especially true for bursty IP communication because the arrival rate of IP packets is just that: bursty. Figure 4-1 shows three users (A, B, and C) sharing a channel by multiplexing information over time.

Figure 4-1   *TDM Demonstrations*

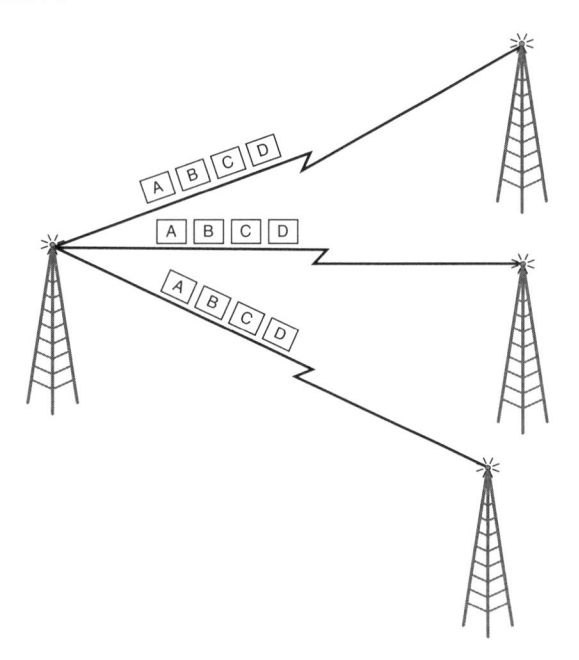

Frequency-Division Multiplexing

FDM is the most fundamental technique used in multiplexing a number of users. With FDM, the multiplexing occurs on a frequency basis. That means that multiple users get access to the channel all the time, but the spectrum is now divided among the multiple users so as to share the frequency band. In FDM, the bandwidth is basically shared among the multiple users, causing each user to get a portion of the bandwidth. This means that each user has access to the network all the time. Although such a scheme does not allow users to have large bandwidth, it still ensures that each user gets a portion of the spectrum all the time.

Figure 4-2 shows multiple hexagonal cells forming a best-fit pattern. The hexagonal shape of the cells results in a best-fit pattern because they are nonoverlapping. Also note that we can create a set of frequencies in each cell that differs from the frequencies in all the adjacent cells.

This way we achieve good frequency reuse, by duplicating the pattern (which looks like honeycombs) in the network. A seminal paper by Kleinrock, et al., "Why Six Is a Magic Number," discusses this theory and highlights the frequency reuse concept in FDM networks.

Figure 4-2 *Frequency-Division Multiplexing and Why Six Is a Magic Number*

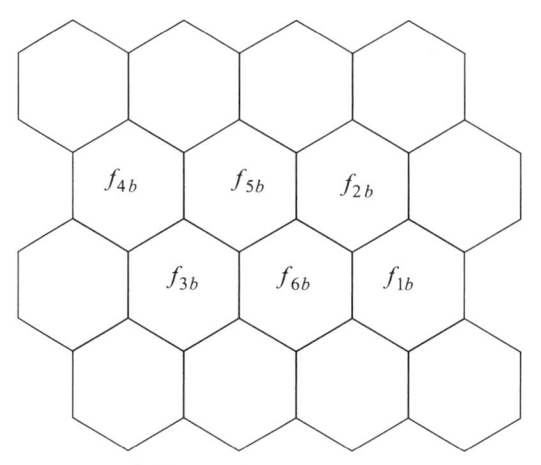

Cell Pattern - Why 6 Is a Magic Number

Code-Division Multiple Access

The third and most important type of multiplexing technique is CDMA. CDMA, or spread spectrum technology, uses the entire bandwidth for every user such that a particular user is selected based on a code specific to that user. In CDMA systems, on many occasions the bandwidth made available to the user is several times the bandwidth needed for actual communication. CDMA is a difficult concept to visualize, but it has tremendous benefits compared to FDM and TDM, among which are the following:

- By using CDMA, the power requirement by wireless antennas is much lesser as compared to the other counterparts. Typical CDMA antenna can consume up to 6 to 7 milliwatts (mw) of power for transmission, which is significantly less than a TDM or FDM system.

- In the other systems, we have to carefully plan the networks, taking into account frequency reuse. Frequency reuse is a basic concept in wireless networks which helps reuse the same frequency for transmission in two areas that are geographically loosely disjoint from one another. By using the same frequency in two disjoint areas (cells) (see Figure 4-2), we are able to have more users in the network. Moreover, the geographical diversity ensures that there is no electromagnetic wave interference of the two signals on the same frequency.

- In a cell-based system, when a mobile user travels from one cell to another, he has to undergo handoff. Handoff is the procedure of the mobile user switching from a channel in the ingress cell to another channel in the egress cell. Power-based or otherwise algorithms are available that allow handoff to happen between cells. Hand off can be a problem if a user moves into a cell that does not have any free (available) frequencies (cannot support the new user). In such a case, the call may be dropped. In CDMA systems, however, the user may go to the next cell, but the call won't be dropped on account of CDMA's code-based multiplexing strategy, whereby the user will continue to use the same channel as in the egress cell. Here we digress to study CDMA systems and benefits.

Classification of CDMA: Frequency Hopping and Direct Sequence

The idea of spreading a signal over a bandwidth greater than the basic signal capacity at baseband, by using some form of coding that spreads the signal, is called spread spectrum (SS) technology. SS systems were first deployed in World War II to deny the enemy access of classified information in the wireless medium. Despite access to the SS signal, it is not possible to decode the signal unless the receiver knows the "code word," which codes the basic baseband signal and spreads it over a higher bandwidth carrier. SS techniques provide good immunity against jamming and against unwanted interference. After the end of World War II, the application of SS technology to wireless communication was natural. There are two basic types of SS technology:

- Frequency-hopping spread spectrum (FHSS)
- Direct-sequence spread spectrum (DSSS)

Frequency-Hopping Spread Spectrum

In frequency hopping (FH), we have a baseband signal (the information bearing signal), and a carrier, which varies its frequency $f_c(t)$ with time. The signal is modulated by the carrier, which itself is varying over time. The variation of the carrier is what produces the term frequency hopping. Note here that the code word per se is not a bit-based code word but is basically the hopping nature of the carrier. This means that a code word in FHSS is basically a carrier that moves its carrier frequency from one frequency to another and so on, in a predictable manner, thus creating the code word. The hopping pattern is the code word in this case, against which the signal is varied. There are two kinds of frequency hopping: a slow hopping and a fast hopping. In slow hopping, the carrier frequency changes in a time period that is slower than the time period of the information signal (1/t=f). In fast hopping, however, the time period of change of the carrier is faster than the time period of a bit in the information bearing signal.

Here we define frequency chirp as the time required for the carrier to change from one frequency to the next one. See Figure 4-3. (Hence in slow hopping the chirp frequency, 1/chirp length, is less than the information signal. In fast hopping, on the other hand, the chirp frequency is greater than the information-bearing signal.

Figure 4-3 *FH-CDMA Shows f_i is the Frequency of Transmission and _ is the Chirp*

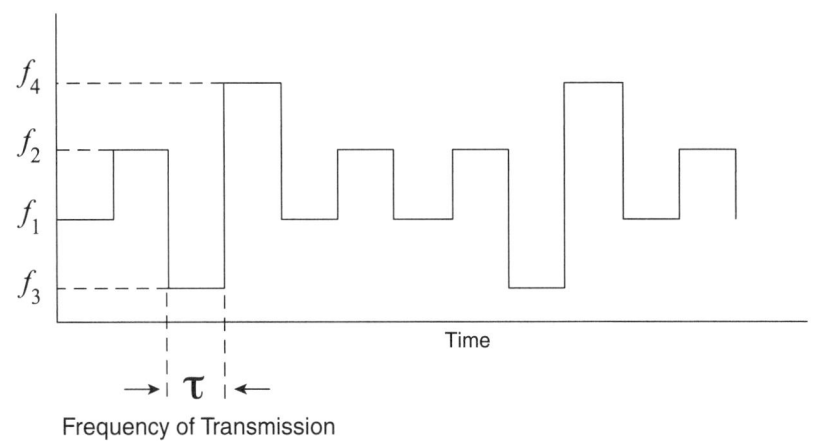

Direct-Sequence Spread Spectrum

In a direct-sequence spread spectrum (DSSS), the information-bearing signal is modulated by a code word (which repeats itself) such that the code bit rate is greater than the bit rate of the information bearing signal. This means that the low-bit-rate information-bearing signal, when modulated by the higher-bit-rate code signal, produces a signal that has higher bit rate as compared to the original information signal. Thus the signal is spread in frequency, resulting in a spreading of the spectrum (and hence SS).

Direct-Sequence Spread Spectrum Theory

A DSSS signal is such that the amplitude of an already modulated signal is further amplitude-modulated by a very-high-rate nonreturn to zero (NRZ) binary stream of digits. Now assume the original signal is j(t) such that

$$j(t) = \sqrt{B}d(t)\cos\omega_0 t$$

and the spread spectrum signal is

$$v(t) = j(t)g(t) = \sqrt{B}d(t)g(t)\cos\omega_0 t$$

where g(t) is the sequence signal (code word) and can also be considered as a pseudo-random noise binary sequence having values +1 or –1. g(t) in other words is a sequence so random that two (time-dispersed/shifted) strings in the sequence have minimal correlation with each other—or, in other words, are perfectly pseudo-random. Second, the bit rate of g(t) is much greater than the bit rate of j(t), and hence d(t). So for every bit in d(t), we have multiple bits in g(t). Therefore, when the multiplication occurs, each bit in d(t) is broken into several sub-bits, also called as *chips*. Hence the rate of g(t) is called the *chip rate* of the system. We note here that g(t) spreads the signal in the frequency domain by this chipping action.

Note here that the basic signal is spread into the spectrum by the fast multiplication of the basic baseband signal by a code word. In other words, we can view the code word as a distributor, which distributes a bit from the baseband signal into multiple bits (hence the frequency gain, which is explained later).

Receiver for DSSS

To recover the DSSS signal, a receiver works as follows: A multiplier multiplies the incoming noise-affected (through the channel) DSSS signal by g(t)—the sequence used to generate the direct sequence code—in the first instance at the transmitter, assuming, of course, the receiver knows this code word. Subsequently the signal is multiplied by the carrier ($\cos\omega_0 t$). The resulting waveform is then integrated for the bit duration and the output of the integrator is sampled, yielding the data d(t). Figure 4-4 shows the DSSS communication methodology.

Figure 4-4 *DSSS System Block Diagram for Transmitter and Receiver*

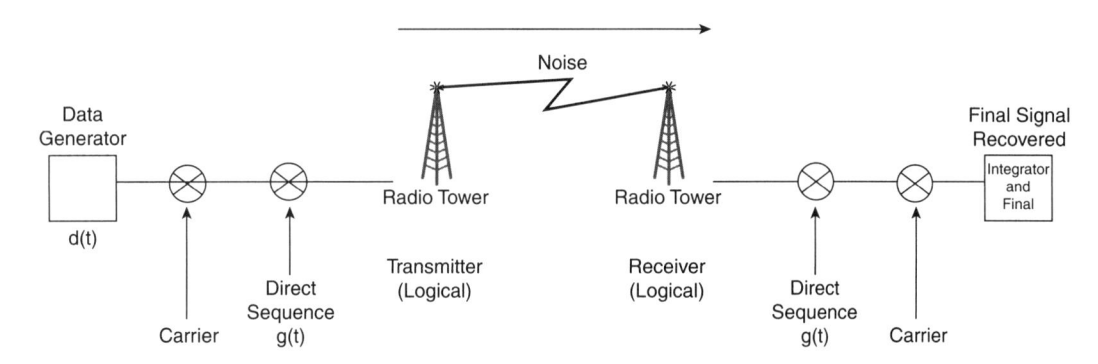

Frequency-Hopping Spread Spectrum Theory

Consider a signal d(t) such that a 0 bit in d(t) is represented by some frequency f_1 while a 1 bit in d(t) is represented by another distinct frequency f_2. Now s(t) can be considered as a Binary Frequency Shift Keyed (BFSK) signal; and this kind of modulation, whereby a baseband digital signal is transmitted by taking the logic levels (0 and 1) on two different frequencies and is then keyed accordingly, is called *Binary Frequency Shift Keying*.

The BFSK signal s(t) is as follows

$$s(t) = \sqrt{B}\cos(\omega_0 t + d(t) + \kappa t + \theta)$$

where k and theta are two phase constants. The frequency modulation is then applied by varying the carrier frequency so that the resulting frequency-hopping spread spectrum is as follows:

$$FH = \sqrt{B}\cos(\omega_i t + d(t)\kappa t + \theta)$$

In the preceding equation, the frequency-hopped signal has a carrier frequency of $\omega_i/2\_$, which changes as the hopping rate f_i (the chirp rate of the system). The frequency chosen for each chirp, is pseudo-random from a batch of 30 to 500 frequencies.

The Advantage of Spread Spectrum

So far we have discussed SS technology from both frequency-hopping and direct-sequence perspectives. We understand the direct advantage of SS technology in terms of security. Obviously in the present scenario this is a big advantage, but not the only one. The biggest advantage of using spread spectrum technology is the ability to relax the stringencies on signal to noise ratio of a transmitted signal. Based on Shannon's theory, we can quantify the performance gain achieved in terms of signal to noise ratio as follows:

$$B_w = \frac{C}{1.44} \cdot \left[\frac{S}{N}\right]^{-1}$$

Here, B_w is the signal bandwidth, C is the capacity of channel, and S/N is the signal to noise ratio in dB.

Now as S/N decreases, to support this decrease we need to increase the bandwidth B_w of the information signal. Conversely, by spreading a signal in the frequency domain we allow the S/N ratio to be decreased further, but still are able to achieve successful transmission. Note that by spreading a signal, the transmission is successful because the noise in the system is throughout the transmission frequency band (additive white Gaussian noise [AWGN]) and this means that probability of detecting a bit increases with the frequency spread of the bit. This indicates that transmission is successful through SS. Therefore, a low power signal can be transmitted successfully through a lossy (fading) channel such as air if we spread this signal sufficiently over the frequency domain. Hence SS communication is an effective strategy for wireless communication. Moreover, because wireless devices are generally portables (such as Bluetooth applications), power in such devices is always a scarce commodity.

Another interesting point about SS systems is the processing gain realized by spreading the signal, a gain that basically quantifies the capability of the signal to avoid interference. If B_w is the information bandwidth, and R is the data rate of the original signal, then

$$G = \frac{B_w}{R}$$

signifies that G is the gain due to spreading the original signal that occupied R, but which now occupies B_w bandwidth. This is also known as *processing gain* of an SS system. Typically, its value is between 10 and 75 dB.

Antenna Theory: Gain and Half-Power Beamwidth

The fundamental characteristics of an antenna are its gain and half-power beamwidth. According to the Reciprocity theorem (a generic theorem for transmission processes), the transmitting and receiving patterns of an antenna are identical at a given radio frequency. The gain quantifies the amount of input power that is concentrated in a particular direction. It is expressed with respect to a hypothetical isotropic antenna, which radiates equally in all directions. (Isotropic means radiating in all directions.)

Thus, in a given direction (q , f), the gain is as follows:

$$G(q, f) = (dP/dW)/(P_{in}/4\pi)$$

where P_{in} is the total input power, and dP is the increment of radiated output power in solid angle dW. The gain is maximized along the bore-sight direction. The meaning of the word *isotropic* refers to the radiation pattern of the antenna, which is said to be in all directions (360 degrees).

The input power is this

$$P_{in} = E_a^2 \, A/h \, Z_0$$

where E_a is the average electric field over the area A of the aperture, Z_0 is the impedance of free space, and h is the net antenna efficiency.

The output power over solid angle dW is as follows

$$dP = (E^2 \, r^2 \, dW)/ Z_0,$$

where E is the electric field at distance r. By the Fraunhofer theory of diffraction

$$E = E_a \, A/r \, l$$

along the bore-sight direction, where l is the wavelength.

Thus the bore-sight gain is given in terms of the size of the antenna by the following important relation:

$$G = h(4\pi/1^2)A$$

This equation determines the required antenna area for the specified gain at a given wavelength.

The net efficiency h is the product of the aperture taper efficiency h_a, which depends on the electric field distribution over the antenna aperture, and the total radiation efficiency (h * = P/P$_{in}$) associated with various losses. For a typical antenna, h = 0.55.

For a reflector antenna, the area is just the projected area. Therefore, for a circular reflector of diameter D, the area is A = π D²/4, and the gain is this:

$$G = h(\pi D/1)^2$$

which can also be written as:

$$G = h(\pi D f/c)^2$$

Because c = l.f, where c is the speed of electromagnetic waves in free space (3 * 10⁸ m/s), l is the wavelength, and f is the frequency. As a corollary, the gain increases as the frequency increases.

The half-power beamwidth is the angular separation between the half-power points on the antenna radiation pattern, where the gain is one-half the maximum value. For a reflector antenna, it may be expressed as follows:

$$HPBW = a = k1/D$$

where k is a factor that depends on the shape of the reflector and the method of illumination. For a typical antenna, k = 66°. Thus the half-power beamwidth decreases with decreasing wavelength and increasing diameter.

Effective Isotropic Radiated Power (EIRP) AND G/T

For the radio-frequency (RF) link budget, the two required antenna properties are the equivalent/effective isotropic radiated power (EIRP) and the "figure of merit" (G/T). These quantities are the properties of the transmit antenna and receive antenna that appear in the RF link equation and are calculated at the transmit and receive frequencies, respectively.

The equivalent (or effective) isotropic radiated power is the power radiated equally in all directions, which would produce a power-flux density equivalent to the power-flux density of the actual antenna. The power-flux density F is defined as the radiated power P per unit area S, or F = P/S. But P = h * P_{in}, where P_{in} is the input power and h * is the radiation efficiency, and S = d^2 W_A, where d is the slant range to the center of coverage and W_A is the solid angle containing the total power.

$$F = h*(4\pi/W_A)(P_{in}/4\pi d^2) = GP_{in}/4\pi d^2$$

Because the surface area of a sphere of radius d is 4p d^2, the flux density in terms of the EIRP is as follows:

$$F = EIRP/4\pi d^2$$

Equating these two expressions, one obtains the following:

$$EIRP = GP_{in}$$

Therefore, the equivalent isotropic radiated power is the product of the antenna gain of the transmitter and the power applied to the input terminals of the antenna.

The "figure of merit" is the ratio of the antenna gain of the receiver (G) and the system temperature (T). The system temperature is a measure of the total noise power and includes contributions from the antenna and the receiver. Both the gain and the system temperature must be referenced to the same point in the chain of components in the receiver system. The ratio G/T is important because it is an invariant that is independent of the reference point where it is calculated, even though the gain and the system temperature individually are different at different points.

Smart Antennas

The use of the spectrum in multiple cells, each of which is geographically diverse from one another, creates a situation in which the number of wireless users in a given area is bounded on the maximum number. To increase the number of wireless users, and to provide a much better reception to individual users, the concept of smart antennas was proposed. A smart antenna is an antenna with some degree of intelligence, such that the antenna can perform functions in an adaptable manner. By using digital signal processing (DSP), for example, a smart antenna can increase its beam's diversity in such a way that the beam focuses on a particular reference plane, thereby allowing more users in that geographic location.

Smart antennas are categorized according to their transmission strategy, as follows:

- **Switched beam**—A finite number of fixed, predefined patterns or combining strategies (sectors)
- **Adaptive array**—An infinite number of patterns (scenario based) that are adjusted on a real-time basis

Smart antennas provide the following benefits:

- **Better coverage and range**—Because smart antennas can cover a particular area better than other areas by changing the pattern of their directivity (signifying the direction of general propagation of the signal), smart antennas can cover a wider range and hence incorporate even weakly linked users.
- **Capacity (interference benefit)**—Because of this better (more directed) coverage just mentioned, wireless users experience less interference (especially in densely populated areas that may have multiple antenna sites, which otherwise could lead to severe interference).
- **Power efficiency (economic benefit)**—Because the directivity patterns of the smart antennas can be intelligently changed depending on traffic conditions and other requirements, power use can be minimized. If the signal in a particular area becomes weak, but there are a lot of users in this area, for example, the smart antenna changes its directional pattern in such a way that the major lobe falls better covers this area (see Figure 4-5). In such a scenario, users get more signal power (enhanced reception) for less of a power strain on their resources (batteries).

Figure 4-5 *Major Lobe of an Antenna*

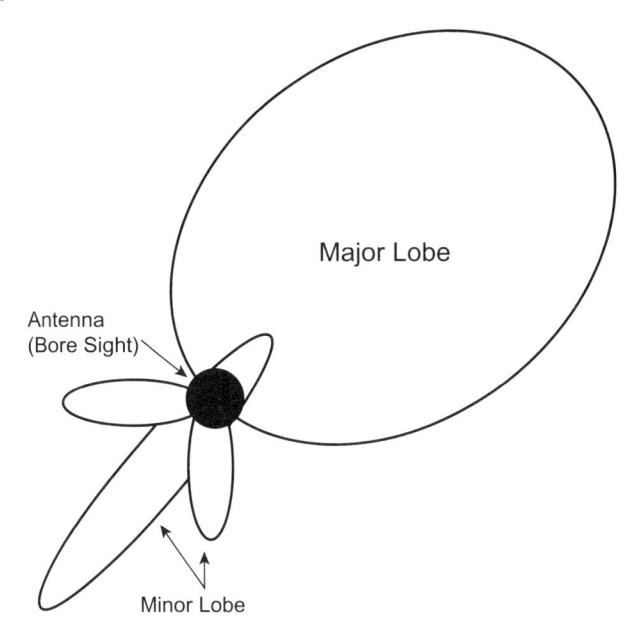

Major lobe and minor lobe determination of an antenna (note: the antenna is perpendicular to the plane of the page going into it)

- **Multipath rejection**—Consider Figure 4-6, A is a major cell site where a base station antenna is located, C is a wireless user, and B is a local cell site minor antenna. Wireless user C gets two copies of the signal from A and B. However, the two paths differ (AC and ABC), meaning that the distance the two signals covers differs (thus inducing a phase shift between the two signals, as shown in the lower part of Figure 4-6). This causes problems as the signal to noise ratio degrades. In addition, the wireless user requires extra power to read the signal (differentiating between the multipath faded pattern). If antenna B were a smart antenna, it could reduce its beam-propagation pattern intelligently so that the user at C gets only a substantial chunk of the signal from A. In such a scenario, user C could more easily receive the data and multipath-fading losses would be minimized. The principle here is to transmit the power of the signal in an optimum path, such that only those users to whom the antenna directs its transmission get the maximum benefit of the antenna. Figure 4-7 shows an example of a multipath phase problem.

Figure 4-6 *Multipath Rejection*

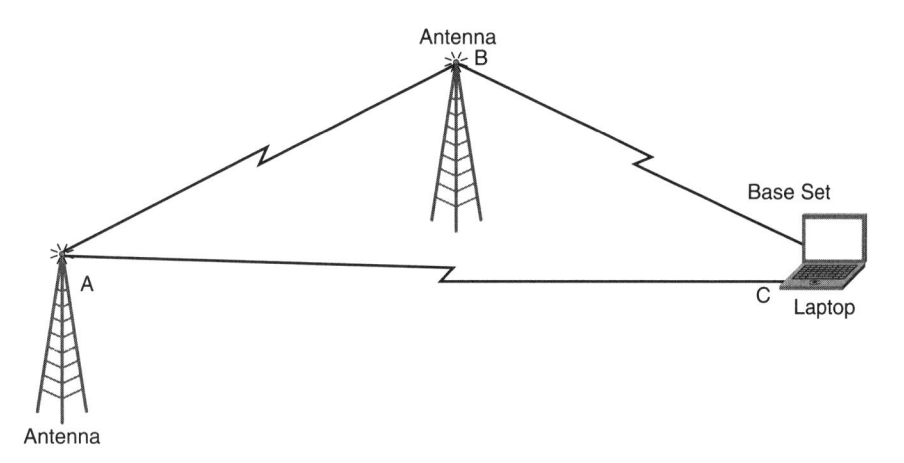

Figure 4-7 *Multipath Effect-Phase Rejection*

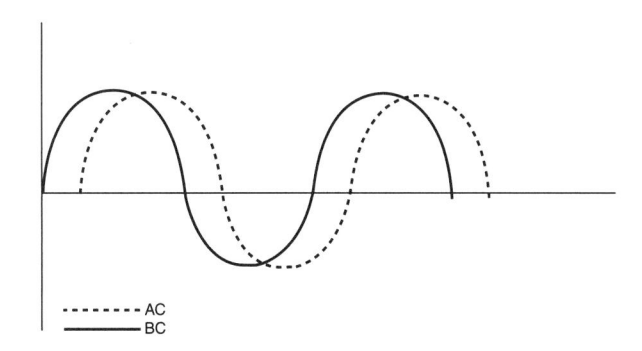

Other Wireless Effects: Fading and Delay Spread

Fading is a natural phenomenon in wireless systems. As a signal propagates through free space, it experiences small zones of fading where the signal loses some of its power or gradually fades. The phenomenon is quite severe for multipath signals, because of the phase discrepancies in them as discussed previously. Fading in this case is called *Raleigh fading* and is inversely proportional to the fourth power of wavelength.

Another negative effect in wireless communication is *delay spread*, which is particularly pronounced in TDM systems. Consider, for instance, a user getting data in a particular slot. If the user is receiving data from two or three different antennas, he is bound to get two or three copies of the same signal. However, each copy may differ in phase (lagging/leading) from the earlier copy because of the multipath concept—that is, because each antenna is at a different distance from the user. This creates smudging of adjacent bits, creating what is called delay spread. This phenomenon is particularly harmful for high-speed wireless systems because it creates severe problems of intersymbol interference (ISI)—that is, two adjacent bits may smear into one another.

Wireless Data: The Evolution of the 802.11 Series of Standards

Xerox implemented Ethernet in the early 1970s as a rudimentary network protocol. Chlamtac and Franta's seminal book on Ethernet, *Local Area Networks*, conceptualized the first approach to local-area networking in 1979. Since then, Ethernet has evolved from just a data-link layer protocol to a universally acceptable protocol that integrates and works as a method of communication over multiple network types. The concept of Ethernet is based on carrier sense multiple access—that is, a user sends its frame of data into a network only after listening to the network and making sure that no one is transmitting data at that time. This means that two users may be listening to a busy channel, and as soon as the channel becomes free, both transmit their data frames, resulting in collision. The collision may result in some different signal, which can be detected by the transmitting stations. That is, because of this different signal, the transmitting stations understand that a collision has occurred. This scheme is known as collision detection (CD), and the strategy is called *carrier sense multiple access/collision detect (CSMA/CD)*. Transmitting stations can be intelligently modified so that there is no collision at all, or can be modified to state that they have a collision-avoidance mechanism. The strategy then is called *CSMA/CA*, CA meaning *collision avoidance*. This discussion has so far focused on LANs implemented on wired media such as copper, coaxial cable, and fiber. We can also extend this approach to wireless media, in which case these LANS are called *wireless LANs* (WLANs). The transmission process—that is, the sending of frames—remains the same. In other words, we can now send Ethernet frames into the network. The following two factors brought about this paradigm shift in communication from wired networks to wireless networks:

- IP mobility
- IEEE 802.11 standardization process

IP Mobility

Mobile IP is intended to enable nodes to move from one IP subnet to another in a network domain. It is just as suitable for mobility across homogeneous media (single physical domain) as it is for mobility across heterogeneous media (multiple physical and logical domains). That is, mobile IP facilitates node movement from one Ethernet segment to another and it accommodates node movement from an Ethernet segment to a wireless LAN, as long as the mobile node's IP address remains the same after such a movement (RFC 2002, "IP Mobility"). The important factor here is that the mobile node's IP address has to remain the same even after moving from its geographic location to another one. This is the fundamental concept of mobile IP and is enhanced in IPv6.

Consider, for instance, a campus environment in which students on a university campus have access to laptops. When outside of classrooms, the students may still surf the network, through a wireless LAN, and thereby create for themselves virtual classrooms outside the physical ones. This makes learning a much more interesting and simpler concept. Extend this concept to a conference room that is about to host a session of salespeople to discuss the previous-quarter sales. If this room is equipped with a WLAN, the salespeople just need to bring their laptops and connect to this WLAN through a wireless interface. The salespeople can then communicate with each other through the wireless network. In such a network, the users may come and go abruptly (in what's called an *ad hoc manner*). Such networks, which can be set up and torn down on demand, are known as *ad hoc networks*. Ad hoc networking is a very important aspect of wireless data networking because of its direct relationship with applications that exhibit dynamic bandwidth demands.

One issue remains, however: how to implement a network that enables *user* mobility. That is, users move from one base station area to another base station area, while maintaining the same IP address. Ipv4 has some rudimentary provisions for mobility. However, Ipv6 takes care of the drawbacks that we face when the users are mobile, primarily the issue of mobile agents informing the base agents of their mobility. Consider an example. User X is communicating to a backbone network through base station A. If mobile user X wants to move to a physical location that is under the influence of base station B, base station A must learn of this change. Mobile agents inform the base agents of such changes, so that the new base agent is the recipient of the forwarded information for user X. In some cases when the user moves from one base station area to another base station area, the mobile user informs the base station about the next hop, so that the data to be transmitted through the wireless channel is automatically sent to the new base station more proximate to the user.

The 802.11b High-Rate Wireless LAN

Two reasons compel us to study the 802.11b wireless LAN standard, as standardized by the IEEE for broadband data communication in wireless medium:

- 802.11b is the most practical implementation (providing a new concept in wireless network altogether).
- 802.11b's application to first mile access networks is almost immediate.

The IEEE released the first of the 802.11 standards (just plain 802.11) in June 1997. Subsequently, in 1999, the 802.11b standard was released for broadband wireless communication. The 802.11b operates in the 2.4-GHz band, typically from 2.4 GHz to 2.483 GHz. This band is the unlicensed band also previously known as the *Industrial, Scientific, and Medical* (ISM) *band*. The 802.11b standard supports operations from 1 Mbps to 2 Mbps to 5.5 Mbps and up to a maximum of 11 Mbps. The 802.11b standard basically defines two operating levels among the seven OSI layers: the data-link control (DLC) or media access control (MAC) and the physical (PHY) layers.

By definition (according to the IEEE's "802.11b 1999 Supplement to 802.11 Standard") IEEE 802.11b defines two units as part of the standard for wireless LAN: a wireless network interface card (NIC), which is attached to a portable device such as a laptop or a personal digital assistant (PDA), and an access point (AP), which is attached to the wired network or to a network point through which data transfer takes place. The AP acts as a bridge to support these wireless devices (NICs).

The standard defines two methods of operation for the wireless LAN:

- The infrastructure method
- The ad hoc method

Infrastructure Method

In the infrastructure embodiment of the 802.11b standard, an AP is defined, which is connected to the wired distribution network. Multiple wireless users are also defined; these users communicate with the AP over the wireless channel. In this scenario, if there are multiple APs, at least one should be connected to the wired distribution network. Two subtypes are also defined:

- **Base service set (BSS)** — The AP in this case forms a logical bridge for any connection in the network. Even if two wireless users just want to communicate with each other, they may do so, but not directly. They must communicate with each other though the AP, thus resulting in multihop communication. In such a case, we may view the network as a bidirectional star with the AP as the center of the star communicating to the multiple wireless network elements.

- **Extended service set (ESS)** — In this embodiment of the infrastructure method for the 802.11b standard, multiple APs communicate among themselves, thus easing the pressure on a single AP. In this manner, wireless users can talk to other wireless users, even those spread across larger distances, making use of the distribution wired network. Also, this ensures mobility of the wireless users themselves. Therefore, wireless users can move from a region covered by one AP to a region covered by another AP but still access the same network. In such cases, however, the procedure involves soft and hard handoffs (as discussed later in this chapter).

Ad Hoc Mode

In the ad hoc mode, all wireless users can communicate to each other directly. This means that there are no APs. The implication is that stations may not be able to communicate with each other because of distance limitations, and hence all stations must be proximate to one another. This mode is also known as *independent basic service set* (IBSS) or *peer-to-peer* because of the one-to-one relationship the users have with each other.

The 802.11b Physical Layer

The physical layer of the 802.11b is responsible for transmission (successful) of raw bits between multiple users, at variable line rates. In addition to 1-Mbps and 2-Mbps line rates, the 802.11 standard supports 5.5-Mbps and 11-Mbps line rates. It does so using coding techniques such as Complementary Code Keying (CCK) and Quadrature Phase Shift Keying (QPSK), which are forms of digital modulation. In addition to these modulation methods, 802.11b uses DSSS, which significantly contributes to the efficient use of the spectrum (minimizing error possibilities while not requiring too much power for transmission). The interchannel frequency difference for 802.11b is specified at 25 MHz between two adjacent transmitting channels. As mentioned before, however, the standard also defines a variable rate. In other words, depending on factors such as the channel condition and the number of users, the transmission rate of an 802.11b system can be changed. In a clear-channel scenario, for instance, an 11-Mbps transmission rate may be achieved.

Note on QPSK

Phase shift keying techniques provide a way to modulate a digital signal into an analog carrier that propagates through free space. The idea is to send a digital signal, primarily a binary signal of 1s and 0s, such that it modulates a carrier, one that is sinusoidal, so that the receiver can detect the original signal at the end of a lossy channel by just comparing the phases of the sinusoid that is broadcast. This technique is more effective than sending a digital signal directly into free space or a lossy channel (because severe interference, attenuation, and, above all, power loss of the signal occurs as it propagates through space). Moreover, the bandwidth required for a digital signal to be sent in free space is much more than a sinusoid that is sent. For wireless networks, sending a digital signal directly means coupling the data in digital format to an antenna. If the signal is directly coupled to an antenna, the antenna—to transfer all the components of the square digital wave—needs far more power than to transfer any sinusoids. Therefore, we prefer to have analog modulation of the digital signal, by a phase or quadrature amplitude method. The simplest technique for such a scheme is Binary Phase Shift Keying (BPSK). In BPSK, the signal is a pure sinusoid such that it has one fixed phase when the data is at a logical 1 and another fixed phase (shifted in this case by 180 degrees) when the data is at a logical 0. The inherent problem with the simplicity of BPSK is the bandwidth required for transmission. To transmit a bit stream with each bit occupying T seconds, the bandwidth required in BPSK is 2F, where F=1/T. To overcome this problem and to make transmission more efficient, QPSK is introduced as a technique for digital modulation.

QPSK consists of a QPSK transmitter and receiver, such that the QPSK transmitter creates a QPSK waveform from a binary data stream. The QPSK transmitter consists of a local clock, two D-type delay flip flops, two multipliers, and a signal adder. The binary signal is sent to the two flip flops, each of which has the characteristic of delaying the signal by the amount specified by another driving signal. A local clock generates a clock signal through a toggle flip flop, such that the period of this clock is twice the period of the data signal. (If T_b is the data period, then $2T_b$ is the clock period.) The toggle flip flop has two outputs for the clock, such that one output is the actual clock, whereas the other output is the inverted clock. The local and inverted clocks Q and $\overline{Q}$ are then fed to the delay flip flops. Because these flip flops function such that they delay the incoming signal depending on the variation of their clocks, Q and $\overline{Q}$, they create two versions of the incoming signal in each arm. The flip flop that runs by the triggering of the Q waveform only allows even bits of the data signal d(t) to go through, whereas the other flip flop that runs by the triggering of the $\overline{Q}$ signal allows only the odd bits to go through. Let's call these two signals in the two arms $b_e(t)$ and $b_o(t)$, where e and o mean even and odd. Further, each of the two signals, is sent through a multiplier, whereby $b_e(t)$ is multiplied by the following:

$$\sqrt{P_s}\cos\omega_o t$$

whereas $b_o(t)$ is multiplied by the following:

$$\sqrt{P_s}\sin\omega_o t$$

Now $b_o(t)$ and $b_e(t)$ both can assume any of the values (0 and 1). So now we have four possible combinations—(0,0), (0,1), (1,0), and (1,1)—giving rise to four phases each in quadrature to one another.

The final equation then is as follows:

$$v(t) = \sqrt{P_s}b_e(t)\cos\omega_o t + \sqrt{P_s}b_o(t)\sin\omega_0 t$$

And this signifies the QPSK modulation technique.

Complementary Code Keying

Complementary Code Keying (CCK) is used to increase the peak data rate for 802.11b transmission. This is justified for the quantum increase from 2 Mbps to 11 Mbps. CCK helps to increase the data rate by increasing the data clock rate from 1 Mbps to 1.375 Mbps and then taking data in 8-bit blocks (in parallel, and thus achieving 8 * 1.375 = 11 Mbps). Of the 8 bits, 6 are used to choose 1 of 64 complementary codes, which are each 8 chips long (48) and clocked at 11 MHz. All 8 chips are used in one 8-bit sequence. The other 2 bits are combined in the code in the QPSK modulator (explained previously).

Dynamic Rate Shifting in 802.11b Standards

Because the channel under consideration in a wireless system is a lossy air channel, such that the channel conditions change with time, 802.11b defines the data rate to be variable. Under excellent conditions (low fading and minimum multipath effects), the rate available is 11 Mbps. However, if the channel is affected by fading, for example, we cannot maintain the same data rate. In other words, we have to drop the data rate. 802.11b devices allow four transmission rates for wireless LANs: 11 Mbps, 5.5 Mbps, 2 Mbps, and 1 Mbps. Typically, DSSS is used for transmission of wireless LANs. The reason FHSS is not used is because of a limitation set by the FCC.

802.11b Physical Layer Functioning

The 802.11b physical layer is split into two parts: the Physical Layer Convergence Protocol (PLCP) and the physical medium dependent (PMD). The PMD takes care of the wireless coding and so on. Note that we mentioned CCK and QPSK techniques previously, which are deployed by the PMD. The PLCP forms an interface for the higher MAC sublayer. See Figure 4-8. The interface is important because of the distributed nature of the system, whereby Ethernet frames have to be sensed in the medium to use CSMA. In other words, the physical layer has to sense the carrier and inform the MAC sublayer of the same so that the MAC sublayer can initiate or stop transmission depending on the status of the medium.

Figure 4-8 *PLCP in 802.11b (IEEE 802.11b Standard: Courtesy* IEEE Network Magazine*)*

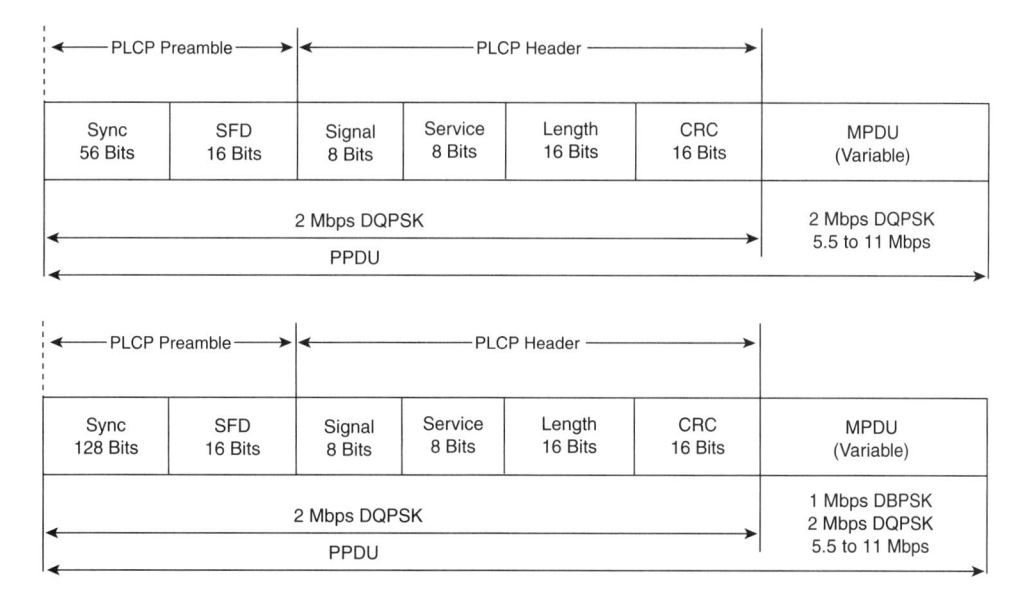

802.11b MAC Sublayer

The next most important aspect is the MAC sublayer, which serves as an interface between the physical layer and the 802.11b device. The MAC supports both the infrastructure and the ad hoc modes as explained earlier. The most important attributes of the MAC layer for wireless networking are error correction and ensuring successful delivery of packets. For the former, the MAC layer uses cyclic redundancy check (CRC) for every packet it transmits. CRC is latched on to a string of bits, calculated depending on the data, so as to prevent the data from being corrupted and recover the data if the bits are ever corrupted. The latter technique, that of ensuring successful frame delivery, is done by packet fragmentation. Packets are fragmented to avoid the need for retransmission of entire packets in case the channel is noise prone. It's better to lose a few fragments and retransmit them than to lose the entire packet and have the cumbersome job of retransmitting the whole packet. Analytically, if we study the probabilistic behavior of packet corruption through a noisy channel, it is found that the probability of a packet corrupting decreases with the length of the packet; hence, fragmentation is justified.

Accessing the 802.11b Medium and Interframe Space

Interframe space (IFS) defines the time between any two medium access activities carried out by the 802.11b device (for instance, the sending of two frames, the receiving of a frame, or the sending of an acknowledgement). Four kinds of IFS have been defined in the standard. Short IFS is the time period between the completion of a transmission of a complete packet and the start of an acknowledgement frame. Extended IFS means when a node receives a packet that is corrupt or that it cannot understand. Point-coordination IFS is one time slot greater than short IFS. Distributed IFS is one more time slot plus the time required to wait for point-coordination IFS.

Management Frames for 802.11b

Eleven management frame types are used for link management. They provide the means for establishing and terminating a link—for instance, a beacon transmission and probe requests.

The 802.11b standard mentions another seven major kinds of control frames, as follows:

- Request to send
- Clear to send
- Acknowledge
- Power save
- Contention free
- Contention free-end
- Contention free-acknowledge

WLAN Device

The most common WLAN device is a PC adapted with a PCMCIA card. This is suitable for direct connection to a PC or a laptop or any other application. The WLAN card is divided into two major sections: the PHY layer and the MAC processing area (see Figure 4-9). An antenna receives the 802.11b signal, and this is often sent through frequency doublers (squaring loops) so that the signal can be properly extracted from the carrier. If DSSS is used, a local synthesizer generates the code word and subsequently demodulates the bit stream. Typically, intermediate frequency gain modulation is used for extracting the bit stream. In such a case, a local frequency generator generates the frequency (carrier), and it is mixed with the incoming signal. The added and subtracted parts of the signal with the local frequency are processed, to make sure there are minimum effects of phase change and multipath transmission. The signal is then decoded using the gain modulator, and then converted in parallel to a digital signal from its analog state. For transmission, the reverse operation is carried out (see Figure 4-9).

Figure 4-9   *WLAN Card*

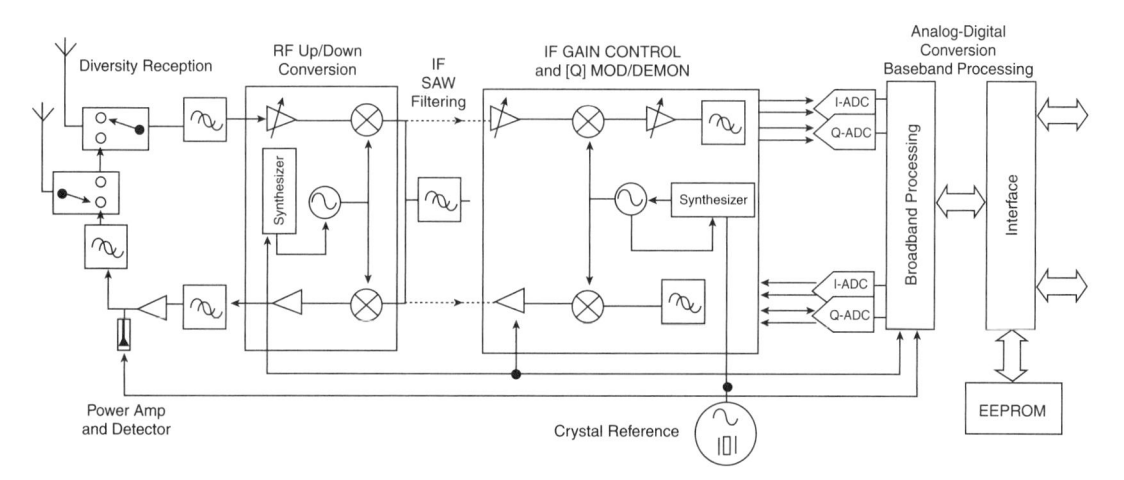

Operation of the IEEE 802.11 Standard

We now look at two key operations of 802.11 LAN networks from a first mile access perspective. First we discuss how a wireless station joins an AP and begins communication. The second operation is the process of handoff from one AP to another. In the first case, there are three periods of time when a WLAN user can seek to join an existing BSS. The three periods are as follows: when a WLAN user starts its process (powering up), when a user enters the sleep or dormant mode (not having any active transmission), and when it enters an area that is under the influence (e-m influence) of a BSS. The WLAN user now, while attempting to access the BSS, needs basic synchronization information. This information can be obtained from the APs or from fellow WLAN users already in the network (ad hoc mode). The 802.11b standard specifies two ways a user can obtain this synchronization information: active and passive scanning. In active scanning, WLAN users send request frames to an AP and repeat the procedure until the AP replies with response frames, thus completing the synchronization schedule. In passive scanning, a WLAN user scans the channels and tries to capture a beacon that is transmitted by the AP to some other user. This beacon frame has within it data pertaining to synchronization.

After finding the most accessible AP, a WLAN user then has to start communication with this AP. Before doing so, however, the AP must authenticate the new WLAN user. This ensures good security. The WLAN user gives the AP its network key (a shared key), which enables the AP to determine whether the WLAN user is authentic. Once authenticated, communication can proceed. After authentication, the next process is association. Association inherently implies the sharing of the WLAN user's information with multiple APs in the vicinity so as to create the necessary information base that pertains to the new WLAN user.

The very nature of a wireless LAN means that WLAN users are able to preserve network connectivity even when they are mobile. This means that when a WLAN user moves from one place to another, the user's network connection is maintained as long as it is within the range covered by the WLAN. However, we know that the WLAN itself covers a large area by creating multiple APs and BSSs. When a WLAN user moves from one BSS to another BSS without dropping the connection, this process is called *roaming*.

In the process of roaming, a WLAN user, while moving away from its primary AP (through which it is communicating to the network), realizes that the power on this link (to the incumbent AP) is getting weaker and weaker. When the user realizes this, the user begins to scan the channels to try to find a link (and hence another AP) with stronger power. When the user finds a new AP that can serve stronger power, it does an act of re-association with the new AP and hence completes the process of roaming.

The roaming process requires that the old AP be informed that it is no longer associated with the WLAN user and hence is free to add another user who wants association with the old AP. This is done through the distribution system (DS). The new AP that welcomes the WLAN user informs the DS of this, and then the DS informs the old AP of the transfer and successful roaming operation.

Performance of 802.11 Standards in First Mile Access

The first mile area is touted to grow tremendously in the next decade or so, because of its direct proximity to consumers and end users. Considering this, optical fiber would be the best solution in the first/last mile. However, the costs of laying an optical fiber combined with the difficulties of laying fiber at certain locations (for instance, in areas requiring right of way) have created a tremendous potential for wireless networking in the last mile. Because here we are talking about wireless networks to the end users, we are basically assuming that a wired network exists to the curb, and its from the curb to the end user that we need this wireless network platform. It saves us the deployment efforts required for a conventional network and accords us a fair degree of mobility, one that is necessary but also "deluxe" for home and enterprise networking.

The advent of the 802.11 standards for wireless LANs began almost a decade ago and has so far culminated in the release of four standards—namely, 802.11a, -b, -g and -h. Of them, the one that most affects wireless LAN networking is the 802.11b standard (discussed previously in this chapter). Of the many salient features proposed in the standard, two are worth mentioning here: the use of Ethernet frames, and ad hoc networking. The former enables us to create a universal method for data transport, one that allows the end-to-end use of Ethernet frames. This feature greatly simplifies wireless networking in the access area and makes it universally acceptable. The second feature, ad hoc networking, enables users to avail themselves of plug-and-play schemes, which bring to the fore the absolute power and essence of wireless networking—namely, its distributed approach and random behavior. A group of users could directly, set up a wireless LAN without the need for any kind of prior adjustments or settings at any location.

This kind of behavior, termed ad hoc, creates the necessary motivation for wireless networking, bringing to the fore the random and distributed nature so essential for deployment of wireless networks.

The 802.11 standards allow the use of high bandwidth to the tune of 1 Mbps, 2 Mbps, 5.5 Mbps, and 11 Mbps, and create a secure system, which prevents the network from being exposed to fraudulent use. The variable bit rates supported allow the system to be universally deployed, especially in areas where there are severe channel disturbances. The variable nature also is an excellent adaptation for channels that fade. The fading behavior of channels is countered by the variation in data rate, so as to create an efficient system. If the channel fades rapidly, for example, the data rate is rolled back, hence removing the pressure on retransmissions of lost frames and instead allowing higher (better/more) coding of the Ethernet frame so that the higher-encoded frame is successfully transmitted.

From the first mile access point of view, wireless networking represents an area of potential high revenue growth. The demand for high bandwidth is universal among end users, and laying fiber or providing a copper-based solution is not always practical. This is where wireless networking steps: Wireless networking provides a reliable, robust, and efficient solution, both in terms of cost and performance.

802.11g

The 802.11g standard operates at the same 2.4-GHz frequency band. It is same as 802.11b, but data transfer can happen up to 54 Mbps. It is backward compatible with 802.11b. This means that the users can in-service upgrade (gradual) from 802.11b to g. Due to high data transfer rates and because of the overall user friendliness, 802.11g is expected to become the de facto standard for 802.11 networks.

Wireless Data Applications: Hotspot Concept

A hotspot is a public location that offers high-speed WiFi connectivity, a system of unified hardware and operation centers that offers quality access to end users.

Summary

In this chapter we outlined the effects of wireless networking in the access area. Wireless networking was discussed as it relates to wireless LANs, and this chapter introduced TDM, FDM, and spread spectrum technologies. We also covered the QPSK modulation format for spread spectrum and discussed frequency hopping. In this chapter, you also learned about antennas, mathematics behind the antennas, and WLAN behavior.

The chapter concluded by considering WLANs from a first mile access network perspective. The next chapter discusses the complete picture for wireless networks in the first mile by considering metro access (the 802.16 IEEE standard).

Review Questions

1. Describe the main drawbacks of wireless LANs.
2. Compare wireless LANs with optical networks.
3. How does a WLAN function?
4. How many channels can be incorporated in 2.4 GHz to 2.483 GHz at 11, 5.5, and 2 Mbps?
5. Describe the advantages and disadvantages of DSSS over FHSS.
6. What is the primary deployment issue regarding FHSS?
7. Mathematically prove that 16-phase shift keying has a higher probability of error than 4-phase shift keying. Also, describe your understanding of this concept.
8. What are the advantages of ad hoc networking?
9. Describe how authentication occurs for a new wireless user in an 802.11b area.
10. How does an upgrade from 802.11b to 802.11g happen (software or hardware)?
11. Describe how smart antennas can accommodate more users, or can provide a higher level (bit rate) of service.

References

Application Note on 802.11 Physical Layer. Agilent Inc.

Chlamtac, I., and J. Lin. *Wireless Data Communication*. Hoboken, New Jersy: John Wiley & Sons, 1999.

Chlamtac, I. and W. Franta. "Local Area Networks," Minnesota, 1979.

Garg, Vijay Kumar. *Applications of CDMA in Wireless/Personal Communications*. Upper Saddle River, New Jersey: Prentice Hall PTR, 1996.

Keshav, Srinivsan. *An Engineering Approach to Computer Networking*. Boston: Addison-Wesley, 1997.

Kleinrock, L., et al. "Why Six Is a Magic Number." *IEEE Proc of Globecom Conference*, 1983.

Taub, Herbert and Donald L. Schilling. *Principles of Communication Systems*. New York: McGraw-Hill Science/Engineering/Math, 1986.

CHAPTER 5

Broadband Wireless Access (IEEE 802.16: WirelessMAN)

Chapter 4, "Data Wireless Communication," discussed the need for wireless data networking from the end-user perspective (especially the consumer). However, the 802.11b standard covered in Chapter 4 is suited only for applications confined to a small area—namely, LANs. The next issue we face in first mile access is to deliver higher-data-rate bandwidth to customers in their premises and to customers who are more geographically dispersed. 802.16, is a recent IEEE standard for wireless MANs. This standard addresses the issue of providing broadband access to end users at higher data rates and at comparatively longer distances. The standard is a true metropolitan-area application appropriately suited for customers to whom fiber cannot be laid and for whom copper solutions are not effective (in terms of bandwidth). We consider this standard to be appropriate for both metropolitan and access applications—specifically to enable first mile access users high-quality broadband access at significant distances within the last mile. In this chapter we discuss the implementation of wireless MANs, 802.16, from a first mile access network perspective. This chapter describes this powerful technology and discusses its benefits for broadband networking.

IEEE 802.16—The WirelessMAN Standard

IEEE 802.16 specifies the WirelessMAN air interface for wireless MANs. This standard defines the Media Access Control (MAC) layer and the physical layer (particularly operating at 10 GHz to 66 GHz) specifications of a fixed point-to-multipoint broadband wireless access system. The 10- to 66-GHz air interface is based on single-carrier modulation and is known as the *wirelessMAN-SC air interface* (SC standing for single carrier). An amendment to this standard is also defined: IEEE-SA-802.16a, to support 2 to 11 GHz using an enhanced version of the same basic MAC layer along with new physical layer.

The 10- to 66-GHz band provides a physical environment for the formation of fixed broadband wireless access (BWA) that supports end-user network access at speeds comparable to those offered by high-speed fiber-optic networks. Channels used in this physical environment are typically wide (25 or 28 MHz) with high-capacity uplinks and downlinks. The setup is ideal for point-to-multipoint access—serving large corporations, as well as medium and small office applications and enterprises—with data rates in excess of 120 Mbps.

Figure 5-1 shows typical wireless MAN network architecture. It consists of a base station (BS), which is connecting to a public switched network; and subscriber stations (SSs), which typically serve large buildings or residences and small offices. A single BS serves multiple SSs. The BS along with the SSs provide for the first/last mile access solution. Both BS and SSs are fixed (stationary devices) and can support multiple services and different quality of service (QoS) options simultaneously.

Figure 5-1 *WirelessMAN Architecture*

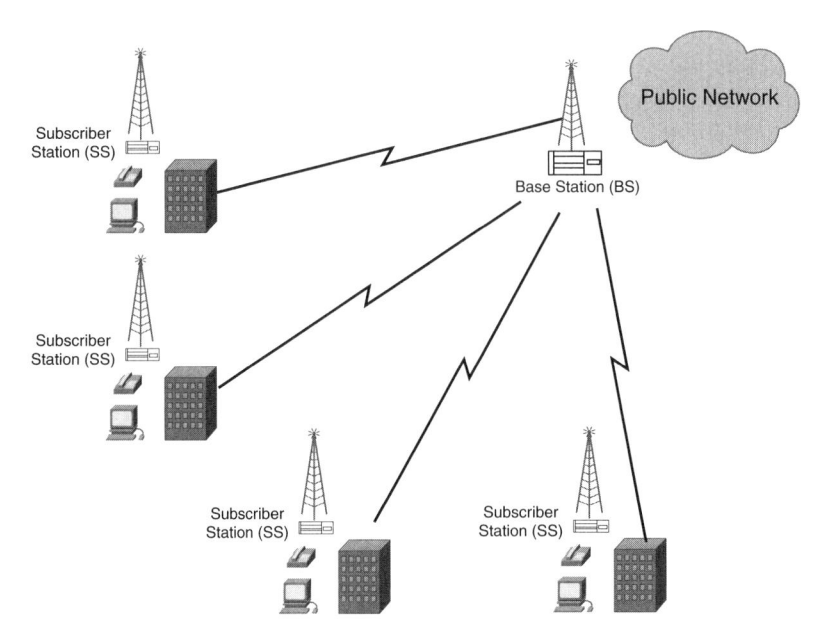

Reference Model

Figure 5-2 shows the reference protocol model described in IEEE 802.16. The MAC encompasses three sublayers:

- The service-specific convergence sublayer (CS)
- MAC common part sublayer (CPS)
- Privacy sublayer

Figure 5-2 *Reference Model*

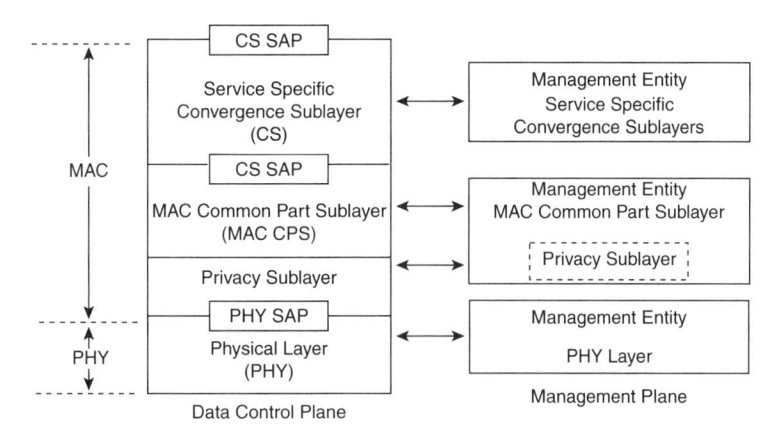

```
                    ┌─ CS SAP ─┐
                    │ Service Specific │ ←→  Management Entity
                    │ Convergence Sublayer │   Service Specific
                    │       (CS)       │      Convergence Sublayers
         MAC        ├─ CS SAP ─┤
                    │ MAC Common Part Sublayer │ ←→  Management Entity
                    │      (MAC CPS)      │       MAC Common Part Sublayer
                    │ Privacy Sublayer │ ←→   Privacy Sublayer
                    ├─ PHY SAP ─┤
                    │ Physical Layer │ ←→   Management Entity
         PHY        │     (PHY)     │         PHY Layer
                    Data Control Plane       Management Plane
```

The service-specific CS provides mapping of external network data, received through the service access point (SAP) into the MAC service data units (SDUs). MAC SDUs are received by MAC CPS through the MAC SAP. Classifying external network SDUs and associating them to the appropriate service flows and connection identifiers (CIDs) is done by this part of the MAC sublayer. Multiple specifications are provided for interfacing with various protocols. MAC CPS is responsible for the core MAC functions, such as system access, bandwidth allocation, connection establishment, and connection maintenance. MAC CPS receives data from the various service-specific CSs through the MAC SAP (physical access point) assigned to a particular connection. QoS is applied to the transmission and scheduling of data over the physical layer (PHY), in this way maintaining adequate QoS. The MAC also contains a separate privacy sublayer, responsible for authentication, secure key exchange, encryption, and other security-related activities. The physical layer addresses multiple specifications, each appropriate to a particular frequency range and application.

The IEEE 802.16 MAC is designed to support high-bit-rate, point-to-multipoint broadband wireless solutions and connectivity. The MAC is capable of delivering high bit rates, both in the upstream and downstream directions. The design also accommodates access to hundreds of terminals per channel, with these terminals being shared by multiple end users. The IEEE 802.16 MAC also supports both continues (voice) and bursty traffic (IP centric), which makes it ideal for both voice and data communication as well as high-QoS applications.

Physical Layer

The communication is typically limited to the line of sight in the 10- to 66-GHz band due to small wavelength (mm range), severe attenuation, fading, and scattering of electromagnetic waves in free space at these frequencies. SC modulation was selected as the default modulation format, and the air interface was designated as *WirelessMAN-SC*. In the point-to-multipoint network, the transmission from BS to the SSs is based on time-division multiplexing (TDM), with each SS having allocated time slots. Transmissions from SSs to the BS are by TDM access (TDMA)—that is, by using statistical TDM, meaning time slots are not of fixed duration.

Several duplex-oriented schemes are supported by the MAC protocol. The recommended burst design supports both time-division duplexing (TDD) and frequency-division duplexing (FDD). The choice of duplexing schemes influences the physical layer design and the features associated with the design. In a TDD scheme, the uplink and downlink share a single channel and do not transmit simultaneously. In the FDD, the uplink and downlink transmit on separate channels; therefore, it can be simultaneous. Both the TDD and FDD support adaptive burst profiles in which modulation and coding options are dynamically assigned on a per-burst basis. The adaptive burst profiling technique used in WirelessMAN has the capability to adjust transmission parameters, including the modulation and coding schemes. The adjustments can be done to each SS on a per-frame basis; these parameters and ranges are supported by MAC layer signaling. Both TDD and FDD support adaptive burst profiling.

The forward error correction method used is Reed-Solomon GF(256) (which is inherently Reed-Solomon code [RS code]) with variable block sizes and error correction/parity check modes. To maximize air link use, the physical layer supports a multilevel modulation scheme. The modulation scheme can be selected based on the quality of the radio-frequency (RF) channel on a per-subscriber basis. More complex schemes can be used to maximize the air link throughput. If the air link degrades due to rain, fog, or other environmental factors, the system reverts to less-complex schemes to allow reliable data transfer. The system inherently supports Quadrature Phase Shift Keying (QPSK), 16-QAM (Quadrature Amplitude Modulation), and 64-QAM modulation schemes. The date in the transmitter is encoded using forward error correction (FEC) and then is mapped to a QPSK, 16-QAM, or 64-QAM constellation. Figure 5-3 shows the effects of QPSK, 16 QAM, and 64 QAM.

Figure 5-3 *Adaptive Modulation Schemes*

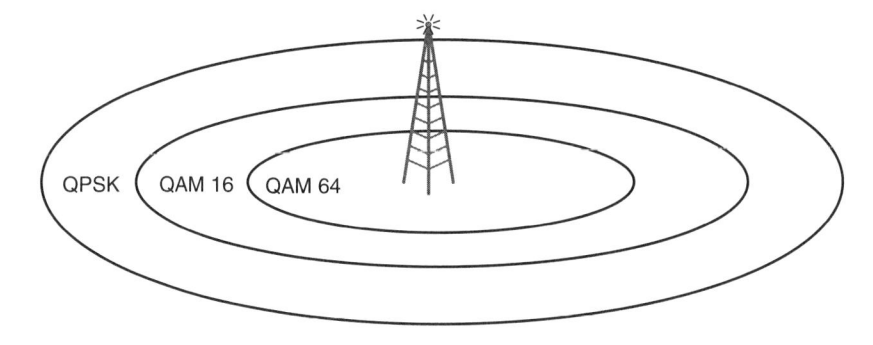

The 802.16a project addresses the 2 to 11-GHz band for both licensed and unlicensed spectrums. Because the 2 to 11-GHz band is not limited to line-of-sight communication, this band is best suited for residential applications, where rooftop antennas are not preferred.

Burst FDD and TDD Operations

In a burst (framed) FDD system, separate channels are allocated for upstream and downstream transmission of the frame. A fixed-duration time slot (for transmission) is used for both uplink and downlink transmission, which allows simultaneous use of both full- and half-duplex SSs. Figure 5-4 shows basic burst FDD mode operation.

Figure 5-4 *Example of FDD Frame Transmission (Courtesy and Adapted from IEEE Standard 802.16)*

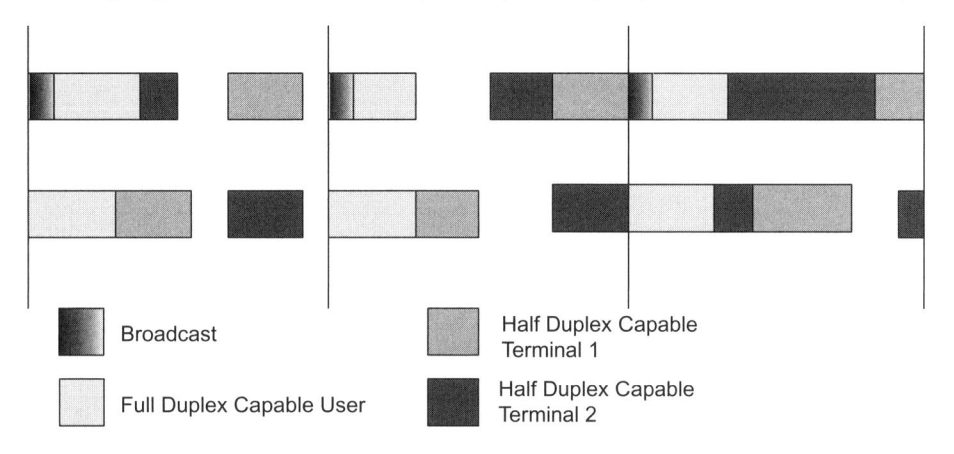

A full-duplex SS is capable of continuously listening to the downlink channel. A half-duplex SS can listen to the downlink channel only when it is not transmitting in the uplink channel.

In the TDD system, the uplink and downlink channel transmission occurs at different time slots while sharing the same frequency. A TDD frame contains a downlink and an uplink subframe that has a fixed duration. The frame is divided into an integer number of physical slots (PSs). Figure 5-5 shows the basic TDD operation. The bandwidth allocated to the downlink versus the bandwidth accorded to the uplink can differ (adaptive). The higher layer in the network controls the difference in the uplink and downlink parameters.

Figure 5-5 *TDD Operation (Courtesy IEEE WirelessMAN Standard 802.16)*

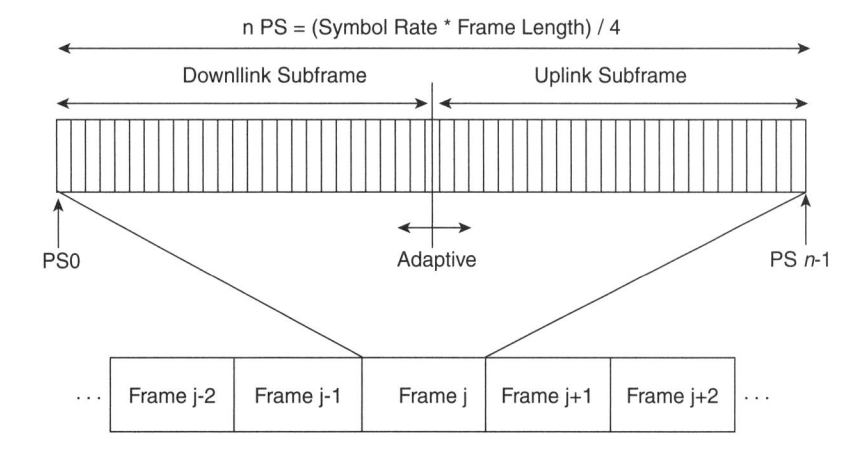

Several management messages are defined, out of which downlink map (DL-MAP) and uplink map (UL-MAP) messages are the most important. The DL-MAP message defines the usage of the downlink intervals for a burst mode. The DL-MAP message defines the access to the downlink information. Table 5-1 shows the DL-MAP message format.

Table 5-1 *DL-MAP Message Format*

Syntax	Size
DL-MAP Message Format	
Management message type =2	8 bits
PHY synchronization field	Variable
DCD count	8 bits
Base station ID	48 bits
Number of DL_MAP elements	16 bits
Begin PHY-Specific Section	

The following are definitions of the contents of the DL-MAP message format:

- **PHY synchronization**—Depends on the PHY specification used. The encoding is given in each PHY specification separately.
- **DCD count**—Describes the downlink burst profile that applies to this map.
- **Base station ID**—An externally programmable 48-bit-long field identifying the BS. The most significant 24 bits are used as the operator ID.
- **Number of elements**—Specifies the number of information elements that follows after the base station ID.

The UL-MAP message defines the usage for the uplink channel. Table 5-2 shows the UL-MAP message format.

Table 5-2 *UL-MAP Message Format*

Syntax	Size
UL-MAP Message Format	
Management message type =3	8 bits
Uplink channel ID	8 bits
UCD count	8 bits
Number of UL-MAP elements	16 bits
Allocation start time	32 bits
Begin PHY-Specific Section	

The following are definitions of the contents of the UL-MAP message format:

- **Uplink channel ID**—Identifier of the uplink channel to which this message refers.
- **UCD count**—Describes the uplink burst profiles that apply to this map.
- **Number of elements**—Number of information elements in the map.
- **Allocation start time**—Effective start time of the uplink allocation defined by the UL-MAP in units of minislots.

Map Information Elements

Each information element (IE) consists of the following three fields:

- Connection Identifier (CID)
- Uplink Interval Usage Code (UIUC)
- Offset

IEs define uplink bandwidth allocations. Each UL-MAP message contains at least one IE that marks the end of the last allocated burst. The CID represents the assignment of the IE and can be either a basic CID of the SS or the transport CID for one of the connections of the SS. A UIUC defines the type of uplink access and the uplink burst profile associated with that access.

Physical Layer Support for Subframing

Within each frame (defined in the section "Burst FDD and TDD Operations") are downlink and uplink subframes. The system uses a frame of 0.5, 0.1, or 2 ms and is divided into slots for bandwidth allocation. In TDD transmission, the uplink follows the downlink (in other words, the downlink comes first), whereas in FDD operation, both uplink and downlink frame transmission occurs simultaneously. In the downlink direction, the available bandwidth is defined with one PS granularity.

NOTE A PS is a unit of time, depending on the physical layer specification for allocating bandwidth. A minislot is a unit of uplink bandwidth, allocating the equivalent to N PSs where $N = 2^m$ and m is an integer within the range 0 to 7.

In the uplink direction, the available bandwidth is defined with a granularity of one minislot. One minislot length is given as $2^m * PS$ (m = 0 through 7), and the number of PSs within each frame is a function of the symbol rate. Note here that in QPSK and QAM types of modulation formats, we use symbols for transmission rather than just logical 0s and 1s.

Figure 5-6 shows a TDD downlink subframe structure. The downlink frame begins with a preamble (start of frame) used by the PHY medium for synchronization and equalization. The frame control section contains the downlink map and the uplink map messages. Following the frame control section, a downlink subframe typically contains a TDM section, which carries the data. The data is organized into bursts, with different burst profiles and different levels of transmission robustness (transmission characteristic types). The bursts are transmitted in order of decreasing robustness—that is, data begins with QPSK modulation, followed by 16-QAM, and finally 64-QAM. Each SS decodes the control information of the downlink frame, and MAC headers indicate the locality of the data (whether for itself or not) for a particular SS in the remainder of the subframe.

Figure 5-6   *TDD Subframe (Courtesy IEEE 802.16)*

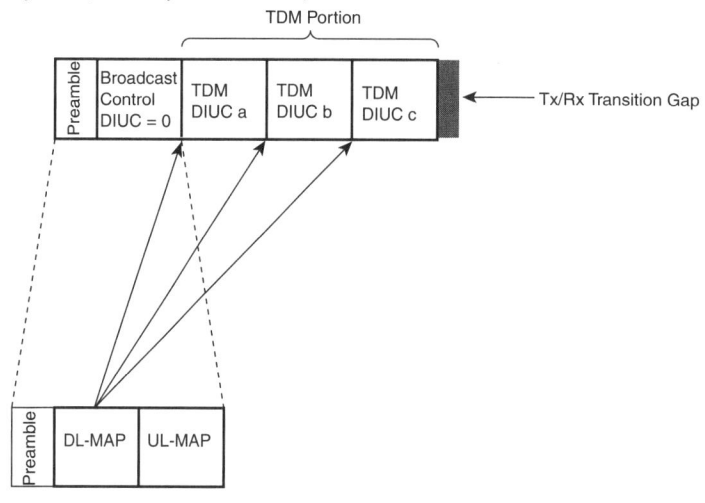

In FDD transmission, as in the TDD transmission, the downlink subframe begins with a preamble followed by a frame control section and a TDM portion. The TDM portion of the downlink subframe contains data for full-duplex and half-duplex SSs. The FDD downlink subframe carries a TDMA portion used to transmit data to half-duplex SSs. This setup allows an individual SS to decode a specific portion of the downlink without decoding the entire subframe. A TDMA burst preamble is transmitted within each TDMA portion for synchronization purposes. The frame control section contains a map for both the TDM and TDMA bursts. Figure 5-7 depicts the FDD subframe structure.

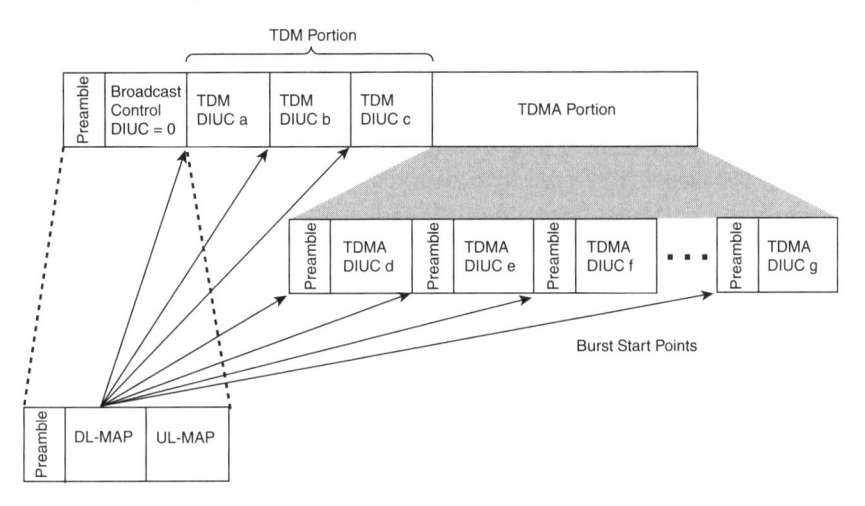

Figure 5-7 *FDD Downlink Subframe Structure (Courtesy IEEE 802.16)*

Between the physical layer and the MAC layer is a transmission convergence (TC) sublayer. The function of this layer is to transform variable-length MAC protocol data units (PDUs) into fixed-length FEC blocks of each burst size suitable for transmission. The PDU starts with a pointer indicating where the next MAC PDU header begins (see Figure 5-8). The TC PDU format allows synchronization of the next MAC PDU in the event a fatal error occurs to a previous block during its transmission.

Figure 5-8 *MAC PDU Header*

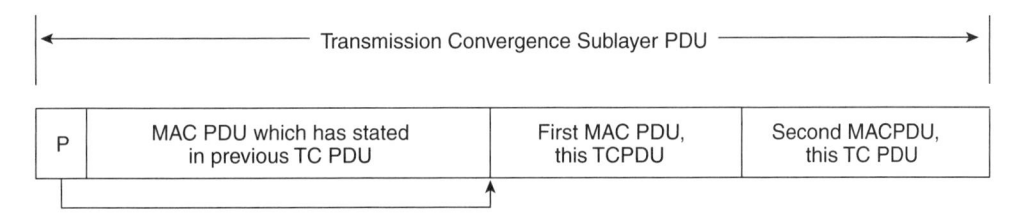

Without the TC layer, there is a potential for the entire burst to be lost in case of an irrecoverable bit error.

Modulation Schemes

The system uses a multilevel modulation scheme to maximize the use of the air link. If the link condition allows, a more complex modulation scheme can be used to maximize throughput while providing a reliable data communication path. Because of environmental factors, if the air link degrades, the system reverts to less-complex modulation schemes, thus retaining reliable data communication. In the downlink, the BS supports QPSK, 16-QAM, and 64-QAM; and in the uplink direction, modulation used by the SS is variable and is in turn set by the BS. Even though QPSK is the recommended modulation for the SS, 16-QAM and 64-QAM are also supported if the link condition permits. The sequence of modulation bits that are mapped onto a sequence of modulation symbols is given by M(x), where x is the corresponding symbol number. The number of bits per symbol depends on the modulation type: For QPSK, $n = 2$; for 16-QAM, $n = 4$; and for 64-QAM, $n = 64$, where $n = \max(x)$. The BS uses two basic power adjustment modes (constant peak power and constant mean power) when changing from one burst profile to another. The power adjustment rule is configurable through channel-encoding parameters.

Figure 5-9 depicts bit mapping for QPSK modulation, and 16-QAM constellation. Phasors are vectoral representations of sinusoids, which represent the strength of the signal in vectoral form—that is, in terms of amplitude and direction.

Figure 5-9 *Bit Mapping for QPSK Modulation (a) and 16-QAM Constellation (b)*

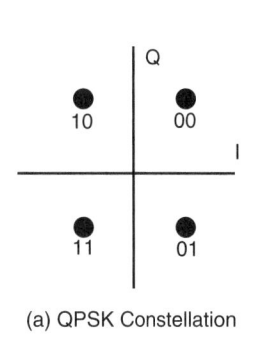

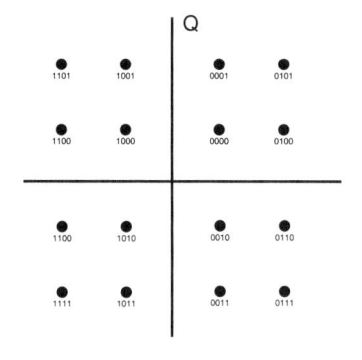

(a) QPSK Constellation

(b) 16-QAM Constellation

Note the amplitude dependence on the phasors.

Privacy Sublayer

The privacy protocol is based on the Privacy Key Management (PKM) protocol defined by the DOCSIS specification. The DOCSIS specification has been modified to fit into the MAC protocol defined by 802.16 and also includes strong cryptographic methods. Privacy is achieved by encrypting connections between the SSs and the BS. Privacy provides protection against theft of services and unauthorized access to the data transport network. A client/server key management protocol (with digital certificate) is used, where the BS is the server that controls and distributes the key to the clients (SSs). Privacy has two component protocols: Packet Encapsulation Protocol for encrypting packets across the wireless MAN access network, and a key management protocol for providing the secure distribution of keying data from the BS to SSs.

Packet Data Encryption

Encryption services are defined within the MAC privacy sublayer. MAC header information that is specific to encryption is located in the generic MAC header. MAC PDU payload is always encrypted, but the generic MAC header is not always encrypted.

Key Management Protocol

The key management protocol uses X.509 digital certificates, the RSA public key encryption algorithm, and strong symmetric algorithms are used to perform key exchanges between SSs and the BS. The Privacy Key Management (PKM) protocol uses public key cryptography to set up a shared secret key (such as an authorization key) between SSs and the BS. The shared secret key is then used to secure succeeding PKM exchanges of traffic-encryption keys. During the initial authorization exchange, a BS authenticates a client. Every SS is provided with a unique X.509 digital certificate, which is issued by the SS's manufacturer. The public key and MAC address of the SS is encoded in the digital certificate. During the authorization phase, an SS presents its digital certificate to the BS. The BS then verifies the digital certificate and associates a SS's authenticated identity to a valid customer. When requesting an authorization key, an SS presents its digital certificate to the BS. The BS associates an SS's authenticated identity to a valid paying customer, and hence to the authorized data services that the subscriber is allowed to access.

Security Association

Security associations (SAs) are identified using security association ID (SAID). The SA is a set of security information that a BS and SS share to support secure communication in an 802.16 wireless network. Three types of SAs are defined:

- Primary
- Static
- Dynamic

An SS during the initialization process establishes a *primary* SA. *Static* SAs are provisioned within the BS. In response to the initiation and termination requests, *dynamic* requests are established and torn down dynamically. Both static and dynamic SAs are shared by the multiple SSs. Using the PKM protocol, the BS ensures that each client only has access to the SA it is authorized to access and also specifies how the SSs and BS maintain key synchronization. Detailed analysis of the PKM protocol is beyond the scope of this book; for more information, refer to http://www.wirelessman.org.

MAC Common Part Sublayer

The MAC common part sublayer is responsible for key MAC functions such as addressing, service definitions, service requests, and service access grants.

MAC Addressing

Each SS uses a 48-bit unique MAC address defined by IEEE 802.16. This address is used during the registration process to establish connection and during the authentication process to authorize access to the network. A 16-bit CID identifies the connections. During initialization, three management connections in both directions are established between the SS and BS. To exchange short urgent MAC management messages, the BS MAC and SS MAC use the basic connection. To exchange longer management messages, the BS MAC and SS MAC use the primary connection. The secondary management connection is used to transfer configuration files, Dynamic Host Configuration Protocol (DHCP) information, Trivial File Transfer Protocol (TFTP), Simple Network Management Protocol (SNMP), and so on. These messages are carried in IP packets. For data services, the higher layers initiate connections based on the provisioning information. The registration or modification of the service contract at an SS stimulates the BS to initiate the set of connections.

Figure 5-10 shows a typical MAC PDU. Each PDU begins with a generic MAC header, followed by the variable payload. The variable MAC PDU allows the MAC to tunnel various higher-layer traffic flows transparently.

Figure 5-10 *MAC PDU*

Generic MAC Header	Payload (Optional)	CRC (Optional)

Two direct types of MAC header formats are defined:

- **Generic header**—Contains data or MAC management messages
- **Bandwidth Request header**—Used to request additional bandwidth

A single bit Header Type (HT) field distinguishes the Generic and Bandwidth Request header formats; the HT field is set to zero for a Generic header and set to one for a Bandwidth Request header.

A Generic header is shown in Figure 5-11, and a Bandwidth Request header is shown in Figure 5-12.

Figure 5-11 *Generic Header*

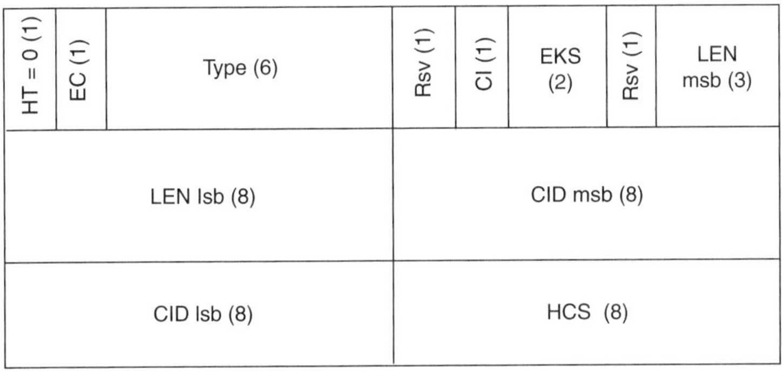

HT = 0 (1)	EC (1)	Type (6)	Rsv (1)	CI (1)	EKS (2)	Rsv (1)	LEN msb (3)
LEN lsb (8)				CID msb (8)			
CID lsb (8)				HCS (8)			

CI(1)-CRC Indicator
1 = CRC is appended to the PDU.
0 = No CRC is appended.

CID(16)-Connection Identifier

EC(1)-Encryption Control.
0 = Payload is not encrypted.
1 = Payload is encrypted.

EKS(2)-Encryption Key Sequence
The index of the Traffic Encryption Key and Initialization Vector used to encrypt the payload. This field is only meaningful if the Encryption Control field is set to 1.

HCS(8)-Header Check Sequence
An 8-bit field used to detect errors in the header. The generator polynomial is $g(D)=D^8 + D^2 + D + 1$.

HT (1)-Header Type. Set to zero.

LEN(11)-Length
The length in bytes of the MAC PDU including the MAC header.

Type(6)-This field indicates the payload type, including presence of subheaders.

Figure 5-12 *Bandwidth Request Header*

HT = 1 (1)	EC=0(1)	Type (6)	BR msb (8)
BR lsb (8)			CID msb (8)
CID lsb (8)			HCS (8)

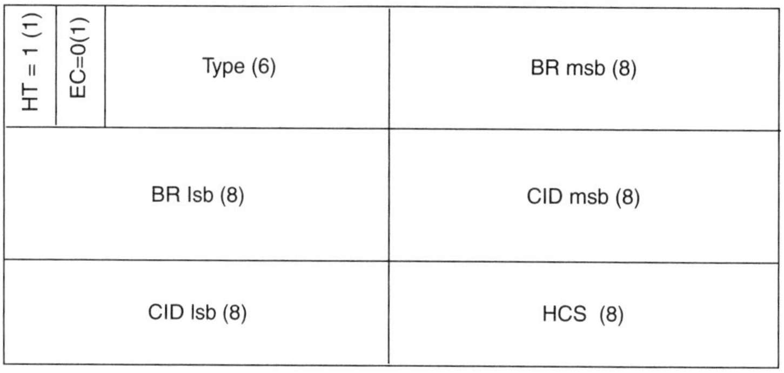

BR(16)-Bandwidth Request. The number of bytes of uplink bandwidth requested by the subscriber station. The Bandwidth Request is for the CID. The request does not include any PHY overhead.

CID(16)-Connection Identifier.

EC(1)-Always set to zero.

HCS(8)-Header Check Sequence. An 8-bit field used to detect errors in the header. The generator polynomial is $g(D)=D^8 + D^2 + D + 1$.

HT(1)-Header Type = 1.

Type(6)-Indicates the type of Bandwidth Request header.

Three types of MAC subheader may be present:

- **Grand Management subheader**—Used by an SS to convey bandwidth-management messages
- **Fragmentation subheader**—Contains information that indicates the presence of fragmented payload
- **Packing subheader**—Indicates that the MAC PDU is made up of multiple MAC service data unit(s) (SDUs). For more information on the MAC header and subheaders, refer to the IEEE 802.16 standard.

Scheduling Services

Scheduling services improve efficiency of the system. A BS can predict the throughput and latency needs by specifying scheduling services and QoS parameters associated. The scheduling services in WirelessMAN (802.16) reuse the services defined by cable modem DOCSIS standards. Table 5-3 summarizes the services. Each of these services is fine-tuned for specific data flow.

Table 5-3 *Services*

Scheduling Type	Piggyback Request	Bandwidth Stealing	Polling
Unsolicited Grant Service (UGS)	Not allowed	Not allowed	PM bit is used to request a unicast poll for bandwidth needs of non-UGS connections.
Real-time Polling Service (rtPS)	Allowed	Allowed for GPSS	Scheduling only allows unicast polling.
Non-Real-Time Polling Service (nrtPS)	Allowed	Allowed for GPSS	Scheduling may restrict a service flow to unicast polling via the transmission/request policy; otherwise, all forms of polling are allowed
Best-Effort (BE) service	Allowed	Allowed for GPSS	All forms of polling allowed.

The UGS is designed to support real-time service flows that generate fixed-size data packets on a periodic basis (T1/E1 and Voice over IP [VoIP] without silence suppression). The rtPS is designed to support real-time traffic flows with variable-size data packets on a periodic basis (MPEG video). The nrtPS is designed to support non-real-time traffic flows of variable sizes on a regular basis (huge file transfers). BE service provides efficient service to best-effort traffic (e-mail applications/the Internet).

Request: Grant per Subscriber Mode and Grant per Connection Mode

Request is defined as the mechanism that the SS uses to ask for needed uplink bandwidth. The request may come as standalone Bandwidth Request header or Piggyback and can request incremental or aggregated bandwidth. IEEE 802.16 defines two modes of operation to enable proper granting of bandwidth request:

- Grant per Connection (GPC) mode
- Grant per Subscriber Station (GPSS) mode

In GPC mode, the BS grants bandwidth connection explicitly to each SS. In GPSS mode, the bandwidth is granted to all the connections associated to the SS. GPSS mode is more useful for real-time applications that require smaller uplink (UL) maps and allows intelligent SSs to make last-moment decisions and use the bandwidth originally granted by the BS more efficiently. Both of these SSs request bandwidth per connection to allow the BS scheduling to consider QoS when allocating bandwidth. In GPC mode, the bandwidth requests are addressed explicitly to individual CIDs. In GPSS mode, the bandwidth requests are addressed to the individual CID, whereas the bandwidth grant is addressed to the SS's basic CIDs and not explicitly to individual CIDs. The GPSS SS needs more intelligence to handle multiple QoS connections. If the QoS requirement at the SS has changed since the last request, the SS has the option of sending the higher-QoS data along with a request to modify the bandwidth borrowed from a lower-QoS connection.

The 802.16 MAC uses a self-correcting protocol rather than an acknowledgment protocol for both classes of grants. A self-correcting protocol has lesser delay and requires lesser bandwidth compared to an acknowledgement-based protocol. Bandwidth request may not be available due to several reasons, including the following:

- The BS or the SS did not see the request due to PHY errors or collision of a contention-based reservation,
- The BS did not have enough bandwidth available to grant.

The self-correcting protocol treats all these conditions the same. After a timeout period for the QoS, the SS re-requests the grant. There are numerous ways to request bandwidth; for detailed analysis, refer to the 802.16 standard.

Contention Resolution

The BS controls assignments on the uplink channel through the UL-MAP messages and resolves which minislots are subjected to collisions. Collisions occur during initial maintenance and request intervals defined by their respective IEs. The contention resolution is based on a truncated binary exponential backoff scheme, where the BS controls the initial backoff window (how much do we retransmit based on the available window) and maximum backoff window. Backoff means the BS backs off from sending more data when it sees a collision in the channel. When an SS is ready to send data, it enters the contention resolution process. It sets its initial backoff window equal to the request backoff (initial window set up for communication). The SS selects a random number within its backoff window. This random number indicates the number of contention transmission opportunities that the SS will defer before transmitting.

The SS then waits for a data grant from the BS; once received, the contention transmission is complete and the SS is ready to transmit. The contention transmission is considered lost if no data grant has been received within the response time (range). The SS repeats the process with a new backoff window, and the retransmission continues until the maximum number of retries has been reached. The PDU is discarded when the SS reaches the maximum retransmission limit. (The maximum number of retries is independent of the initial and maximum backoff windows that are defined by the BS.) The BS has the flexibility to control the contention-resolution process by using request and backoff and emulating an Ethernet style of contention resolution.

Initialization

Figure 5-13 shows the initialization procedure of an SS.

Initialization   **129**

Figure 5-13 *Subscriber Station Initialization Process*

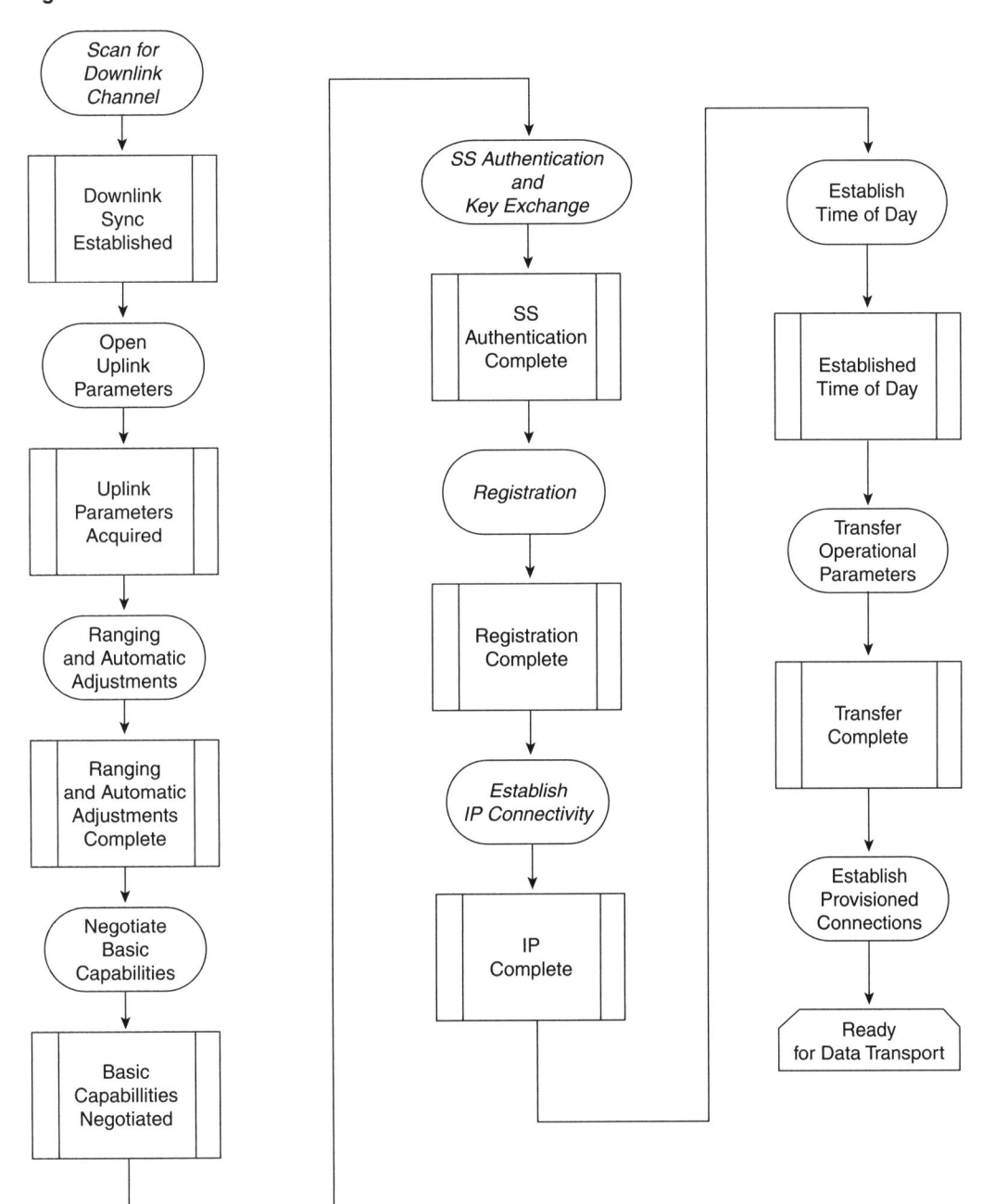

The initialization can be divided into several phases, as follows:

- **Scanning and synchronization for downlink channel**—During initialization or after signal loss, the SS tries to acquire a downlink channel. For a signal-loss condition, the SS tries to reacquire the original channel from the information stored in the nonvolatile memory. If it fails to acquire the original channel, it scans for possible channels of the available downlink frequency band until it finds one. When the physical layer has achieved synchronization, the MAC layer initiates the channel-control parameters.

- **Obtain transmit parameters**—When synchronization is achieved, the SS waits for a control message from the BS in order to retrieve transmission parameters. These messages are transmitted periodically for possible uplink channels by the BS and are addressed to the MAC broadcast address.

- **Perform ranging**—Ranging is the process by which the SS acquires correct timing offset parameters. The SS's transmissions are aligned to a symbol that marks the start of a minislot. The delay through the physical layer is kept constant and accounted for by the guard time (safe time to minimize substation transmission overlap). Automatic ranging adjustments are also defined by IEEE 802.16.

- **Negotiate basic capabilities**—After the ranging procedure is complete, the SS and BS exchange a basic capabilities message. The SS transmits a subscriber basic capability request (SBC-REQ) with its capabilities set to on. The BS responds with the intersection message capabilities set to on.

- **Authorize subscriber station and perform key exchange**—Authorization and key exchange are done as explained in the section "Privacy Layer."

- **Perform registration**—Registration is the process by which the SS receives its secondary management CID and thus becomes manageable. To register, the SS has to send a register request (REG-REQ) message to the BS. The BS responds with a register response (REG-RSP) message, which includes the secondary management CID.

The next step is to establish IP connectivity, followed by establishing time of day, and then transferring of operational parameters. During the IP connectivity phase, the BS also negotiates which version (4 or 6) of IP to use during communication.

Finally, connection is set up using all the QoS parameters and the SS is ready for data transmission. The preceding stages do not take into consideration error situations. Each stage has an error-handling function that gets initiated if an error occurs, but it is beyond the scope of this book to discuss error handling.

Propagation Conditions

To obtain high quality and high availability of the service offered by 802.16, line-of-sight radio propagation is required. The SS requires highly directional antennas to minimize the multipath and interferences. Multipath can cause intersymbol interference (ISI) and is not desirable.

The predominant fade mechanism deals with rain attenuation in the 10- to 66-GHz frequency region. Fade depth is a function of rain rate and dependent on the frequency of operation and distance. An internationally accepted method for computation of rain-fade attenuation probability is that defined by ITU-R P.530-8. For any given SS and BS, the maximum cell radius depends upon the rain-fade factor.

Design Considerations

The range of the system for given availability with given rain fading is calculated by estimating the link budget. Worst-case rain fade is considered in system design. The level of the desired received signal decreases until it equals the receiver thermal noise plus the specified signal to noise ratio specified at the receiver end. The amount of interference is calculated by taking the ratio of carrier level (C_L) to the sum of noise (N) and interference (i), given by $C_L/(N + i)$. The link-budget degradation is related to received power-flux density tolerance, for a given receiver noise figure (NF) and antenna gain in any given direction. This tolerance can be turned into separation of distances for various scenarios.

Electromagnetic propagation over frequency ranges of 10 to 66 GHz is comparatively nondispersive. The absorption of emissions by structures and buildings is significant, which limits the requirement to line-of-sight communications for reasonable performance. The operation range is within a few kilometers due to larger free space loss and is also limited by thermal and interference noise. The radio systems are susceptible to interference from emissions well beyond their operational range. The rain cells producing the most severe rain losses are not uniformly distributed over the operational area. This creates the potential situations in which the desired signal is severely attenuated but the interfering signal is not.

Design Issues

Radio waves infuse through boundaries (national and regulated), and emissions (electromagnetic) spill outside the spectrum allocations. This poses a challenge for two operators to coexist in the same geographical location.

Two recommended approaches are as follows:

- **Co-channel scenario**—Two operators are in either adjacent territories or territories within radio line of sight of each other and have the same spectrum allocation.
- **Adjacent channel scenario**—The licensed territories of two operators overlap and they are assigned adjacent spectrum allocations.

Table 5-4 shows guideline for the most extreme case among the several mechanisms that apply.

Table 5-4 *Guideline*

Dominant Interference Path	Scenario	Spacing (Interference Is Generally 6 dB Below Receiver Noise Floor)
Point to multipoint (PMP) (BS-BS)	Adjacent area Same channel	60 km
Mesh SSs to PMP BS	Adjacent area Same Channel	12 km
PMP BS to PMP BS	Same area Adjacent channel	1 guard channel
Mesh SSs to PMP SS	Same area Adjacent channel	1 guard channel

Services Offered and Relation to the First Mile Problem

The network operates in conjunction with the basic network backbone, and the users are not made aware that the services are delivered through radio networks. A typical 802.16 network supports hundreds of user/premises within a radio coverage area. The traffic demand is often statistically multiplexed, allowing the network to deliver significant bandwidth on demand with a high level of spectral efficiency. The range of applications includes voice, video, data, and entertainment services. The network supports traffic flows that are unidirectional, asymmetrical or symmetrical, and changes with time with a different mix of services. The mix of services changes rapidly as connections are established and terminated. These systems compete with other wired and wireless delivery mechanisms for the first/last mile access. Use of this technique results in a number of benefits, including rapid deployment and service creation. It is our recommendation that along with the 802.11 standard, the 802.16 standard is a viable approach to provide wireless connectivity—namely, in the broadband category—to end users in an access network. The 802.16 standard must be also considered with other contemporary solutions to create an effective and user-friendly first mile access network.

Summary

In this chapter we covered an emerging standard for wireless metropolitan access networks. The need for which is evident in broadband first mile access networks for the transport of traffic to the end user on high-speed backbone links. As we discussed in Chapter 4, the emergence of the 802.11b standard has greatly thrust wireless data networking into the LAN area; this chapter concentrated on the entire metro access area, which is a prelude to first mile access. The emergence of the 802.16 standard, and the strong indications of its warm acceptance by the industry, has created a need for wireless MANs. These networks are typically beneficial for enterprise and small business solutions, where fiber cannot be laid and where copper does not provide sufficient enough bandwidth. Despite the transmission limitations seen in the wireless area, the wireless MAN is an excellent candidate for best-effort IP kinds of traffic.

Review Questions

1. Differentiate between 802.11 and 802.16 in terms of performance, coverage, and implementation.
2. Is Ethernet a right choice for frame transmission in wireless MANs?
3. Differentiate between multiple modulation formats for WMANs.
4. Develop an algorithm to place antennas, taking line of sight into consideration to create the following:

    a) A ring with 5 nodes

    b) A ring with 8 nodes and diverse protection (The work and protection path cannot be on the same geographical path.)

5. How does fading affect transmission?
6. Describe the 3 sublayers in the MAC for 802.16.
7. Describe the performance benefit of adaptive modulation. What is backoff?
8. How does FEC help in bettering system gain?
9. How does registration of subscriber stations occur in 802.16?
10. How is security achieved for multiple subscriber stations in the same network?
11. Describe the two approaches for assigning spectrums in the same geographic area. Compare these and identify the better one.
12. Design an end-to-end wireless network with N number of WLANs that are fed by a backbone of WMAN (802.16). Each of these N WLANs supports K users. These users run some QoS-sensitive applications.

    Develop this system and describe how to provide end-to-end QoS in such a network.

    Develop an algorithm to enhance the system performance, taking into consideration end-to-end operation.

    Would there be any adverse effect, such as intersymbol interference or channel smudging, due to the interaction between 802.16 and 802.11? If so, why?

Discuss scalability of end users K * N if more users are randomly added to WLANs. Will bandwidth per user increase or decrease? What should we do to maintain the same bandwidth? Is there any effect on the 802.16 network in this case?

How would the system perform if we knew the traffic pattern—that is, we could predict the burstiness profile of traffic beforehand?

References

IEEE. *IEEE Communications Magazine*. June 2002

IEEE Standard 802.16. IEEE, 2001

IEEE Standard 802.16a. IEEE, 2003

IEEE Standard 802.16c. IEEE, 2002

http://www.wirelessman.org

CHAPTER 6

Free Space Optics Solutions in the First Mile

Free Space Optics: Complement to Radio-Frequency Wireless

Of the many solutions seen so far in this book for solving the first mile access problem, we can conclusively map out a three-pronged approach. This trident of solutions is focused on wireless-, optical-, and copper-based mediums. Each of these three is discussed in details in Chapters 4, "Data Wireless Communication," 2, "Passive Optical Networks in the First Mile," and 8, "Power-Line Communication," respectively. They have their own advantages and disadvantages. Optical solutions such as coarse wavelength-division multiplexing (CWDM) and passive optical networking (PON) are the optical implementations that have been proposed for the access area. PON and CWDM are two technologies that can cater to excellent high-bandwidth applications. Both PON and CWDM exemplify high-quality connectivity, hence high bandwidth. However, both mean the deployment and use of fiber—an expensive medium to deploy everywhere. The second technology of focus here is the wireless solution, among which we have wireless LAN standards (such as the 802.11 series) and the wireless MAN standard (802.16), which have been covered in detail in Chapters 4 and 5, "Broadband Wireless Access (IEEE 802.16: WirelessMAN)."

Wireless technology exemplifies a ubiquitous geographic reach. This means that where fiber cannot go, wireless can. This kind of universal connectivity achieved through wireless communication creates a system that easily facilitates both mobile computing and wireless networking. However, such a system is quite limited in bandwidth. The reason for that is the limited bandwidth spectrum of wireless. Therefore, we have a trade-off: By using fiber, we get more bandwidth, but fiber cannot reach everywhere; on the other hand, wireless means universal reach, but the spectrum that we get through wireless itself is limited.

Considering this, we need a solution that can give high bandwidth, and this high bandwidth should be available globally. Free space optics represents a technological solution that allows universal (line-of-sight-limited) access with high bandwidth for the end user and thus combines the salient features of both optical and wireless technology.

By deploying a free space optical network in the first mile, we get connectivity where there is no fiber, and the bandwidth that we get through free space optics is much more than wireless. Hence we replicate the universal connectivity feature that is characteristic of wireless, and at the same time get performance that can be compared to a fiber solution.

What Is Free Space Optics?

Free space optics (FSO) is a nascent technology currently considered a strong contender to solve the first mile access problem. The FSO system consists of a transmitter and a receiver separated in distance and by atmosphere (free space). Communication occurs over air, the medium of communication. A laser first tracks a receiver and then emits light. Data is modulated onto the light. The receiver gets the beam of light and detects the beam, and, hence, there is one-way communication. Note here that there is loss in the medium, and the transmission is exposed to natural impediments, such as rain and fog (or even humans and animals), blocking the laser beam. The other issue in FSO is to track the receiver. Sometimes the receiver may be mobile or may not be aligned with the transmitter—that is, line of sight may not be achieved. In such cases, the transmitter may direct a beam, but the receiver may not be able to receive the beam correctly, resulting is significant data loss due to errors in transmission. To solve this issue, we need tracking. Key components of an FSO system are the transmitter, the receiver, and the tracking device. We study these and their operation in this chapter.

FSO Operation

FSO systems use wavelengths around 850 nm and 1550 nm because of low loss of signals (atmospheric absorption) in that operating frequency window and also because of the abundant availability of industry standard components (transmitters and receivers). Unlike low-frequency microwave systems, FSO does not require operating licenses and is not regulated by regulatory bodies (such as the Federal Communications Commission [FCC] in the U.S.).

A rudimentary FSO system consists of an FSO transmitter and a receiver. FSO systems require line-of-sight communication (without any obstacles between the transmitter and receiver), and these systems are typically deployed in a point-to-point network. Figure 6-1 shows a typical FSO system.

The transmitter is essentially a light source (LED/laser) complemented with a telescope assembly. The telescopic lens narrows the beam and projects the converged beam toward the receiver. The receiver subsystem is made of a photodetector and lenses or a mirror assembly (to collect the focused beam). The typical diameter of the receiver-side telescope is about 10 to 20 cm.

During transmission, the beam gets diverged (due to Huygen's wavelet principle), and the divergence varies between a few hundred microradians to a few hundred milliradians. The beam at the receiving end is much larger than the size of the lenses; therefore, part of the transmitted energy is lost. This loss, due to the disparity between the beam size and receiver subsystem's lenses is known as *geometric path loss*. The geometric path loss can be reduced by use of a narrow beam. The narrow beam requires a very complicated stable beam-tracking system, which can be very expensive and difficult to implement (analogous to a guided missile tracking an interceptor aircraft). Typical distances between transmitter and receiver are not more than 1 km. Beyond that distance, the beam fades, the light undergoes dispersion, and it is very difficult to track the receiver and maintain alignment.

Figure 6-1 *Schematic of FSO Transmission*

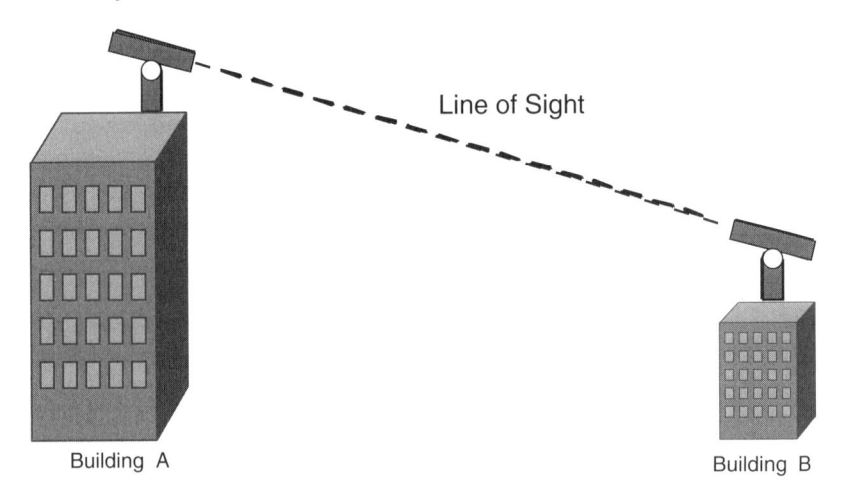

Line of Sight

Building A

Building B

Transmitters: Lasers

Lasers are used as optical sources for emitting modulated data into an optical fiber. Lasers have a distinct property whereby they can emit a narrow beam of light having a small optical spectrum (line width), while having a high-output optical power (a concentrated beam of photons of approximately the same phase and frequency). A laser is a semiconductor device (for optical purposes at least, although different forms of lasers do exist) whose operation is governed by the population inversion condition. The population inversion condition specifies the numeric superiority in volume of the electrons in the excited state (formed by absorption of energy by normal state electrons) over the electrons in the ground state in a semiconductor junction device. Optimally, a laser used in typical optical networking operation has a narrow spectral line width, in addition to fast response and the capability to couple a significant amount of optical power into the fiber, which in our case is replaced by the atmosphere as the medium.

Lasers used in optical communications are generally of two types: semiconductor lasers and fiber lasers. Semiconductor lasers are most commonly used in networking applications and are discussed here in detail. Fiber lasers are not so commonly used and are therefore not discussed. Semiconductor lasers are based on the optical properties of a p-n junction. Semiconductors as such have intermediate properties as compared to a conductor or an insulator. Silicon and germanium have been traditionally used as semiconductor materials. Indium phosphide (InP) and gallium arsenide have also recently found applications in lasers. A semiconductor material can be made p-type or n-type by doping the material (adding an impurity) with electrons (n-type) or extracting the material of some of its free electrons (p-type). An electron is a fundamental atomic particle of unit-negative charge and negligible mass. Its addition creates an n-type material, whereas its removal creates a p-type substrate. Removal of the electron is also called *addition of holes*, which are considered to be positively charged particles in theory but do not exist in reality. A semiconductor material that contains a region of p-type and a region of n-type with a shared boundary between the p and n types is called a p-n junction, as seen in Figure 6-2.

Figure 6-2 *P-N Junction Diode*

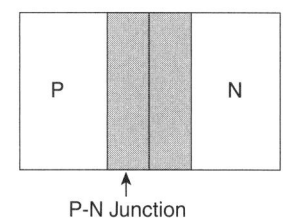

Free electrons in a semiconductor can flow when a voltage difference is applied linearly across the semiconductor, and this state is called the *conduction state* or, more appropriately, the electrons are considered to exist in the conduction band. An energy level is associated with the conduction band (that is, the energy of electrons in the conduction band is predetermined) and is given by Fermi-Dirac distribution equations. On passing a current across a semiconductor material, free electrons absorb a quanta (an integral multiple of $h\upsilon$; h being Planck's constant, and υ being frequency of absorbed radiation) of energy and jump into an excited state. After a period of time, these electrons (excited), which have absorbed the excess energy and have risen to a higher excited state, drop back to the original state by emitting the excess absorbed energy in the form of photons at frequency υ.

This random oscillation of electrons from a lower energy level to a higher energy level and subsequently emitting photonic radiation (light) is called *spontaneous emission*. The set of output optical frequencies is proportional to the energy between the stable and excited states, called the *bandgap energy*. In spontaneous emission, there is no frequency or phase matching between consecutively emitted photons. In other words, every emitted photon has random phase and frequency distribution. Spontaneous emission cannot sustain optical communication for the simple reason of low power and wide spectrums of emission. For lasers to function for FSO, the emission should be stimulated (that is, externally controlled).

Consider a case when a few electrons have adsorbed energy and risen to an excited state. Now assume that while these electrons are in the excited state an external photon is bombarded onto theses electrons. These electrons fall from the excited state to the ground state emitting photons, which have the same frequency as well as phase as the incident-bombarded photon—in other words, we would get a powerful beam of light at a controllable frequency, predetermined by the bombarded photon. To sustain such a source for a long period of time, however, it is necessary to ensure that at any given time there is an abundance of photons in the excited state. This kind of emission is called *stimulated emission* on account of the external stimulus involved in the emissive process. To sustain emission of this kind, a condition called *population inversion* needs to be accomplished. In other words, the number of electrons in the higher state (excited) should be greater than the number of electrons in the lower (stable) state. If this is not achieved, the emission is spontaneous (random phase and frequency distribution).

One possible way to achieve population inversion is by having multiple energy levels. The cut-in point, at which stimulated emission is the dominant emission in the system, is called the *lasing threshold*. When population inversion is established, the system would exhibit an optical gain on account of the feedback achieved due to the bombardment of photons on the excited electrons. This gain would amplify an optical signal exponentially. The optical gain is coupled by one more factor for good laser operation: the optical feedback. By placing the p-n junction inside a cavity consisting of reflecting walls, optical feedback can be achieved. Optical gain initiates the stimulated emission into a gain profile analogous to an electrical amplifier, while optical feedback ensures the oscillatory function of the p-n junctions.

Distributed Feedback (DFB) Lasers

By ensuring feedback, an oscillation function is realized in the previously discussed laser.

Feedback is realized by placing the p-n junction in a cavity that has full-reflecting walls on all but one side and a partial reflector on the remaining side. By inserting a grating (corrugated surface) within the cavity, as shown in Figure 6-3, optical feedback is achieved.

Figure 6-3   *Distributed Feedback Lasers*

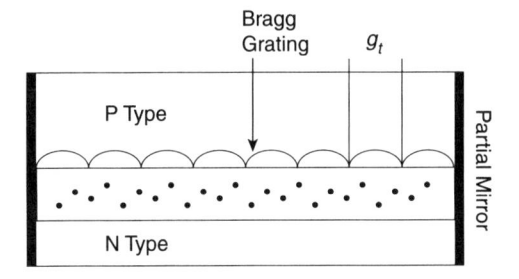

This optical feedback is called *distributed feedback* due to its diverse occurrence in the cavity. The feedback is essential for maintaining the lasing threshold, and is due to Bragg diffraction. When clusters of wavelengths hit upon a grating, only the wavelengths that correspond to the Bragg's wavelength are reflected. Only wavelengths corresponding to Bragg's condition are reflected. Bragg's condition is as follows (Agrawal, pp. 105, 301):

$$g_t = a\left(\frac{\lambda_B}{2n}\right)$$

In this equation, n is the refractive index, g_T is the period of grating, *a* is the order of Bragg diffraction, and λ_B is the Bragg's wavelength of our interest. These types of lasers are called DFB lasers (see Figure 6-3).

The feedback wave adds in phase to the emitted radiation, which is due to the drop of electrons from the excited state to the ground state. The grating formed is by methods of holography. The grating gives best performance for the Bragg's wavelength. The grating can be so formed that the periodic perturbations can be varied, giving optimum performance for a number of different wavelengths finding solid-application FSO networks.

Distributed Bragg Reflector Lasers

Distributed Bragg reflector (DBR) lasers are a conceptual extension to DFB lasers. In other words, the principle is very much the same; however, the feedback associated through the grating is extended through the entire region of the cavity (see Figure 6-4).

Figure 6-4   *Distributed Bragg Reflector (DBR) Laser*

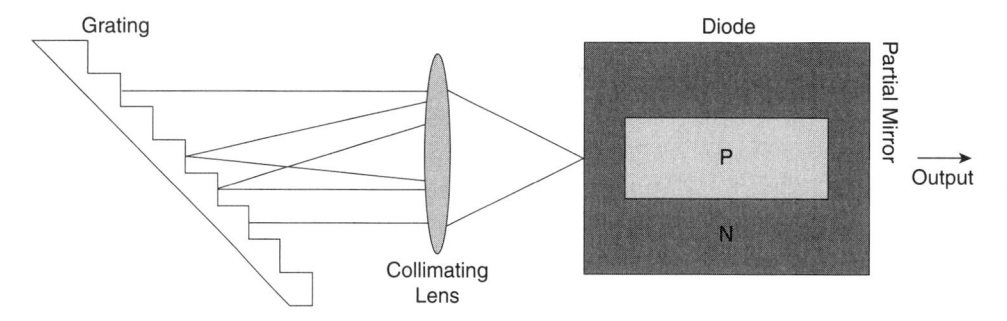

The corrugation or grating now extends to the mirrored walls, thus enhancing tunability. Wavelength tunability can be achieved by varying the grating periods outside the gain medium (p-n junction). For a DBR laser, Bragg's condition needs to be satisfied:

$$g_t = a\left(\frac{\lambda_B}{2n}\right)$$

DBR lasers are good candidates for tunable optical sources. DBR and DFB lasers are temperature dependent and, therefore, need temperature-controlling elements for stable uniform operation.

Receivers: Photodetectors

The function of an optical receiver is to decode and interpret the optical signals. The main component of an optical receiver is a photodetector, which converts the optical power into an electrical current. Photodetectors need to meet very stringent requirements to achieve desirable performance. Requirements include good responsivity (sensitivity) to a wide range of wavelengths used for transmission (usually in the 850-nm, 1300-nm, and 1500-nm region), low noise characteristics, low or no sensitivity to temperature variation, low cost, and extended operating life. Even though several types of photodetectors are available, semiconductor-based photodetectors (also known as *photodiodes*) are used exclusively for optical free space communications. The most common photodiodes used in FSO systems are the PIN photodetector and the Avalanche PhotoDetector (APD) due to their small size, fast response, high sensitivity, and comparably low cost.

The PIN Photodetector

The PIN diode is an extension of the p-n junction diode, where slightly doped intrinsic material (i-type; i stands for intrinsic) is inserted between the p-n junction, thus increasing the depletion width (region) of the p-n junction. The depletion region is the region between the p-n regions that is formed by some of the electrons from the n-type moving over and depleting the holes in the p-type, thus creating a region of neutral charge, upon condition of reverse bias. Figure 6-5 shows a PIN photodiode.

Figure 6-5   *PIN Photodiode*

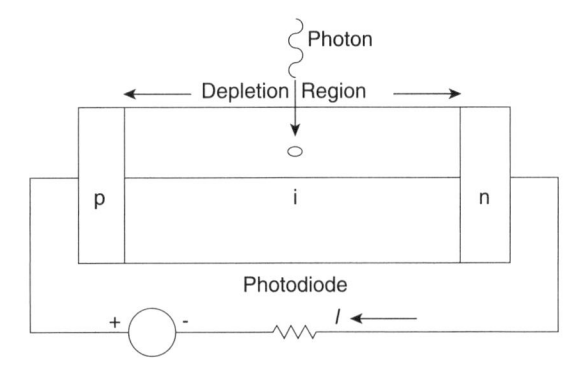

A high reverse-biased voltage is applied across the PIN diode so that the intrinsic region is completely depleted. Figure 6-5 represents the normal operation of PIN, with reverse bias applied across the PIN junction. When light (photons) is incident on a semiconductor material, it gets absorbed by electrons in the valence band. As a result of this absorption, the photons transfer their energy and excite electrons from the valance band to the conduction band, leaving holes in the valence band. The design of the PIN photodiode is optimized in such a way that electron-hole pairs are generated, mostly in the depletion region. See Figure 6-6. On the application of voltage across the depletion region, the formed electron-hole pairs induce an electric current flow (also known as *photocurrent*) in an external circuit. Each electron-hole pair generates one electron flow (Keiser).

Figure 6-6   *Energy Band Diagram for a Photodetector*

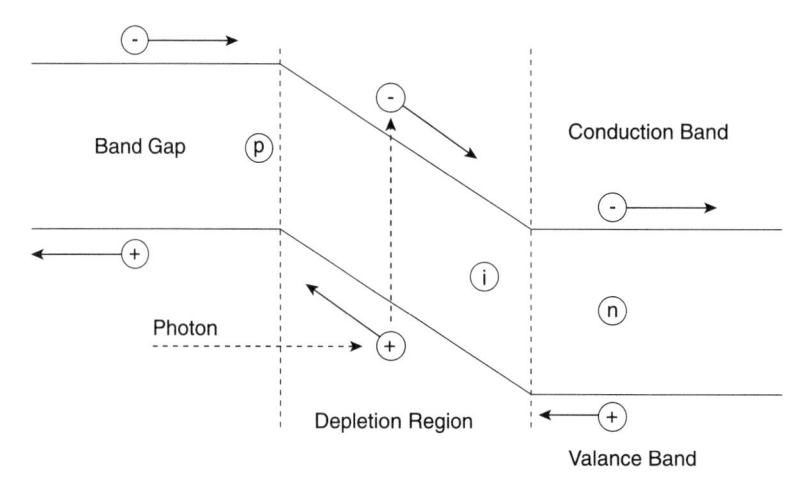

The energy of the incident photon must be equal to or greater than the band gap energy for the photocurrent to be generated.

The energy is calculated with the following equation:

$$hf_c = \frac{hc}{\lambda} \geq eE_g$$

In this equation, λ is the wavelength, E_g is the energy of the bandgap, c is the velocity of light, and e is the charge of the electron. λ is the wavelength at which the semiconductor material will function as a photodetector. There is an upper limit of wavelength (λ_{cutoff}) above which any particular semiconductor material will not generate a photocurrent. The cutoff wavelength is about 1.06μ m for silicon. Note that h here is Planck's constant.

An analysis of the photocurrent generated is beyond the scope of this book. For more information, refer to the resources listed in the "References" section at the end of this chapter.

The photocurrent (I_p) resulting from power absorption of photons is shown in the following equation:

$$I_p = \frac{eP_0(1 - EXP(-\alpha_s w))(1 - R_f)}{hv}$$

In this equation, P_0 is the optical power, e is the electron charge, hv is the photon energy, α_s is the absorption coefficient, and R_f is the reflectivity of silicon.

Important characteristics of photodetectors are quantum efficiency and responsivity. Quantum efficiency (η) is the number of electron-hole pairs generated per incident photon. Responsivity is the amount of current produced at a particular input optical power.

Responsivity of a photodetector is expressed in terms of λ, and is given by the following equation (Ramaswami, Ch. 2):

$$R = \frac{e\eta\lambda}{hc} = \frac{\eta\lambda}{1.24} Ampers/Watts$$

Typical PIN photodiode responsivity values are 0.65 A/w for silicon at 850 nm and 0.45 A/W for germanium at 1300 nm.

Avalanche Photodiodes

When light is absorbed by a PIN photodetector, only one electron-hole pair is generated per photon. The sensitivity of the detectors can be increased if more electrons are generated, which means less power is needed for photodetection; therefore, the signal can travel longer.

If a high electric field is applied to the generated electrons, it procures enough energy to excite more electrons from the valence band to the conduction band, which results in the generation of more electron-hole pairs. These secondary electron-hole pairs that are generated by this process can produce more electron holes if subjected to a high electric field.

This process of multiplication of electron-hole pairs is called *Avalanche multiplication* (or the *Avalanche effect*) (shown in Figure 6-7) and the photodiode that is designed to achieve this electron-hole pair multiplication is known as an *Avalanche photodetector (APD)*.

Figure 6-7 *Avalanche Multiplication Process*

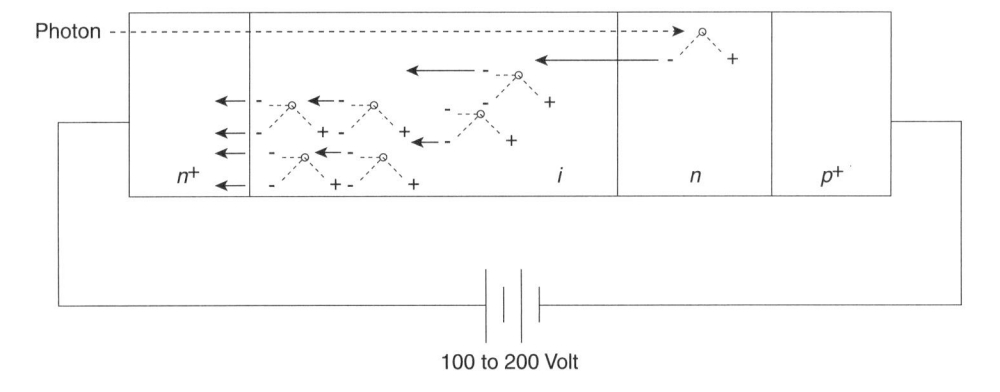

In practice, the Avalanche effect is a statistical phenomenon—that is, electron-hole pairs generated by the primary electron are randomly distributed. The statistical value is called the *multiplicative factor* or *multiplicative gain* (M_f) and is as follows:

$$M_f = \frac{I_a}{I_p}$$

In this equation, I_a is the average value of the total current (including the current generated due to Avalanche effect), and I_p is the current due to initial electrons. An APD can be designed with a multiplicative factor equal to infinity, a condition called *Avalanche breakdown*. However, large values of M will produce unpredictable effects in the generation of the photocurrent, which affect the noise performance of APDs (see Figure 6-8).

Figure 6-8 *Functional Diagram of an Optical Receiver*

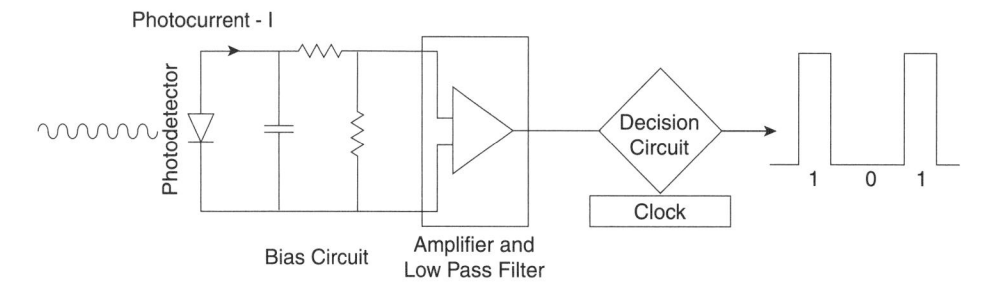

Optical Receivers

An optical receiver consists of a photodetector followed by a preamplifier. The function of a preamplifier is to amplify the photocurrent for further processing. The next stage consists of a high-gain amplifier and a low-pass filter. An amplifier gain control circuit automatically limits the amplified output to a fixed level, regardless of the optical power incident on the photodetector. The low-pass filter reduces the noise level and shapes the pulses. The low-pass filter is designed in such a way that the intersymbol interference (ISI) is minimized. Receiver noise is proportional to receiver bandwidth, and noise can be reduced by loss-pass filters with bandwidth (BW) less than the signal bit rate (B). The electric pulse spreads beyond the bit slot for BW < B and results in ISI, which interferes with proper detection of nearby bits.

The final stage of an optical receiver consists of a decision circuit and clock recovery circuit. The decision circuit compares the output to a threshold level at sampling times defined by the clocking circuit and decides whether the signal is a 1 bit or a 0 bit. Due to the noise associated in receivers, there is a probability that the decision circuits will detect bits incorrectly. The receivers are usually designed in such a way that the error probability of detecting a 1 for a 0 and a 0 for a 1 is quite small (from 10^{-9} to 10^{-12}).

Receiver Noise

Noise is a serious problem in detection of optical signals at the receiver. This electrical noise due to current fluctuations affects the receiver performance. There are two major contributions to noise:

- Shot noise
- Thermal noise

Shot Noise

In simplistic terms, we can define the interarrival rate between electrons flowing as a random phenomenon, hence contributing to immense fluctuations in an electrical circuit (and therefore called *shot noise*). For a photodiode of responsivity R, the current induced is as follows:

Current induced = $I(t) = RP_{input} + I_s(t)$
$I(t) = <I_{pd}> + I_s(t)$
(because RP_{input} = average value of I_{pd})

In this equation, $<I_{pd}>$ signifies the average value. For information on each of the terms, refer to *DWDM Network Designs and Engineering Solutions* (Cisco Press).

Shot noise can be statistically depicted as a Gaussian function (mean = 0, variance = 1).

Thermal Noise

The self-random motion of electrons due to the possession of kinetic energy based on the temperature gives rise to uneven fluctuations or thermal noise. This noise does not need any voltage to sustain itself.

Thermal noise, like shot noise, can be approximated as a Gaussian process. The final equation for induced current now is as follows:

$$I(t) = \langle I_{pd} \rangle + I_s(t) + I_{thermal}(t)$$

Noise Figure

The noise figure (NF) is a figure of merit associated with a device. In the receiver, the photodetector is followed by the front-end amplifier, and the noise figure gives the amplification ratio of input noise to output noise across the amplifier:

$$\sigma^2_{thermal} = \frac{4k_B T}{R} NFB$$

In this equation, B represents the bandwidth of the receiver, k_B is the Boltzmann constant, R is responsivity, T is absolute temperature, and NF is the noise figure. $\sigma^2_{thermal}$ represents the variance of the thermal coefficient of the noise throughout the entire operating frequency spectrums.

Receiver Performance

Receiver performance is an important factor in the FSO system design. The FSO design depends on the performance of the receiver's capability to detect 1s and 0s from the incoming optical signal. Bit error rate (BER) is a figure of merit to measure receiver performance. Receiver sensitivity is another performance-measuring standard for optical detectors and is important for FSO design. Finally, the signal to noise ratio (SNR) can be regarded as the absolute quantitative measure of the signal.

Bit Error Rate

During the transmission of data through an optical channel, it is desired that the receiver be able to receive individual bits without any errors. Errors occur when a receiver fails to detect an incoming bit correctly. Causes for errors generally stem from impairments associated with the transmission channel. A receiver fails to detect a bit correctly when it detects a 1 bit when a 0 bit is transmitted or a 0 bit when a 1 bit is transmitted. The receiver is also bit-rate sensitive. For different bit rates, a receiver has different magnitudes of errors; therefore, the BER is a figure of quality in an FSO network. Typically, optical end systems should have a BER of 10^{-9} to 10^{-12}—in other words, for every 10^9 to 10^{12} bits transmitted, one corrupted bit is allowed.

Mathematically, BER is the sum of probabilities, such that when a 0 bit is transmitted a 1 bit is received, and when a 1 bit is transmitted a 0 bit is received. This summation of conditional probability gives the BER of the system statistically. The equation to calculate BER is as follows:

Receiver Performance

BER = P (1) P (0 received for 1 transmitted) + P (0) P (1 received for 0 transmitted)

BER = P (1) P (0/1) + P (0) P (1/0)

P (0) = Probability of a zero bit transmitted = 1/2
P (1) = Probability of a one bit transmitted = 1/2
P (0) = P (1) = 1/2 (because a 1 or 0 is equally likely to be transmitted)

P(0/1) and P(1/0) depend on the distribution of the current over time while detecting the signal. P(0/1) signifies the probability that a 0 bit is received for a 1 bit transmitted, and P(1/0) signifies the probability that a 1 bit is received for a 0 bit transmitted. In other words, the probability density of the noise associated with the system affects the final waveform of the current. That is, if we consider noise being superimposed on the signal, this superimposed waveform is what will determine how many wrong decisions were made at the receiver. Figures 6-9 through 6-11 cover BER.

Figure 6-9 *Original Signal*

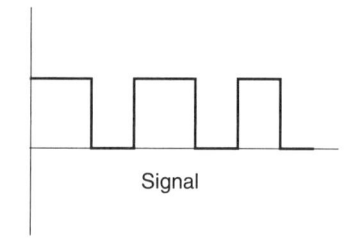

Figure 6-10 *Noise*

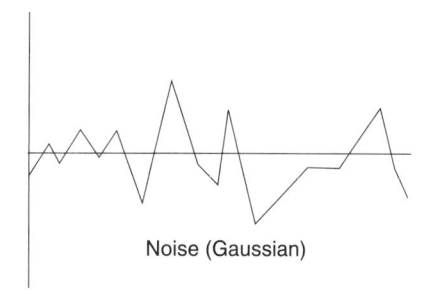

Figure 6-11 *Final Signal*

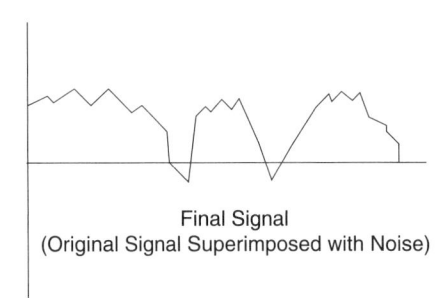

Final Signal
(Original Signal Superimposed with Noise)

The noise spectrum can be analytically given as a summation of the probability density function (PDF). PDF is defined as the first-order derivative of the distribution function F(x), where F(x) is the distribution function. The distribution is given in the following equation:

$$F(x) = P(x \leq x)$$

In this equation, x is a random variable; therefore PDF is shown in the following equation:

$$f(x) = \frac{d}{dx}F(x)$$

Further, the noise is classified into shot and thermal noise. Both shot noise and thermal noise can be approximated as Gaussian density functions.

A Gaussian distribution is defined as follows:

$$f(x) = \frac{1}{\sqrt{2\pi\sigma^2}} e^{-\frac{(x-m)^2}{2\sigma^2}}$$

In this equation, m is the mean of f(x), and σ^2 is the variance of f(x).

Signal to Noise Ratio

The SNR of a receiver is defined as the ratio of signal power to noise power in the electrical domain:

$$SNR = \frac{(SignalPower)Electrical}{(NoisePower)Electrical}$$

The SNR is proportional to input power squared. SNR can be enhanced by increasing the load resistance:

$$SNR = \frac{R_L R^2 P_{in}^2}{4 K_B TFB}$$

In this equation, F is a proportionality factor for increasing thermal noise content of a receiver. B is bandwidth, P_{in} is input power, R_L is load resistance, and T is temperature. K_B is Boltzman's constant.

Lenses and Mirrors in FSO

A lens is a transparent device made of glass or plastic that can refract light in such a way that the rays form an image. A mirror, on the other hand, forms images by using the principle of reflection as opposed to refraction in a lens. Both lenses and mirrors are used in FSO in assembly of transmitters and receivers.

There are basically two types of lenses: convex and concave (see Figure 6-12). A convex lens directs the light beam toward the central axis of the lens, whereas a concave lens directs the light beam away from the axis of the lens. By looking at the shape of the lens, one can easily determine whether the lens is concave or convex.

Figure 6-12 *Concave and Convex Lenses*

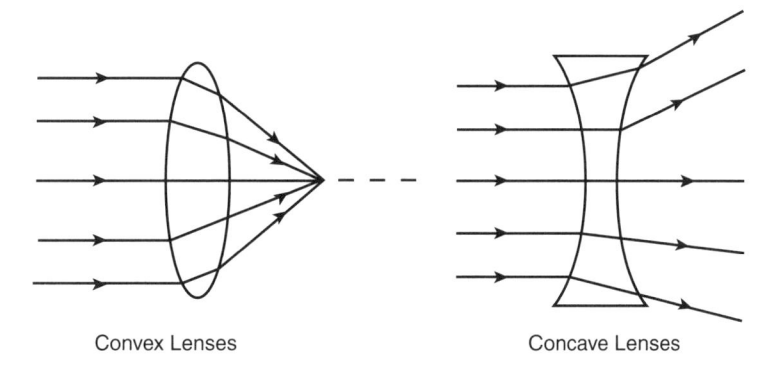

Convex Lenses   Concave Lenses

Note here that convex lenses are used to collect light at the receiver end. The light is already scattered and spread out, hence the lens gathers as many optical photons as possible. In FSO, concave lenses have limited application due to their divergence property of light, which means that they scatter a light beam and hence are not conducive to FSO communication (which needs a more concentrated light beam to track a receiver and ensure good communication).

Tracking and Acquisition

To reduce geometric path loss, the receiver needs to align the beam in the transmitter path to collect maximum beam power. Even though FSO operations are limited to within a few kilometers (line of sight), it is not easy to align the beam perfectly due to multiple external unbalanced forces acting on the transmitter and receiver. Tracking and acquisition techniques are used to align the signals perfectly and reduce path loss. The methods for tracking include the use of servo motors, stepper motors, mirrors, and micro-electromechanical systems (MEMSs).

An auto-tracking feature automatically realigns the receiver if the equipment premises shift slightly due to environmental factors—such as swaying of equipment premises due to wind and such. Tracking systems usually use a beacon beam that is different from the data beam to enable tracking and acquisition. The data and beacon beams must be lined up in the same direction for the tracking system to function. Tracking systems have been proposed for satellite and military applications for several years. The challenge still is in quick tracking and homing in on a movable object whose movement is random. The idea is to home on to a moving device and keep track of its movement, thus ensuring a steady flow of data from the transmitter to receiver and back.

Link Margin and Design Considerations

Like fiber-optic network design, link-margin analysis is important in FSO networks. To determine the link budget, we need to know the transmit power, the receiver sensitivity, and the path loss. Transmit power and receiver sensitivity are obtained from the vendor data sheets. Loss can be calculated. Unlike fiber-optic networks, FSO communication systems suffer losses due to weather (rain and fog conditions). A proper margin for weather conditions is necessary for satisfactory system performance.

Power budget is calculated using the following equation:

> **Power budget** = transmit power (dbm) – receiver sensitivity – total Loss (dB) – margin (dB)

Total loss is calculated as follows:

> Total loss = optical loss + geometric path loss + pointing loss + atmospheric loss (loss due to weather conditions)

Optical loss results because of imperfect lenses and other optical elements. A lens typically transmits 96 percent of the light, and the remaining 4 percent gets reflected or absorbed. This loss is typically accounted for as optical insertion loss for the link-budget calculation.

Absorption loss due to water vapor is the primary absorption factor near the infrared region. Use of a wavelength in the 850-nm and 1550-nm region helps to minimize the loss due to absorption. The next section discusses losses due to fog, rain, and other particles in more detail.

Factors Affecting FSO Operations

When light is transmitted through the atmosphere the strength of the light (power) is attenuated due to scattering and absorption. Attenuation due to absorption and scattering affects light as it travels through the atmosphere. This is given through the following equation:

$$\frac{I_0}{I_i} = e^{-\gamma d}$$

In this equation I_0 is the detected intensity at distance d, I_i is the initially launched intensity, and γ is the attenuation coefficient. The attenuation coefficient, a function of the wavelength, is composed of molecular and scattering coefficients and molecular and aerosol absorption coefficients. Here the aerosol absorption coefficient means the absorption of photons by an atmosphere on a volumetric basis.

Mie Scattering

Mie scattering occurs for particles that are about the size of the wavelength. Therefore, FSO communication that deploys wavelengths near the infrared range, fog, haze, and aerosol particles are the major contributors to the Mie scattering effect. This process of scattering due to the wavelength semblance near the infrared band and its direct proximity to fog, haze, and aerosol particles is called *Mie scattering*.

A simplified formula to calculate Mie scattering is as follows (for more information, refer to *Free Space Optics* [Sams, 2001]); this formula is also used to calculate the attenuation coefficient due to Mie scattering:

$$\gamma \equiv \frac{3.9}{v}\left(\frac{w}{550}\right)^{-\delta}$$

where δ is proportional to the third root of v. And v signifies visibility in the transmission wavelength. Lambda is wavelength.

Based on empirical observation, researchers have concluded that Mie scattering caused by fog characterizes the primary source of beam attenuation, and that this effect increases with distance. For design purposes, visibility conditions of a typical geography must be studied and factored in. The link power budget should accommodate the effect of weather for proper operations of the FSO systems.

Rayleigh Scattering

Light scatters due to dense fluctuations of molecules in the atmosphere, leading to a phenomenon known as *Rayleigh scattering*. This phenomenon results from the collision of light quanta with molecules, causing scattering in more than one direction. Depending on the incident angle, some portion of the light propagates forward and the other part deviates out of the propagation path. Raleigh scattering is inversely proportional to the fourth power of the wavelength:

$$\left(R \propto \frac{1}{\lambda^4}\right)$$

Therefore, short wavelengths are scattered more than longer wavelengths. Any wavelength that is below 800 nm is unusable for optical communication because attenuation due to Raleigh scattering is high.

NOTE The impact of Rayleigh scattering on the transmission signal can be ignored in FSO systems, because FSO systems usually operate in the longer wavelength (near-infrared wavelength) region. On sunny days, the sky appears blue because of Rayleigh scattering.

Beam Spreading

If it were an ideal world and there were no environmental attenuation, the only distance limitation would be because of the beam spreading. FSO systems typically have approximately 1 m of beam spread per kilometer of distance. The loss due to beam spreading is also considered in link-budget analysis.

Impact Due to Rain, Fog, and Snow

Rain, snow, fog, and minute particles in the atmosphere can cause attenuation and hence limit the distance that FSO can support. However, the effect of rain is significantly less than that of the fog, because the radius of raindrops is much larger than the operating wavelength of typical FSO lasers (and, therefore, rain does not significantly affect FSO communication). Snow is crystallized water, and attenuation varies with the density of snow (light snow or heavy snow). Scattering is also a minor issue with snow, because of its relatively large size of the flakes.

NOTE Radio-frequency (RF) wireless technologies that use frequencies above 10 GHz are adversely impacted by rain but are little impacted by fog. The unlicensed RF frequencies in the 2.4-GHz and 5.5-GHz range are relatively unaffected by both rain and fog.

Link Design Example

Customer ABC wants to have OC-3 (155 Mbps) worth of bandwidth between his primary and secondary site to back up. His critical data and these two sites are separated by 1 km distance in a straight line. He knows that no fiber is available between those sites. FSO was chosen as an ideal technology. Design the given network using the following parameters specified by the vendor:

- Transmitter power, 7 dBm.
- Receiver sensitivity, –40 dBm for OC-3 (155 Mbps) digital transmission rate and a BER $1 * 10^{-10}$.
- Assume a geometric loss of 25 dB for 1 km.
- Assume an 18-dB loss due to fog and other unpredictable events in this region.

The answer is as follows:

Transmitter power (7 dBm) minus total loss (calculated as follows) should be greater than the receiver sensitivity (–40 dBm) for the system to function.

Total loss is calculated as follows:

Total loss = geometric loss + optical loss + loss due to weather (link margin)
Total loss = 25 + 4 + 18 = 46
Transmit power – total loss = 7 dBm — 46 dB = –39 dBm

This is well within the receiver sensitivity range; therefore, the system works fine.

Laser Safety for FSO Networks

Exposure to a laser beam of high power can cause damage to human skin and eyes. Certain wavelengths (400 nm to 1600 nm) can penetrate the eye with intensity severe enough to damage the retina and permanently damage the eye. To help avoid potential eye hazards, vendors must comply with several standards and regulations before they launch a product.

The Center for Devices and Radiological Health (CDRH), which is a division of the Food and Drug Administration (FDA), is the main regulatory body responsible for laser safety administration in the United States. Each country has its own standards and regulations. Refer to the CDRH website for additional details (http://www.fda.gov/cdrh).

FSO Communication in First Mile Networks

Because FSO networks have to be precisely aligned and installed for proper functioning, advanced planning is necessary. After adequate planning, we need to conduct a site survey and ensure that line-of-sight communication is possible between transmitter and receiver elements. The next step is to install the infrastructure and verify the link budget. Make sure that we have enough link margin to accommodate the effects from dramatic changes in weather.

FSO systems support all types of topologies and are typically deployed in ring, mesh, or point-to-point architectures using classic multihop models. Figure 6-13 shows a sample FSO network deployed in ring fashion and point-to-multipoint fashion.

Figure 6-13 *A Generic FSO Network*

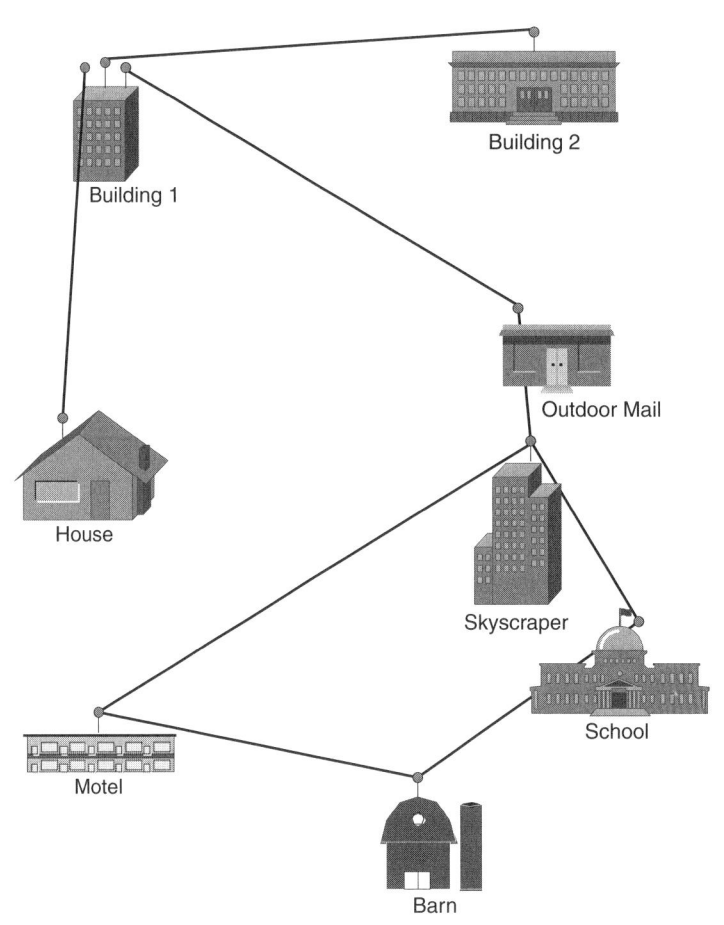

Summary

In this chapter we covered FSOs as a method to solving the first mile advance access bottleneck. This is an amazing technology that needs a lot of work before effective deployment. FSO design is an intriguing subject, consisting of aspects of wireless and optical design and taking into consideration a variety of different issues (such as tracking). This chapter sought to cover these issues.

Review Questions

1. How is FSO different from wireless communication?
2. What is the benefit of FSO as compared to WiFi?
3. What is the key limitation to FSO communication?
4. List application environments (such as enterprise networks) that can deploy FSO communication.
5. List the laser requirements for FSO communication.
6. Under ideal atmospheric conditions, if we have a laser that outputs 50 mw of light at 850 nm, what is the maximum distance at which a receiver can be placed? (Assume receiver sensitivity to be -25 dB and a suitable atmospheric loss.)
7. For the same problem as listed in Question 6, if it starts to rain, and rain accounts for an extra 70 percent attenuation, how will you compensate for the extra attenuation? Describe a compensation technique and mathematically evaluate it.
8. If the receiver is mobile and the transmitter has the task of trying to connect to the receiver using a beacon, list the properties of the beacon that are necessary to ensure good tracking. Does a feedback system help here? Explain your answer. Describe an algorithm that would better the feedback system given that the receiver can ascertain only certain positions in a known topological graph.
9. For a 10-km FSO link based on 10 spans of 1 km each FSO links, with multihopping of the signal, what are the technical requirements of the multihopping device? If the x percent of the data is lost in each span, prove that the data lost in the complete network is more than 10x.
10. Prove the upper limit on distance for FSO communication for the following:

    Assume an ideal system and line-of-sight FSO communication: Show the limit for Earth's radius of 6025 km.

    For a metro network FSO laser on a skyscraper of 890 feet, calculate the maximum allowable distance at which the FSO system can communicate.

References

Agrawal, Govind P. *Fiber-Optic Communication Systems*. Hoboken, New Jersy: John Wiley & Sons, 2002.

Gumaste, Ashwin, and Tony Antony. *DWDM Network Designs and Engineering Solutions*. Indianapolis, Indiana: Cisco Press, 2002.

Heinz, Willebrand, et al. *Free Space Optics*. Indianapolis, Indiana: Sams, 2001.

Keiser, Gerd. *Optical Fiber Communications*. New York: McGraw-Hill Science/Engineering/Math, 1999.

Ramaswami, Rajiv and Kumar N. Sivarajan. O*ptical Networks: A Practical Perspective*, Second Edition. San Diego, California: Morgan Kaufmann, 2001.

CHAPTER 7

DSL Technologies

The phenomenal increase in the amount of data in the Internet and related networks has fuelled a similar demand for bandwidth and faster connection speeds, which has driven the development of several technological approaches to provide broadband access to end users. Bandwidth-killer applications such as videoconferencing, multimedia, and video on demand (VoD) and the ongoing deregulation and privatization of telecommunication networks and the rapid growth of distributed business applications have also contributed to this demand for more and more bandwidth from existing networks. Wavelength-division multiplexing (WDM) and terabit routers have generated a vast switching and fiber capacity for the backbone networks, which have enabled the backbone networks to facilitate high-speed transmission of traffic flowing through the networks. Although the backbone networks have been augmented with terabit-speed capabilities to cater to this huge surge in IP-centric traffic, the area mapped by the first mile, representing the interface between the end users (consumers) of the Internet and these high-speed backbone access networks (that is, the Public Switched telephone Network [PSTN]), is characterized by limited speed capabilities.

There exist different approaches to providing broadband access to business and residential customers (as we discussed in previous chapters). These approaches include direct-broadcast satellite systems; fixed wireless technologies such as local multipoint distribution system (LMDS), multichannel multipoint distribution system (MMDS), and digital electronic messaging services (DEMS); and, of course, fiber in passive optical networking (PON) and coarse WDM (CWDM) implementations. Apart from the natural benefit of mobility, another advantage of wireless technologies is its universal deployment and its cost effectiveness as compared to fiber and conventional copper solutions, due to the absence of a medium (for instance, the fiber that must be deployed). The operating spectrum of wireless technologies is limited. This is because the operable range of frequencies in an electromagnetic medium is severely impaired by signal attenuation, creating only a select narrow band of frequencies suitable for wireless communication, especially under conditions such as attenuation, fading, scattering, and so on. Fiber, on the other hand, generates tremendous bandwidth capabilities but entails a very high cost of deployment, and this deployment is constrained to particular zones. (For example, we cannot lay more fiber in downtown Manhattan due to the space constraint there.) Therefore we need a low-cost solution that uses the existing network infrastructure optimally—thereby alleviating the issue of the high capital cost of deployment—and provides sufficient bandwidth for broadband access to residences or small offices (more popularly known as the small office/home office [SOHO] market) and even extendable to small enterprises. The PSTN has resulted in a very efficient network that ably connects a very large number of end users. It is this movement (PSTN) that we consider the basis for a third approach: a digression from fiber and wireless for low-cost, good-performance broadband access.

The PSTN primarily comprises network lines enabled to carry voice traffic. These lines require only up to 4 kHz bandwidth to provide plain old telephone service (POTS). As traditional voice and newer data networks began to converge in the backbone, there was seen a need to provide a high-speed solution that supports both voice and data traffic using the existing bandwidth in the PSTN. Contemporary copper solutions, such as cable modems and 56-kbps dial-up modems, offer high-speed broadband access but are built to be entirely data centric (and thus need a dedicated medium for just the data transfer). Consider, for example, a dial-up line that uses the RJ-11 jack for connectivity; it uses the voice channel for data purposes and, therefore, the ability for voice communication is lost. Digital subscriber line (DSL) emerged as a technology that offered a dual functionality—supporting data while not doing away with functionality for providing voice—using existing PSTN (which actually carries the voice), thus alleviating the problem of high cost of deployment as in wireless and fiber technologies and providing both high-speed broadband access and POTS.

In other words, DSL is a method of putting voice and data on the same medium, by making of use of the larger spectrum offered by the medium. Thus, DSL allows the successful coexistence of voice and data in the same medium.

With respect to first mile access networks, DSL is an important technology due to its ability to provision multiple services such as voice, data, and video high-bandwidth services at a very low deployment cost. The key user selling point (USP) for DSL is the low cost of initial deployment, because the medium already exists in the form of the PSTN. Second, DSL is able to provide multiple services on the same medium, thus proving to be a good choice for providing broadband connectivity to end users. In first mile networks, the role of DSL is to provide connectivity to consumers who already have access the public telephone system. Here consumers use a DSL modem to tap the multiplexed bandwidth, which is in a different spectrum as compared to voice. The limitation of DSL is primarily due to the physical layer characteristics of the DSL medium—that is, copper loop—and this results in distance and bandwidth limitations. However, a DSL network has multiple regeneration and aggregation points to circumvent the two limitations of distance and speed. The aggregation points enable multiple DSL lines to be coalesced together, aggregating in thicker bandwidth pipes (higher speed).

One of the primary challenges in the first mile network that deploys DSL technology is in providing dynamic provisioning of services.

What Is DSL?

DSL is a generic term for a set of technologies that use the additional 1-MHz bandwidth, apart from the 3-kHz band transmitting voice traffic along the shielded twisted-pair (STP), or copper, connecting a subscriber's telephone and the central office (owned by the provider). The term *DSL* was derived from the basic integrated services digital network, also known as the *ISDN*.

DSL is a dedicated link that establishes a direct connection between the customer and the carrier, providing broadband access to the end user through a DSL modem installed at the customer's premises. One end of the DSL connection terminates at the central office of the telephone company. The DSL connection enters the customer premises through a high-pass filter that recognizes a high-frequency band for data traffic and bypasses a low-frequency band for voice traffic during a traditional phone call. The filter is connected to the DSL modem at the end user's premises, and this DSL connection terminates into a network interface card (NIC) installed in the end user's application device.

In Figure 7-1 we see a typical DSL first mile network. Downstream flow from the service provider to the end user is shown from right to left, whereas upstream flow—that is, from the end user to service provider—is shown from left to right. In the downstream, at the service provider side, data and voice are coalesced and combined together by a POTS splitter, which also reverses functionality as a combiner. The voice comes from a Class 5 switch, which essentially switches voice circuits. The data comes from a DSL access multiplexer (DSLAM) that acts as an aggregation point. The dual signal (voice and data) is sent to the end user. At the end-user site, a filter separates the voice from the data band. The voice is further cleaned through a microfilter. The data is then sent to a DSL modem, which acts as customer premises equipment (CPE) and allows the creation of an interface with the end-user device, such as a PC or laptop.

Figure 7-1 *Typical DSL Network*

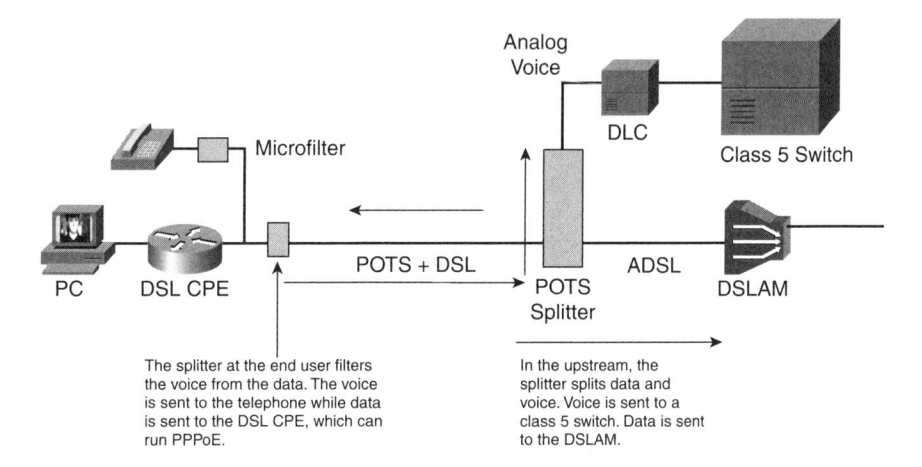

Thus DSL provides higher speeds for Internet access than a standard dial-up network to the customer without interrupting voice calls (as is the case with a dial-up connection) Internally, DSL creates either separate channels for the data traffic and voice traffic by maintaining distinct frequency bands for voice and data by using pass-band techniques as in many DSL approaches, or it uses base-band modulation as in ISDN, where the voice and data frequency bands overlap with each other but maintain individuality through coding.

The various types of DSL services can be categorized broadly into the following two groups:

- Symmetrical
- Asymmetrical

Symmetrical Services

Symmetrical forms of DSL are construed in a duplex mode of communication where the transmission speeds available for upstream and downstream communication between the source and destination nodes are the same. The DSL children in this category are high-bit-rate DSL (HDSL), HDSL2, and symmetrical DSL (SDSL). Because most backbone links are distinctly symmetrical in the physical layout, symmetrical services fit very well with most of the currently used network equipment and the changes, if any, would be minimal. Therefore, they are more attractive if the focus is on reducing costs during installation and simplifying network-level implementation complexities. Symmetrical services are important where there is real-time information exchange between two endpoints (for example, videoconferencing). To honor the QoS-sensitive service-level agreement (SLA), the DSL line must be symmetrical, thus enabling the same bit rate in both upstream and downstream communication.

However, many applications do not require similar bidirectional speed requirements between a source and destination (for instance, web servers, client/server applications, and, most importantly, Internet browsing by consumers). For instance, IP traffic is bursty and is characterized by asymmetrical speed requirements along the upstream and downstream links. This is intuitive in the way we surf the web, for example. In such applications, symmetrical services, which necessitate a bidirectional bandwidth allocation, would result in a significant waste of available bandwidth.

Consider, for instance, a client/server interaction on a web-based interface, where any communication between the client and the server would not require a speed of, say, 1.5 Mbps on both the upstream and downstream links. Such applications motivated the growth of asymmetrical services in the DSL family.

Asymmetrical Services

Asymmetrical services provide different speeds to set up a bidirectional mode of communication between two ends of a network. In our client/server example, the upstream communication from the client to the server can be implemented on a much lower-speed link (around, say, 16 to

640 kbps) when the client requests a certain block of data from the server. Subsequently we can provide very high speeds, such as 1.5 to 8 Mbps, during the downstream communication from the server to the client, yielding the data requested by the client. Various DSL technologies such as asymmetrical DSL (ADSL), rate-adaptive DSL (RADSL), G.Lite, and most strains of very-high-data-rate dsl (VDSL) provide such asymmetrical communication.

DSL Technologies

This section covers some of the DSL implementation methods, highlighting some of the various technologies available, including the following:

- ISDN-DSL (IDSL)
- High-data-rate DSL (HDSL)
- Symmetrical DSL (SDSL)
- HDSL 2 (the next-generation HDSL)
- Asymmetrical DSL (ADSL)
- Rate-adaptive DSL (RADSL)
- Very-high-data-rate DSL (VDSL)

ISDN-DSL

ISDN stands for *Integrated Services Digital Network* and is a method to best use the plain old telephone system (POTS) by dedicating lines for pure data-centric or high-quality hybrid applications. ISDN DSL was the first among the numerous variants of DSL that came to the fore. DSL was conceived from ISDN, which was created to facilitate an environment to integrate voice and data traffic in the PSTN end-to-end network, from one user device to another.

ISDN is an attempt to diversify various bearer services offered by the telephone network. Circuit-switched digital channels, packet-switched virtual circuits, dedicated point-to-point links, and call-control services are segregated at the network layer into different networks supporting the various bearer services and are subsequently brought together at a common ISDN switch and accessed by the user through a common terminal equipment (TE).

As for ISDN functioning, IDSN uses three different channels in various combinations to form the user interfaces. The B (bearer) channel is a 64-kbps channel that supports a circuit-switched connection, a packet-switched virtual circuit, or a dedicated point-to-point link. The D (data) channel is a 16-kbps or 64-kbps control channel that carries signaling information to control calls and for low-bit-rate packet switching. An H channel is analogous to a B channel, providing speeds ranging from 384 kbps, 1.54 Mbps, and 1.92 Mbps for higher-rate services.

ISDN specifies basic access and primary access for users. Two full-duplex 64-kbps B channels are combined with a full-duplex 16-kbps D channel to provide 2B + D basic access. The primary access is a 24-channel 23B + D combination in the U.S, with 23 full-duplex 64-kbps B channels, and is a 31-channel 30B + D combination in Europe, with 30 full-duplex 64-kbps B channels. The D channel is 64 kbps in both combinations. Figure 7-2 shows the basic network configuration of IDSL.

Figure 7-2 *Basic Concept of IDSL*

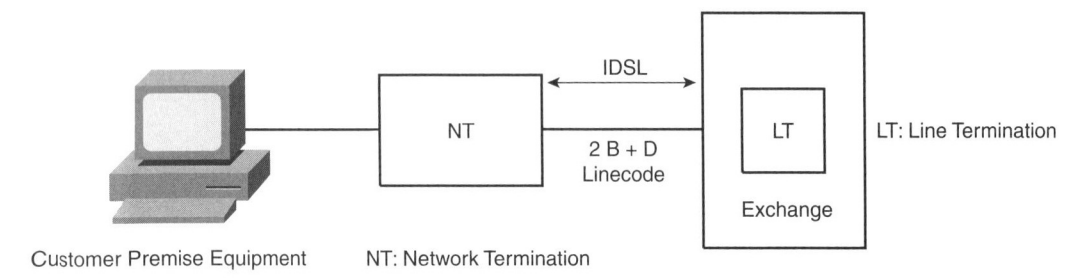

For residential services, IDSL takes the form of a bit-rate interface (BRI) that offers a 144-kbps full-duplex connection organized into a 128-kbps connection comprised of two 64-kbps B channels and one 16-kbps D channel. An extra overhead of 16 kbps is provided for an embedded operations channel (EOC), which exchanges management-related information such as link performance statistics between a line terminal (LT) and a network terminal (NT). A four-level Pulse Amplitude Modulation (PAM) line code known as *2B1Q* is used along with echo cancellation techniques to enable full-duplex transmission.

IDSL is a symmetrical offering whose speeds would be sufficient to provide data communication. However, a 144-kbps connection would be inadequate for applications such as interactive multimedia, video on demand, or remote LAN access. Such applications motivated the development of subsequent DSL offerings.

High-Data-Rate DSL

The need for speed capabilities higher than 144 kbps gave rise to another variant among the SDSL services: high-data-rate DSL (HDSL). HDSL is a bidirectional transmission system that allows transport of signals at a bit rate of 1.5 Mbps (T1) or 2.08 Mbps (E1) on multiple access network wire pairs up to distances of 12,000 feet (as opposed to the 144-kbps speed provided by IDSL). HDSL requires two to three pairs of shielded copper lines between the CPE and the central office. There are several options for the line codes, PAM 2B1Q, Carrier Amplitude/Phase Modulation (CAP), and Discrete MultiTone (DMT). 2B1Q allows duplex transmission on a single pair or parallel transmission on two to three pairs at 2.08 Mbps and parallel transmission for only two pairs at 1.5 Mbps. On the other hand, CAP is applicable for only one or two shielded twisted copper pairs running at 2.08 Mbps. Figure 7-3 shows a conceptual first mile connection of an HDSL network.

Figure 7-3 *Basic Concept of HDSL*

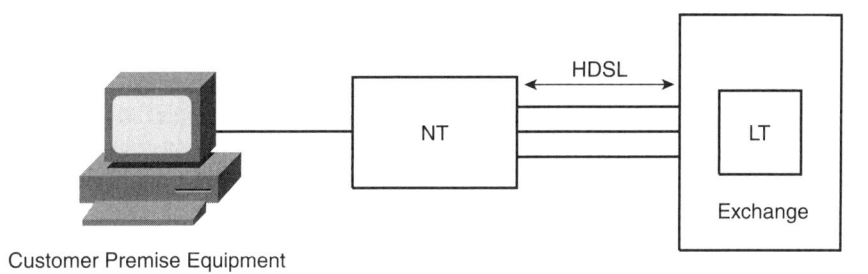

Customer Premise Equipment

The T1 (T carrier, tier-1) multiplexed trunking system forms the basis of the ISDN network in North America and Japan, wherein several ISDN switches are multiplexed on a time-division multiplexed (TDM) basis onto a single 1.5-Mbps T1 line. Analogous to the T1 trunking system, we have the 2.08-Mbps E1 lines providing a TDM multiplexed scheme to the ISDN switches in Europe and other regions outside North America and Japan. To provide residential access, a T1/E1 necessitates two pairs of wires and active line repeaters to regenerate the signal at intermediate points to compensate for the signal loss due to attenuation and electromagnetic interference.

HDSL can be viewed as a repeaterless T1/E1, because HDSL provides a transmission speed of 1.5 Mbps, using two shielded twisted copper pairs. Therefore, the main advantage of HDSL is that it allows service providers to provision T1 services more quickly and cost effectively because no repeaters need to be deployed. Another important benefit is that various management functions of HDSL can be monitored using the same operations system support (OSS) and software as before, because the links outside the HDSL core are still T1/E1 lines.

HDSL is used mainly for Internet access to servers (not just from clients), wireless system base station connections, LAN extensions and connections to fiber rings, videoconferencing, and distance-learning applications. ADSL, discussed shortly, restricts the upstream traffic to a fraction of the downstream traffic because of the different upstream and downstream connection speeds. This would be ideal for servers placed remotely, but not for servers that are onsite, either in homes or small offices. For onsite servers, HDSL is more suitable than ADSL, so that the upstream bandwidth is not restricted to a value below 1.5 to 2 Mbps.

A number of proprietary implementations of HDSL exist, with wide variations in features, which hamper vendor interoperability in a multivendor environment where HDSL products belonging to different vendors are integrated to provide a certain service. Another serious limitation in HDSL is that it disrupts simultaneous functioning of neighboring services, running in the adjacent cable pairs (ranging from other DSL services to analog voice services). In spite of its efficiencies, HDSL can be very slow to deploy due to its requirement for additional copper pairs, which also lessens the availability of T1/E1 services in some areas. As the cost of HDSL products began diving, the cost entailed in implementing the additional one or two pairs of wires became a significant factor in the overall cost considerations.

Symmetrical DSL

Symmetrical DSL (SDSL), also known as *single-pair DSL*, shares much in common with HDSL. It can, however, be seen as an improvement over HDSL because it removes the requirement of an additional pair of wires; therefore, it brings down overall cost of implementation and lessens connection complexity. SDSL evolved to be a vendor-specific DSL scheme running at 784 kbps on a single pair of wires as opposed to the two or three pairs of wires required in HDSL, thus providing greater transmission flexibility. SDSL is therefore a single-line version of HDSL, transmitting T1/E1 signals over a single twisted pair.

Once SDSL made its foray into the DSL market, vendors began customizing their SDSL products to create exclusive vendor-specific versions. Some vendors came up with faster versions of SDSL running at 1.5 to 2.0 Mbps but with limited transmission distance. Other vendors created versions that operated at speeds as low as 384 kbps, intending to support longer transmission distances. SDSL allows a service provider to provision DSL services based on three main parameters: cost, range, and speed of service. Based on an allocated budget, performance requirements, and the transmission distance from the local exchange, customers can choose from a number of available SDSL options.

SDSL speed capabilities are mapped with the corresponding maximum reach of service in Table 7-1.

Table 7-1 *SDSL Speeds and Distances*

SDSL Data Rate (kbps)	Maximum Transmission Distance (km)
128	6.71
256	6.56
384	4.42
768	3.97
1024	3.51

SDSL equipment is vendor specific and, therefore, SDSL products from different vendors are not interoperable. Several SDSL implementations boost the transmitter power at each end of the DSL connection to achieve very high speeds, over very long transmission distances. This surge in the transmitter power produces interference in the form of cross-talk noise with other SDSL services or non-DSL services running in adjacent cables. In a common 50-pair bundle of access lines, one powerful SDSL link could cause problems for all the other services in the same bundle. The problem of cross-talk noise interference has obstructed the standardization process of SDSL, keeping SDSL on the fringes of the DSL world except in Europe. This is the main reason why the next-generation HDSL (also known as *HDSL2*) is expected to replace SDSL in many applications.

HDSL2: The Next-Generation HDSL

The main drawback of HDSL and SDSL is their inability to allow simultaneous functioning of neighboring services within the same cable bundle or adjacent cable pairs in a seamless fashion. A single twisted copper pair between the central office unit and the remote unit at the customer premises forms the basic requirement of providing HDSL2 services. HDSL2 gained relatively higher popularity than HDSL because, unlike the latter, HDSL2 permits simultaneous undisrupted operability of other services in the nearby vicinity and requires fewer twisted copper pairs between the central office and customer equipment. With HDSL2, for instance, we can ensure smooth operability with a couple of access lines bearing SDSL services and others providing ADSL services, grouped together within a common cable bundle.

HDSL2 provides duplex communication at speeds of 1.5 to 2 Mbps with a transmission reach of around 4 to 6 km. The line code used in HDSL2 is PAM. The first of the HDSL2 products based on the International Telecommunication Union (ITU) recommendations (G.shdsl or G.991.2) have begun to appear on the DSL market. To overcome the limitations of HDSL and SDSL, HDSL2 has been developed with a greater focus on power management to ensure that HDSL2 works well with other services on adjacent cable pairs.

HDSL2 monitors the strength of the transmitted signal from the terminal unit at the central office to the remote terminal unit at the customer premises, by actively monitoring both the signal to noise ratio(SNR) and the bit error rate (BER) on the pair of wires running between them. HDSL2 gradually throttles back the transmitted signal power, if the signal strength exceeds a threshold value that represents adequate transmission signal strength for the HDSL2 service provided to the customer, while ensuring that the BER is within acceptable bounds. If the BER rises beyond some critical value (typically 10^{-7}), the signal power level can be increased again to establish the combination of the SNR and BER to the optimum configuration values, so that the errors are within the performance boundaries.

HDSL2 is used widely in Frame Relay access systems, mobile communication systems (as the backbone), LAN and WAN access, and to provide T1/E1 services. In Frame Relay access systems, HDSL2 provides a very cost-effective and high-speed approach for linking the customer's Frame Relay access devices to the service provider's Frame Relay switch. In mobile communication systems, the links between the mobile telephone switching office (MTSO) and the base stations at the various cell sites covered by the MTSO are generally multiple land lines at T1/E1 speeds. HDSL2 requires only one pair of wires for 24 or 30 voice calls to and from a given cell site, and even more calls can be carried by using voice-compression techniques that minimize the bandwidth requirement. Most importantly, HDSL2 is the fastest standard SDSL scheme to link a router on the customer premises to an Internet service provider (ISP) using only one twisted pair.

HDSL2 services, which generally have a span of 12,000 feet, may use line extenders or doublers to reach out to a longer distance of 24,000 feet. Some service providers have tried to provide "extreme" HDSL2 by adding yet another line extender to the previously mentioned

24,000-foot-long HDSL2 service, thereby enabling a transmission distance of 36,000 ft (or 11 km). However, multiple line extenders cannot be powered by the service provider, and, therefore, the additional line extender that traverses the transmission distance from 24,000 ft to 36,000 feet has to be powered at the customer's premises. Therefore, HDSL2 with multiple line extenders does not provide an attractive solution because of these "subject-to-the-client-end" dependencies. Another limitation is that the PAM line code used in HDSL2 barely leaves any room for the 4-kHz voice band; hence there is no possibility of analog voice support in HDSL2, apart from some form of Voice over DSL (VoDSL). Service providers are skeptical about the role of HDSL2 in the residential sector, because of its lack of support for analog voice and also because of its somewhat steep price structure.

Asymmetrical DSL

ADSL is the DSL scheme that transmits two separate data streams with more bandwidth dedicated to the downstream link toward the customer than the upstream link carrying customer requests. In other words, downstream traffic flow far exceeds upstream flow. ADSL is used to transport large amounts of data traffic over existing copper telephone lines, with upstream speeds ranging from 16 to 640 kbps, and downstream speeds ranging from 1.5 to 8 Mbps, while simultaneously supporting POTS for voice. As opposed to SDSL services, in which the signal rate and possible line length are severely limited due to the interference problems between multiple symmetrical signals carried within the copper wires, ADSL succeeds because it takes advantage of the fact that most of its target applications (such as VoD, Internet access, remote LAN access, multimedia, and PC services) function perfectly well with a relatively low upstream data rate. Figure 7-4 shows the ADSL spectrum.

Figure 7-4 *ADSL Spectrum*

In a conventional ADSL setup, there is an ADSL modem at each end of the wire pair, one at the subscriber end called *ADSL terminal unit-remote (ATU-R)* and the other at the central office and called the *ADSL terminal unit-central office (ATU-C)*. The 1-MHz bandwidth of the wire pair running between the subscriber's premises and the central office is divided into three regions. The frequency band from 0 to 4 kHz forms a channel for the voice traffic, whereas the frequencies above 4 kHz are divided into two regions: an upstream channel with a frequency band ranging from 10 to 100 kHz, and a higher-frequency downstream channel with a band ranging from 100 to 1000 kHz. ADSL uses one of the following three modulation technologies: DMT, CAP, or Quadrature Amplitude Modulation (QAM) for its line code to convert the serial bit stream into an electrical signal suitable for transmission.

Figure 7-5 shows a conceptual realization of ADSL.

Figure 7-5 *Conceptual Realization of ADSL*

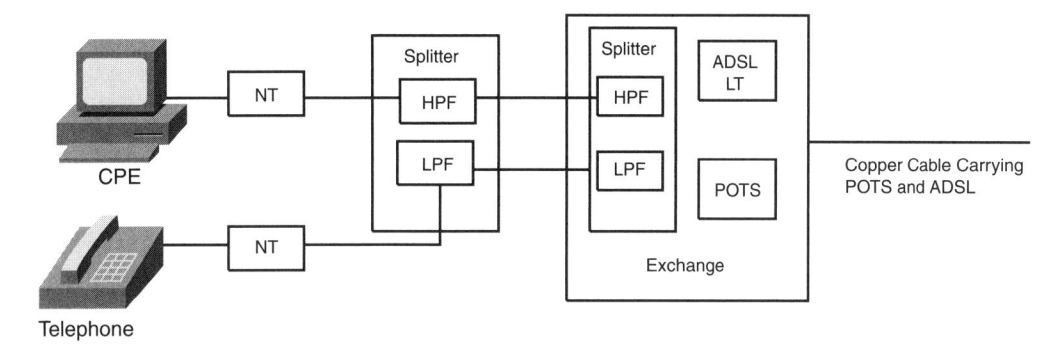

ADSL Concept

LT	:	Line Terminal	HPF	:	High Pass Filter
NT	:	Network Terminal	LPF	:	Low Pass Filter
CPE	:	Customer Premise Equipment	POTS	:	Plain Old Telephone Service

To facilitate both ADSL data access and POTS, POTS splitters need to be implemented both at the ATU-C at the central office end and the ATU-R at the remote subscriber's end to segregate the voice traffic from the ADSL data traffic. The POTS splitter uses a low-pass filter to allow only the voice or low-frequency band to pass through and to block the middle- to high-frequency bands (the data bands). An ADSL service provider allocates a DSLAM in the central office. The voice signal is split off at the ATU-C splitter through the low-pass filter and is transmitted to the telephone switch, whereas the ADSL data stream is filtered through the high-pass filter and is concentrated onto the DSLAM. The DSLAM concentrates a number of ADSL subscriber lines onto a single ATM line connected via a router or a switch to the provider's ATM backbone network, through which a subscriber is connected to a corporate gateway or an ISP.

ADSL solutions are in great demand due to bandwidth-intensive applications such as telecommuting, branch office connectivity, residential Internet access, VoD, distance-learning, and multimedia. ADSL is relatively more popular with network service providers as compared to other DSL schemes because of its ability to offer up to 8-Mbps data access, run over an existing copper-based infrastructure, and support traditional POTS traffic. However, ADSL requires the carrier to install the splitter equipment at the customer site, and, therefore, the equipment is subject to client-end dependencies such as user installation, ownership, powering, and wiring. The splitter has to be powered from the client end and hence ADSL cannot use the power-control solution, as is possible with SDSL services, to reduce any possible interference with neighboring DSL services.

This motivated the development of a new form of "splitterless" ADSL, which does away with the splitter requirement at the remote terminal unit at the ADSL subscriber end. This new version, which grew to be a standard, is also called *ADSL lite* or *G.Lite*. G.Lite is a more cost-effective solution compared to ADSL, and it is far simpler and cheaper to provision a G.Lite connection in the network. G.Lite was aimed at reducing the overall installation costs of implementing DSL at the various client sites. The only difference between the NT of a full ADSL setup and a G.Lite setup is that the NT of a G.Lite setup includes a high-pass filter. G.Lite does away with the integrated splitter at the customer site and distributes to the high-pass and low-pass filters on the customer's premises. The low-pass filter is packaged as a microfilter, which is designed to be customer installable, and one microfilter is needed for each and every telephone or ISDN device on the premises.

Therefore, G.Lite, with its splitterless feature, is ideal for consumers who require low to medium download speeds for their applications. However, G.Lite is unable to offer the speed of a full-rate ADSL device; therefore, it may not be able to provide the high bit rate or speeds required by customers who require huge download rates and are connected to a high-speed backbone. G.Lite is also known as *universal ADSL* and is standardized by ITU as G.992.2.

Rate-Adaptive DSL

Rate-adaptive DSL (RADSL) addresses a possible limitation of some early ADSL products, especially those that used CAP modulation for their line coding. Earlier versions of ADSL products faced serious problems with varying line conditions, especially with the downstream speeds varying from one location to another (typically in increments of 32 to 64 kbps). Line conditions may improve or deteriorate, depending on various factors including seasonal variations, rain conditions, humidity, overall temperature, and solar radiation. For instance, end users may receive traffic that traverses through the lines at 608 kbps in the morning, reaches a high of 640 kbps in the afternoon and drops down to 576 kbps in the evening. Early ADSL installation was hence plagued with problems such as excessive time consumption, complexity, and high expenses, because technicians would have to manually maintain the varying ADSL line speeds onsite.

RADSL is a natural progression of ADSL and can adapt to these changing conditions, eliminating the need for manual maintenance by a technician. Due to the unpredictable variations, one of the ADSL lines would be able to achieve a speed of 640 kbps, for instance, whereas another would manage a mere 582 kbps. However, RADSL-enabled devices determine their own maximum speed level, once powered on without the aid of a technician. Currently, most of the ADSL implementations are essentially RADSL-capable, especially the ADSL products using DMT for line encoding.

The basic architecture of an RADSL link and an ADSL link is quite similar. RADSL devices determine their optimal speeds based on current line conditions at the time of installation. If the line conditions improve, the speed of the RADSL device stays at the predetermined optimum level. If the RADSL devices fail on account of deterioration of line conditions, the RADSL speed setting procedure is repeated.

Currently, a wide range of vendor-specific RADSL products is on the ADSL market.

EZ-DSL, one of the vendor-specific versions of RADSL developed by Cisco Systems, Inc., eliminates the requirement of installing a splitter at the customer premises and the need to rewire at the consumer end (thus maturing RADSL into a consumer-installable product). Because EZ-DSL operates over the existing wiring and its installation procedure at the customer premises is quite similar to that of legacy modems, installation costs for both service providers and consumers are considerably reduced.

Very-High-Data-Rate DSL

Very-High-Data-Rate DSL (VDSL) is an evolutionary step from ADSL. VDSL is capable of providing astounding speeds ranging from 13 Mbps for a copper-line length of 4500 feet (1.4 km) to 52 Mbps for a length of 1000 feet (0.3 km) for downstream broadband communication and a speed of around 1.5 Mbps for upstream communication. VDSL is evolving in two different forms: one symmetrical like HDSL, and another asymmetrical like ADSL. Figure 7-6 shows the frequency bands for VDSL.

Figure 7-6   *VDSL Frequency Bands*

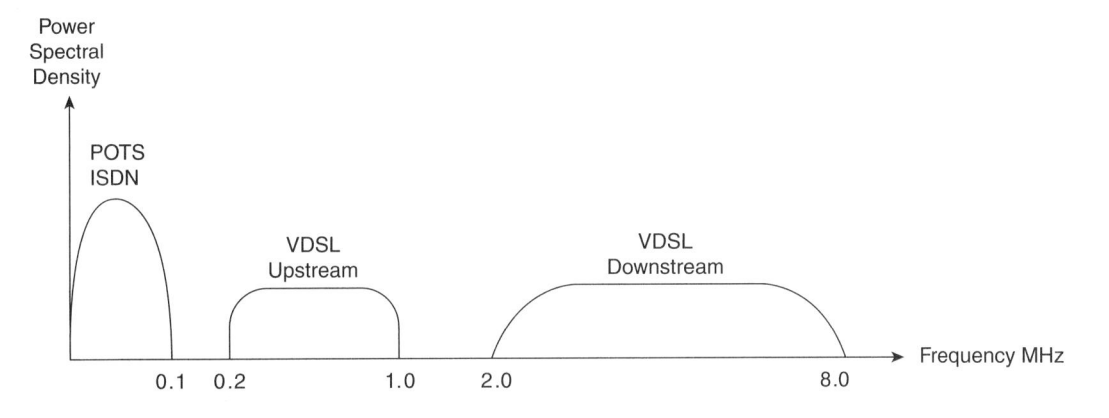

Symmetrical VDSL

VDSL has been designed to deliver a plethora of asymmetrical broadband services, including digital-broadcast TV, VoD, high-speed Internet access, distance learning, and telemedicine.

VDSL is an attempt to address three key issues with current broadband services delivered over the local loop: accommodation of ATM transport, an ever-increasing demand for higher bandwidth, and implementation of high-speed optical fiber feeders in the local loop at the customer end. Extending the optical fiber backbone closer to the customer is quite an expensive proposition. To alleviate that problem, VDSL is poised to implement strategies such as fiber-to-the-neighborhood (FTTN) or fiber-to-the-curb (FTTC), where a high-speed fiber-optic cable runs through a group of customers and high-speed VDSL links provide the final connections from the customer to the end users.

As a result, bandwidth use has increased by two orders of magnitude over the past 10 years or so (from less than 100 kHz for narrowband ISDN to more than 10 MHz for VDSL).

Long-Reach Ethernet

Long-reach Ethernet (LRE) can be considered as Ethernet packet-switch technology with a VDSL physical layer using copper as the physical medium. LRE technology can be used to run Ethernet over single-pair wiring at up to 15 Mbps at distances up to 3500 feet, 10 Mbps at distances up to 4000 feet, and 5 Mbps at distances up to 5000 feet at full-duplex line rate. LRE uses frequency-division duplexing (FDD) to separate the downstream channel, the upstream channel, POTS, ISDN, or private branch exchange (PBX) signaling services in the frequency domain. In addition, service providers can overlay voice traffic, both POTS and VoIP, over the same telephone wiring used for data and video. LRE is essentially Ethernet on both ends (between the switch and CPE is VDSL); therefore, the installation and management is just like Ethernet.

LRE uses QAM, in which two carriers are out of phase by 90 degrees and modulated by separate signals. LRE uses both the signal amplitude and the phase of the signal to define each symbol. Like 802.16 explained in Chapter 5, LRE also uses various QAM modulations (QAM-4 to QAM-256 based upon what distance and speed we want to achieve). By using a modular QAM approach, we can achieve the best performance and the highest possible bandwidth while maintaining the lowest cost and power required.

Ethernet data is encapsulated onto a continuous stream of cells (usually using a proprietary scheme [for example, HDLC]). The system applies a self-synchronizing scrambler mechanism to this continuous nonbursty data-cell stream. A self-synchronizing scrambler mechanism is used with LRE and is initialized to a random value. The scrambling mechanism diffuses burst noise and spreads it out. When that stream is received at the other end, the Ethernet data is reassembled back into the original Ethernet frames. A complex Reed-Solomon (RS) error-correction coding scheme is used for recovery of the data stream, providing both error detection and data recovery.

LRE systems are deployed mostly in multiple-dwelling units (such as hospitals, hotels, and apartment buildings) and large campuses. LRE and CWDM is one possible alternative in the last mile for an end-to-end Cisco solution. LRE is gaining prominence in first mile networks due to cost and performance. LRE is actually in principle EoVDSL (or Ethernet over VDSL). Figure 7-7 shows a typical multiple-dwelling deployment.

Figure 7-7 *LRE Deployment*

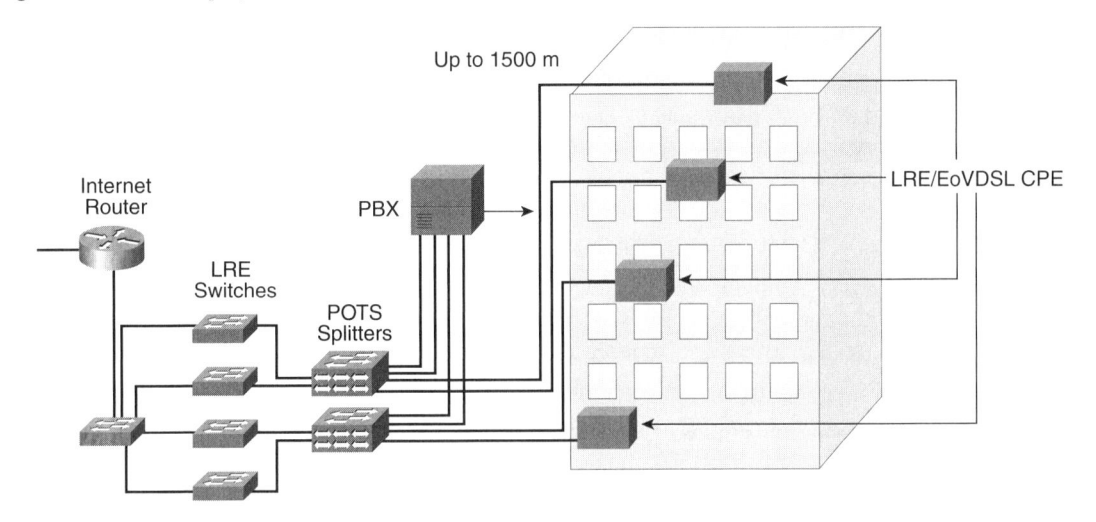

Line-Encoding Standards

Line encoding is the technique used to insert electrical pulses on a medium—in this case, copper twisted pair—so that the receiver can interpret this accurately. The three standard line-coding techniques used in xDSL technologies are as follows:

- Discrete Multi-Tone (DTM)
- Quadrature Amplitude Modulation (QAM) (discussed in Chapters 4 and 5)
- Carrierless Amplitude/Phase (CAP) Modulation

Discrete Multi-Tone

DMT is a method of coding information (line coding) to be transmitted over the physical phone line. The information is split into a number of subchannels, each characterized by the same bandwidth requirements but packaged to be transmitted at different frequencies. This technique has a number of advantages such as channel independence, wherein if a line gets noisy over one range of frequencies, other frequency bands may still be transmitted, so the overall throughput is reduced rather than completely blocked.

Initially, an equal number of tones are transmitted to measure the characteristics of the line. The processing of a signal takes place in the ATU-R. The processed signal is delivered to the ATU-C by using the same phone line at a low speed.

The ANSI specification for ADSL using the DMT line-code technique uses 256 channels for downstream communication and 32 channels for upstream communication, where each channel has a bandwidth of 4.3125 kHz, and an equivalent spacing of 4.3125 kHz between adjacent frequency bands. DMT was chosen for ADSL and G.Lite due to the inherent rate-adaptive nature of DMT devices, which enables them to readily adjust to varying line conditions, such as moisture or interference.

Carrierless Amplitude/Phase Modulation

The Carrierless Amplitude/Phase (CAP) Modulation is similar to QAM in most aspects. Both use amplitude and phase to represent the binary signal. The main difference is that the CAP does not use the carrier signal to represent the phase and amplitude (hence the name Carrierless Amplitude Modulation); instead, it uses waveforms to encode the bit steams.

In QAM, the constellation is fixed, but with CAP the constellation is free to rotate because there is no carrier to hold it to an absolute value. CAP receivers need to be compensated for this free rotation and usually include a rotary function-detector circuit to identify the relative position of the constellation.

Unlike DMT, CAP uses the entire available bandwidth and there are no subchannels as in DMT. CAP-based systems use frequency-division multiplexing (FDM) to separate the frequency of operations of upstream and downstream channels.

A Typical xDSL Network

An xDSL network consists of an xDSL modem or CPE, a DSLAM, and/or an aggregator. In DSL systems, both the DSLAM function and the aggregation function can be in the same platform. The function of the CPE is to act as the customer interface and can be a router, a bridge, or both. DSLAM is used to aggregate individual subscriber lines from the CPE. Figure 7-8 shows a typical xDSL network.

xDSL supports a variety of services and applications. Some of the services that are supported include the following:

- Digital video
- VoIP
- High-speed Internet
- Video on demand
- Interactive games
- Interactive shopping
- Videoconferencing

Figure 7-8 *A Typical xDSL network*

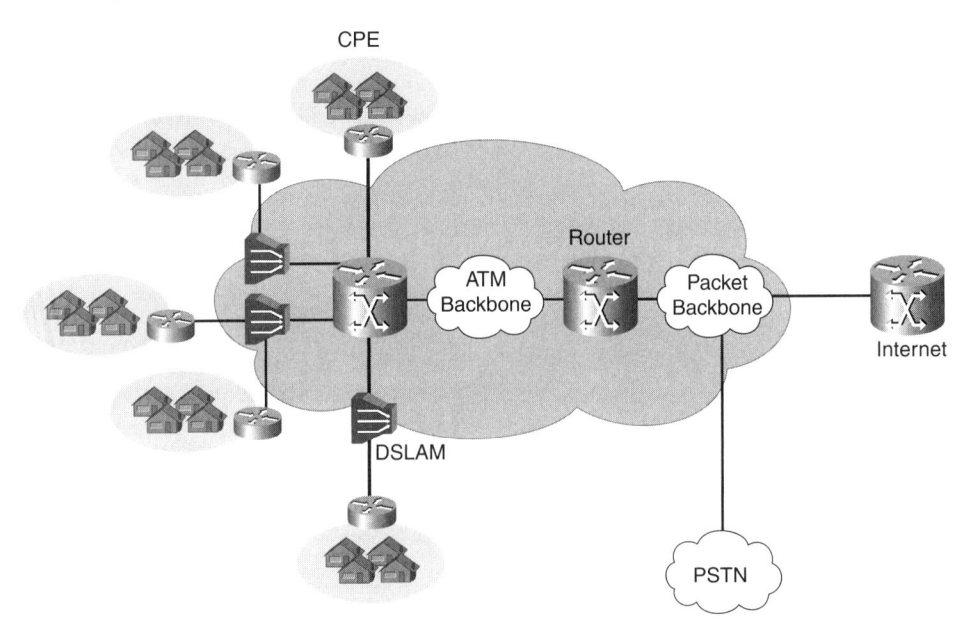

Issues with xDSL

The PSTN networks were originally designed to carry analog voice within a narrowband range of 0 to 4 kHz. There are filters along the lines to condition the signals and eliminate frequencies above 4 kHz. To service multiple customers, service providers deployed bridge taps. Long loops also deploy load coils, which help to flatten 0- to 4-kHz frequency response. Bridge taps, load coils, filters, cross talk, noise, and radio-frequency interference pose significant challenges to the deployment of xDSL in real life. A significant fraction of the DSL downstream bandwidth is unusable because of the distance and the length of the copper cables. (DSL downstream rates are severely restricted due to signal attenuation in the copper lines, bridge taps, cross-coupled interfaces, and cross talk with the services provisioned on adjacent circuits.)

Some of the variants of xDSL are not available in all parts of the world. Because service providers are trying to ramp up the network and support, they are unable to provide adequate attention to customer service and customer satisfaction. Poor customer service is one of the main issues troubling many new broadband access providers.

Other issues prominent with xDSL are management, provisioning, and billing. Figure 7-9 shows a typical first mile DSL-based network. Due to the distance and line speed limitations, we see the number of stages where aggregation is carried out within the first mile itself. Notice here that these aggregation points (versions of DSLAMs) are essentially complex circuit-switched fabrics based primarily on ATM circuit technology. This leads to excess overhead and

178 Chapter 7: DSL Technologies

switching (and hence a network where management is rather difficult). The capacity and performance of these DSLAMs is rather static; therefore, to upgrade or to provision a new service, multiple steps (for example at each stage in Figure 7-9) must be carried out. This inability (or difficulty) in incrementally adding customers is one of the key hindrances behind the increased use of DSL in existing areas. Also, keeping tabs on individual ATM circuits needs excessively large computing power, especially to execute complex functions such as monitoring and billing. This renders xDSL networks rather difficult to operate for large networks.

Figure 7-9 *DSL Network in the First Mile*

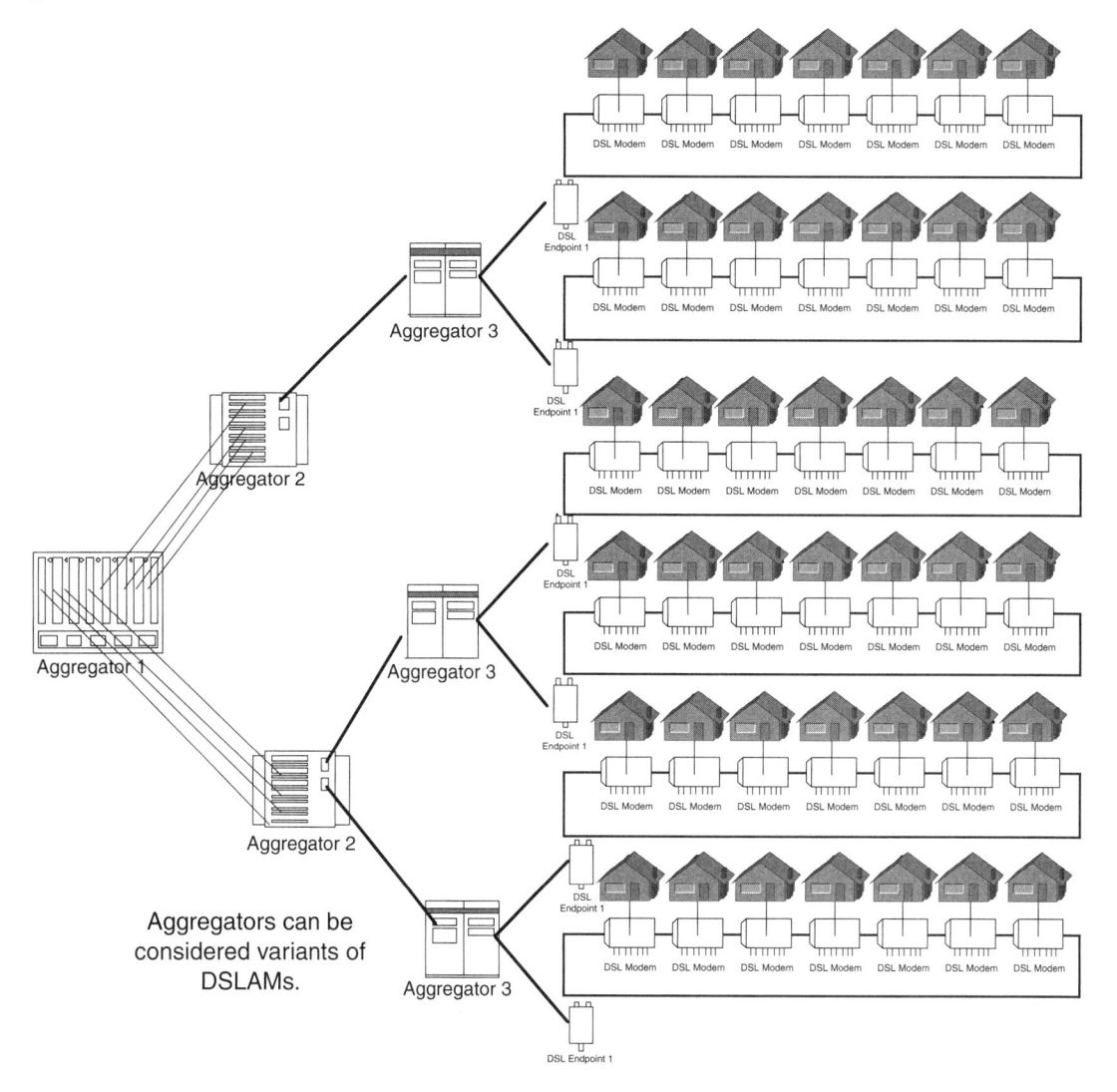

Aggregators can be considered variants of DSLAMs.

Despite these drawbacks, xDSL represents a promising technology for the years to come on account of its low cost and universal acceptability.

DSL in the First Mile

DSL has passed the curve of initial market penetration and is now a viable and available option for most consumers. DSL networking equipment is now available at stores, and connecting to DSL services is no longer considered a high-tech need but is more a necessity (even for home use). DSL, because it can use the existing network infrastructure, offers a low-cost solution and, perhaps more importantly, is publicly well accepted despite its management shortcomings. From the first mile network perspective, DSL can be viewed in two ways: a technological solution, and a viable business proposition for first mile networks.

From a technological perspective, DSL is a TDM solution based essentially on ATM technology. This makes it easier for conventional service providers to provision an end-to-end system that has DSL in the first mile and TDM services such as Synchronous Optical Network (SONET) and so on in the core. This also enables QoS in the first mile and enables service providers to provide multiple services.

From a business perspective, the low cost of deployment is one key reason for deployment of DSL technology. Moreover, the end-to-end solution possibility is the reason why it makes sense to have DSL in first mile networks. With regard to the DSL market, we have to understand that growth in this sector is not exponential; however, we can assume it to be only weakly linear (with a low gradient of growth). Apart from the business-savvy versions such as ADSL and VDSL, we see that native DSL does not offer an order of magnitude increase in connectivity to the end user as compared to 56.6-kbps dial-up modems. This means that the motivation to change from dial-up to DSL is weak, especially when dial-up can be extremely low cost (especially when we use freenets). Therefore, it is expected that service providers will have incremental DSL services that allow the full exploitation of DSL technology to the end user and thereby allow the end user to avail multiple megabits per second of bandwidth.

Summary

In this chapter we discussed DSL in the first mile. DSL technology is an evolutionary technology built on the the POTS infrastructure. Over the past few years, multiple variants of DSL have emerged—namely, ADSL, VDSL, HDSL, and so on. This chapter discussed these variants and their implementation and management issues. We also discussed LRE, a new paradigm for point-to-point communication over copper. LRE shows promise for the first mile—namely, an all-Ethernet approach. This chapter concluded by discussing some of the key limitations of DSL implementation that need to be solved to achieve a complete broadband network.

Review Questions

1. How is DSL implemented over PSTN?
2. Differentiate between asymmetrical and SDSL.
3. Differentiate between ADSL and RADSL. Which one would you use to implement a network in a high rise with bandwidth-killer applications such as video on demand and so on? Which one would you use in a hospital with bidirectional-committed communication? Explain your reasons.
4. Discuss the possibility of a metro access network that has a CWDM ring with 8 nodes each dropping one wavelength of CWDM. Suppose each node services a number of DSL DSLAM points. What form of DSL would you choose for implementation, assuming a very wide range of services needed by the end user? Hint: The CWDM drops Ethernet service, and we prefer to use an end-to-end manageable solution.
5. Diagram the solution of CWDM + xDSL (where x is what you chose in Question 4) and compare the management functions to a generic DSL network.
6. Describe the operation of HDSL? Why did HDSL not grow in popularity? What do you think are the reasons why there was limited interoperability between vendors using HDSL?
7. Compare the three modulation types described, which one would you use and why to attain (a) best throughput (number of bits per symbol), (b) least error, and (c) a combination of throughput and low error?
8. Discuss the issues that affect interoperability between last mile and last inch for the following:
    - DSL in last mile and WiFi in last inch
    - CWDM in last mile and DSL in last inch
    - BPON in last mile and DSL in last inch
    - EPON in last mile and DSL in last inch
9. For each of the following services, how does a DSL network have to be altered for successful implementation?
    - Video on demand
    - Video broadcasting
    - Videoconferencing
    - Telemedicine
    - Telesurgery
    - Internet services
    - Entertainment (gaming)
    - Voice

10. Research the terminal allocation problem, which basically gives an optimum solution that tells you where to keep DSLAMs so that you get a network with the fewest DSLAMs possible. Let there be K DSLAMs each with capacity given by a matrix R (of dimension 1 * K). Write a formulation that gives you the best possible performance for incremental capacity—that is, the network does not change much with an increase in traffic lines (DSL lines) for any given DSLAM. Make suitable assumptions and variables.

CHAPTER 8

Power-Line Communication

Introduction to Power-Line Communication

In the previous chapters we discussed multiple technologies that work as possible solutions to solve the first mile access problems—that of providing bandwidth (preferably on demand) to the end users and alleviating the access bottleneck in the first mile. Multiple solutions have been proposed using different technologies such as optical (passive optical networking [PON]), wireless (WiFi), wireless MANs (WMANs), digital subscriber line (DSL), and free space optics (FSO). Most of these solutions have been structured around a communication technology built on an equipment requirement that is fundamentally unique and diverse from the rest of the technology concepts that directly affect consumers who are the chief exponents of the first mile access technology. This directly translates into a high cost associated with the development and deployment of a communication infrastructure—one that is specifically suited but also must be refurbished or upgraded over time. The capital expenditure required to build a communication infrastructure to the end user is one of the chief inhibitors of first mile access technology deployment.

Although high-speed, end-user networks are appealing to the users and the proponents of such, these high speeds entail a substantial investment and slow returns for most carriers. In most of today's end-user networks, deployment principally relies on the plain old telephone system (POTS). However, the telephone network was designed primarily for voice communication, and the limited bandwidth required for voice presents a problem for data communication, which needs much more bandwidth. Because of this, technologies such as PON and wireless are good deployment alternatives but are limited either by performance or by cost (or both).

The deregulation of the electricity industry paved the way for a new medium for first mile access: power lines (through which digital signals can be sent). *Power-line communication* (PLC) is defined as the delivery of broadband access using the existing electrical power distribution network as a channel for delivery.

The idea behind PLC is as follows: Data signals are coupled over power lines for short-distance communication. The data signals typically occupy a much higher bandwidth compared to electrical signals, which generally occupy lower bandwidth (for instance, 60 Hz in the U.S. and 50 Hz in Europe and Asia). However, the amplitude of the electrical signals is quite high, to the tune of several hundred volts, and, therefore, the electrical signals dwarf the data signals.

As noted in "Power Line Communications: State of the Art and Future Trends" by Pavlidou, et al., that the electrical power infrastructure was not created with data communication in mind. This means that data communication is an add-on onto the electrical power infrastructure; therefore, we must deal with several issues before we can effectively use power lines for data communication.

If we consider the access network as a combination of first mile and first inch networks—that is, the network area from the end user to the curb site is the first inch, and the area from the curb site to the central office is the first mile—PLC technology is invariably relegated to the first/last inch without exception. Significant impediments and massive deployment hurdles prevent PLC from being deployed in the entire access area. However, this is a future technology, and one that will undoubtedly influence the last inch solution.

A Typical Indoor Power-Line Communication Network

Figure 8-1 shows an indoor power-line network. Various appliances, both electric and network related, are plugged into the power-line wiring. The power line—a copper wire—carries both electrical power signals and data signals by frequency differentiating them. In summary, the two signals coexist on the same line but are in different spectrums. Multiple networking appliances are connected to the power line. The data is segregated from the electrical power signal by an end device analogous to a modem. This end device segregates the data by using a demodulation technique.

Figure 8-1 *PLC Inside Homes*

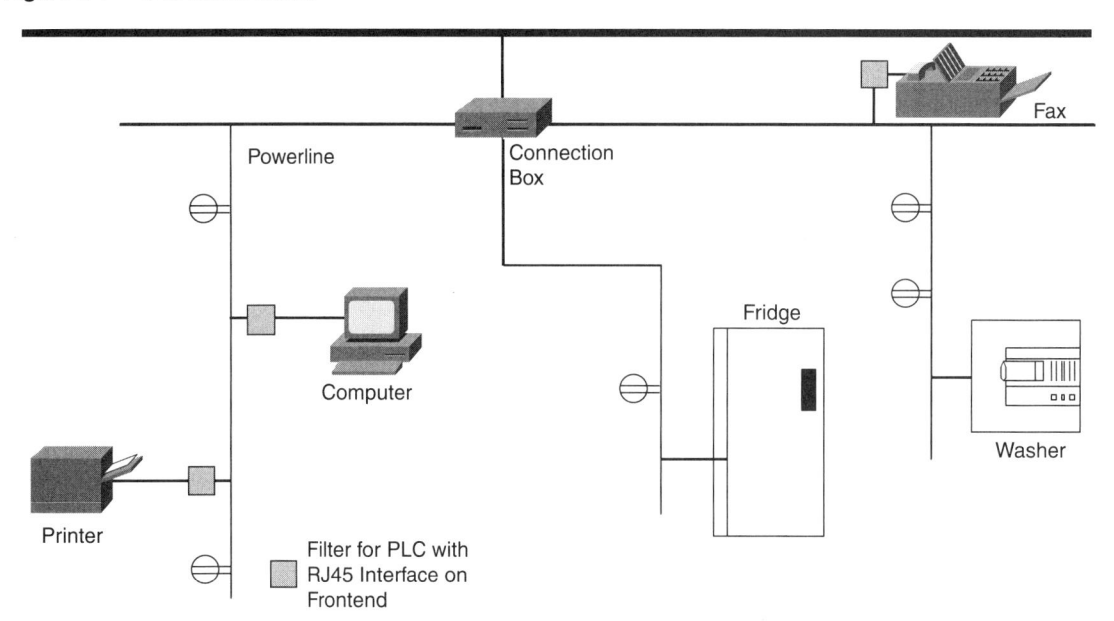

As shown in Figure 8-1, PLC represents a plug-and-play medium—one that does not require much provisioning. Of course, there is a limit to the data rate the PLC medium can sustain, as well as a limit to the distance data signals can travel in a PLC environment. These issues are covered later in this chapter. At this time, it is important for you to understand the direct advantages of PLC networking: the potential to use an existing infrastructure, and power lines as a medium provide more bandwidth than the classic Public Switched Telephone Network (PSTN).

Channel Characteristics That Affect PLC

As discussed in the article "Power Line Communications: State of the Art and Future Trends," power lines are intrinsically a very hostile medium for data transmission (let alone data *communication*). This article identified a number of data-transmission hindrances in power lines, including the following: high noise, extreme attenuation, and variable impedance.

The PLC channel characteristics are time and frequency dependent and are, thus, unstable. Moreover, with joints, crevices, and so on in the channel, it can be described as extremely lossy. The channel quality is usually described with reference to its signal to noise ratio (SNR), which enables us to understand the qualitative capability of a signal. Intense modeling and simulation concludes that the SNR for PLC systems is essentially random in nature and, hence, extremely hard to instantaneously calculate. From this, we understand how unpredictable the channel is. Finally, the PLC channel can be described as having a transfer function that is somewhat bandpass in nature but without a flat top. In fact, the top of this transfer function has several ripples and notches—some of which are very deep and almost serve as transfer-function breaking points.

The channel impedance is highly variable and, therefore, power fluctuations on the data path are obvious. In fact, impedance may be as high as several kilo-ohms ("Personal Computer Communications on Intrabuilding Powerline LANs using CSMA with Priority Acknowledgements" and "Signalling on Low-Voltage Electrical Installations in the Frequency Range 3 kHz to 148.5 kHz—Part 1: General Requirements, Frequency Bands and Electromagnetic Disturbances"). The channel tends to behave like a parallel resonant circuit, often showing capacitive or inductive behavior. For a detailed reference on electrical circuits, refer to *Power Line Communications* by Klaus Dostert. The only important aspect to consider while designing a PLC first mile network is that a severely unbalanced external force from the electrical power makes the power line a capacitive and inductive medium that directly affects the data flow. The qualitative effect on data flow is the random behavior of the channel. The quantitative analysis is not considered here due to its extreme mathematical nature; you can read more about the quantitative analysis in *Power Line Communications*.

The other aspect we need to understand about PLC systems is attenuation. Because of the random channel behavior, as previously discussed, there is severe attenuation in the system. In one PLC system, for instance, the measured attenuation was 15 dB on a 1-km line (without any major equipment breaks making it even more lossy). From this, you can understand the kinds of losses we are looking at when considering PLC for transmission of data signals. Note that this means PLC can find application only in the first mile when distances are short and data rates are low.

Because PLC isn't viable at this point over large geographic areas (because of high loss), it is best to restrict it to first mile access communication. PLC is especially appropriate for first/last inch distribution networks because distances are shorter and so PLC can more effectively deliver high bandwidth. In addition, PLC network management is simpler in the first/last inch because loss is less of an issue.

Noise in PLC Systems

The power line is connected to a large number of appliances and equipment, each radiating its own noise and, hence, creating at different times a different level of noise in the channel. The noise in the channel varies, and this variation results from a number of factors, including frequency, time of day, and load. The article "Power Line Communications: State of the Art and Future Trends" specifies four main types of noise:

- **Colored background noise**—This is the accumulation of the low-power noise sources whose power spectral density is directly proportional to frequency.
- **Narrowband noise**—This is generally due to radio and other electromagnetic interferences that get latched on to the power line.
- **Asynchronous impulsive noise**—This is due to the noise created when circuits are switched on and off and so on.
- **Periodic impulse noise**—This is due to the appliances that produce noise harmonics at 50 GHz and 100 GHz and their integral multiples.

Modulation Requirements in PLC

To transmit a signal over a power line, it is imperative to modulate the signal onto a carrier frequency for two reasons. First, as explained in Chapter 4, "Data Wireless Communication," by modulating a baseband signal onto a higher-frequency carrier, attenuation is substantially reduced. Second, to make good use of the frequency spectrums power lines provide, one must multiplex as many channels as possible in the same power line. This means frequency-division multiplexing (FDM) of multiple frequencies on the same power line. This also serves as an aid for demodulation. If a power-line network has three houses connected, and if each is trying to demodulate the same signal, for example, multiple issues must be dealt with, including granting each of them a good throughput. By sending each of the three data streams on a different carrier, the demodulation scheme becomes far more simple, thus leading to easy access.

The modulation schemes generally used are frequency shift keying (FSK), spread spectrum (SS), and orthogonal FDM (OFDM). Among these, FSK is the simplest form and is accomplished by modulating multiple signals with frequency-separated carriers. The scheme is simple; because of the variation in the channel response and the way the system behaves, however, FSK has inherent limitations in PLC beyond 1-Mbps rates.

The nature of SS as a modulating techniques means that the modulated signal is virtually safe from any interference. This allows transmission beyond the 1-Mbps rate over the power line. One variant that has been studied and implemented is phase-hopped SS, but the limitation here is that of a continuous spectrum. This results because the spectrum does not yield in an even (or predictable) response, resulting in holes throughout the spectrum (due to discrete attenuation in certain bands).

In Japan in 1985, NTT demonstrated (live) PLC using SS technology. Of course the bit rates were of the order of a few kHz, and there were several deployment problems. In SS systems, a viable solution to the main drawback—synchronization—is needed.

Another solution is OFDM. In OFDM, the total bandwidth is divided into N parallel subchannels, and bits are assigned to subchannels in direct proportion to the subchannel SNR. OFDM has the distinct advantage of being able to work despite multipath fading.

Media Access Control Issues in PLC

With regard to Media Access Control (MAC), PLC is quite in conformity with the generic solutions for MAC—namely, token polling, random access, and so on. Most protocols are possible candidates for use on the PLC medium, but they may need some modification. Some examples of these candidate protocols are Ethernet and all its variants, Token Rings, polling protocols, and so on. One of the most important protocols for MAC in PLC is carrier sense multiple access (CSMA). In principle, CSMA is a scheme that allows multiple stations to listen to the channel; however, a station can transmit its data only when the channel is not occupied. CSMA thus became is a direct possible candidature for Ethernet. Another MAC protocol deployed for PLC is Aloha (and Slotted Aloha), a best-effort scheme. However, the throughput achieved (number of successful attempts in accessing the channel) with such is quite low. Therefore, the most successful implementation scheme is CSMA. CSMA schemes deploy CSMA/CA (collision avoidance), and this is typically preferred in PLC systems.

The fundamental advantage of CSMA is its ease of implementation; it is both technologically mature and relatively inexpensive. As explained in the previous chapters, Ethernet is the standard protocol used in first mile access communication, and PLC is no exception. The one problem with implementing CSMA/CA-based Ethernet communication over PLC networks is that the channel is unreliable. This unreliability results from power-frequency-distribution problems. These problems exist because the channel is essentially noisy and because the channel noise varies randomly (making it difficult to approximate, let alone compensate for). PLC networks may need for the MAC layer to have some excess-repetition features, such as automatic repeat request (ARQ), that facilitate the repeated transmission of the lost or corrupted frames. Similarly, another issue that needs to be addressed by the MAC layer is security. Security requires special attention because of the open bus architecture of a PLC system. Such an open system means that anyone can plug in a PLC modem anywhere and access the data by just scanning and tapping the frequency. Therefore, special care must be taken to ensure a secure system, perhaps incorporating special security features such as the 802.1x protocol.

Orthogonal Frequency-Division Multiplexing

OFDM is one of the most promising techniques for data transmission over power lines. OFDM is well documented in the literature and well known in the industry. (See "Personal Computer Communications on Intrabuilding Powerline LANs Using CSMA with Priority Acknowledgements" and "Main.net Communications: Bringing Broadband Internet to Every Electric Socket.") It is currently used in DSL technology and terrestrial wireless distribution of television signals and has been adapted for IEEE's high-rate wireless LAN standards (802.11a and 802.11g). The basic idea of OFDM is to divide the available spectrum into several narrowband, low-data-rate subcarriers. In this respect, it is a type of Discrete Multi-Tone (DMT) modulation (Jeff Norman's PLC presentation). To obtain high spectral efficiency, the frequency responses of the subcarriers are overlapping and orthogonal (hence the name). Each narrowband subcarrier can be modulated using various modulation formats. Because the subcarrier spacing is small, the channel transfer function reduces to a simple constant within the bandwidth of each subcarrier. In this way, a frequency-selective channel is divided into many flat-fading subchannels, which eliminates the need for sophisticated equalizers. OFDM has the following advantages:

- Excellent mitigation of the effects of time dispersion
- Very good at minimizing the effect of in-band narrowband interference
- High bandwidth efficiency
- Scalable to high data rates
- Flexible and can be made adaptive; different modulation schemes for subcarriers, bit loading, adaptable bandwidth/data rates possible
- Excellent interchannel interference (ICI) performance (and so complex channel equalization not required)

For these reasons, it is an excellent candidate for PLC.

This discussion has highlighted some of the intricacies of OFDM. From this discussion, you should realize that OFDM is the preferred modulation style for PLC networks because of its adaptability to such a complex and random channel behavior (noise environment).

Standards and Regulatory Aspects (From "Power Line Communications: State of the Art and Future Trends")

The frequency distribution in the power line and the opening of the power line for multiple vendors to pitch in their broadband solutions creates a need for a hierarchical standardization as well as regulatory authority. The European Union (EU) has defined a set of standards and frequency regulations for use in the power line. Unlike the EU, North America does not have any regulatory standards. Standardization is required primarily because of electromagnetic interference between data and power lines, which needs to be minimized. Standards compliance may minimize this electromagnetic interference.

A Case Study of Power-Line Communications (Courtesy of Main.net)

In this section we analyze the PLC system as sold by a vendor called Main.net. The objective is to examine the subsystems, components, and other requirements of a PLC network first mile access implementation. Management is one of the most important aspects of a PLC network. The management system is responsible for provisioning the network, building up the operational features, billing consumers, and securing the network from unauthorized use. In addition, the management system of PLC networks is part of a higher hierarchical management system that has multiple other network environments within it. Therefore, a PLC network's management plane must seamlessly interface with the management plane of the bigger telecommunication networks. Two generic management protocols need to be considered here: Simple Network Management Protocol (SNMP) and TL-1. The latter, TL-1, is the conventional management protocol for circuit-switched networks such as Synchronous Optical Network/Synchronous Digital Hierarchy (SONET/SDH) and so on. In contrast, SNMP is a more advanced, yet user-friendly scheme for controlling and managing networks that are subject to packet-based traffic—namely, IP communication. It is important for PLC networks to conform to network management standards. Main.net's solution provides an SNMP-manageable example.

The Main.net PLC solution defines the following four elements of the end-to-end PLC network:

- **CuPLUS** is the interface between the PLC network and the traditional backhaul connection. This acts as a gateway of communication between the telco network and the home users who use PLC as a solution for the first mile access.

- **RpPLUS** is a PLC network repeater that expands the reach of the network. The nature of the channel—that is, the power line—is very noisy, and the noise content is randomly variable (as discussed previously). This creates a need for repeaters, which boost the signal level and suppress the noise to extend the reach of PLC systems.

- **NtPLUS** is the network termination device that connects the customer's PC to the Internet. This is quite analogous to a modem in dial-up networks. It acts initially as a filter separating the high-bit-rate data from the low-bit-rate electrical power signal. It then demodulates the data from the carrier frequency as in FSK or by using an appropriate demodulation technique as in phase shift keying (PSK) and so on.
- **PLUSmms** is the PLUS network monitoring and maintenance system. As mentioned previously, this is the brain of the network. It dictates the way the network behaves and how it can be adapted to different traffic and other requirements (provisioning). In addition, call admission and billing features are carried out by this part.

There are two more elements Telplus (an interface to the phone just for voice) and NmPlus (which is responsible for the operations system support (OSS) feature as shown in the diagram).

The four sections previously listed are essential for creating a PLC network for the first mile area.

Figure 8-2 shows the first mile network proposed using PLC. The basic network topology is that of a star, which inherently acts as a distribution network.

Figure 8-2 *PLC System (Courtesy Main.net)*

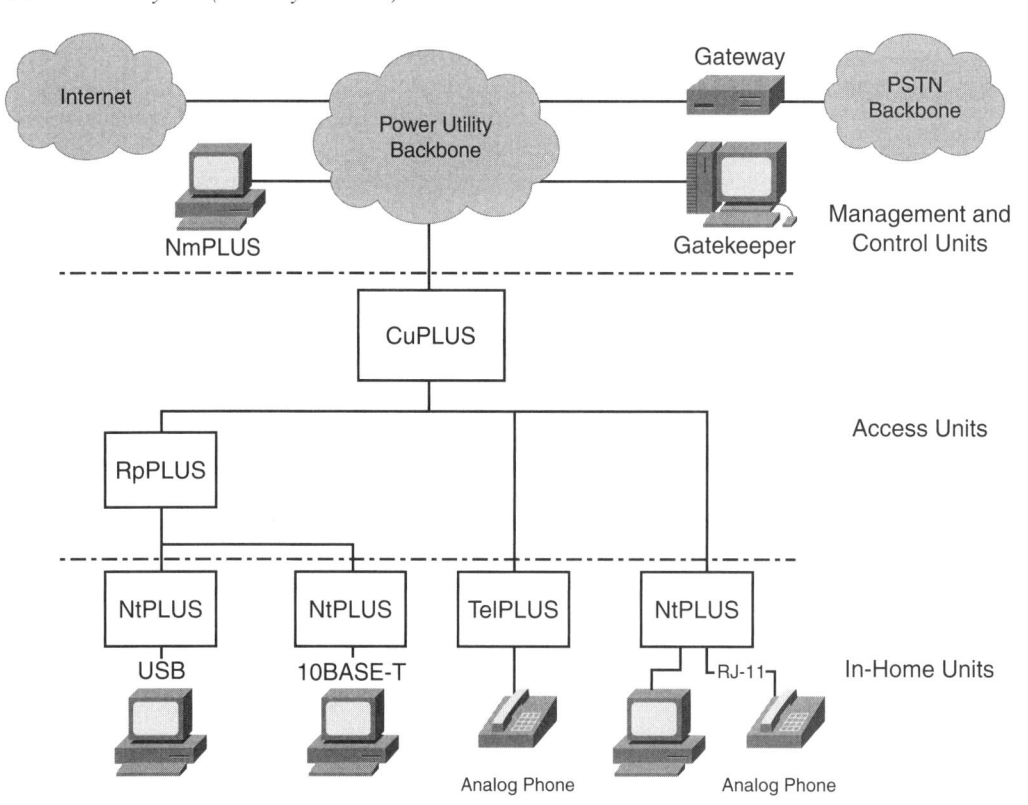

The interface point (A) is placed between the first mile distribution network and the telco access point. It has two main features:

- Regulating data flow through to the distribution network
- Modulating the data, making it power-line compliant

In Figure 8-2, point B is basically a repeater. Two repeater technologies have been defined in industry literature. The first one involves a crude method of direct amplification of both carrier and the modulated signal. In such, the noise may not be suppressed. In the second scheme, the signal is demodulated, and the baseband signal is amplified. In such, the possibility of errors is minimized, and corrupted signals are cleaned up (from the noise). This latter technique is more expensive than the former but is nevertheless very effective.

In Figure 8-2, point C is the direct access point for the consumers. The PC or whichever bandwidth-savvy applications need data access are connected to the PLC network through point C. These are primarily analogous to modems (as mentioned previously).

Some of the specifications for Main.net's model for point C are as follows:

- **Transmission method**—Direct-sequence spread spectrum (DSSS), proprietary direct-sequence spread spectrum; however, the system is PHY-independent and other PHY modulations can be integrated.
- **Transmission**—1.7 to 30 MHz; this means a bandwidth of 29.3 MHz capable of supporting a minimum of 4 to 10 Mb per user.
- **Bandwidth**—2.5 Mbps, 15 Mbps (Q3-2003), 56 Mbps (Q4-2004).
- **Smart repetition**—Allows for expanded network reach with minimal impact on bandwidth.
- **SNMP compliant**—Makes network management (operation, administration, maintenance, and provisioning [OAM and P]) possible.
- **Supports**—Software download and maintenance data upload, via TCP/IP.
- **Interfaces**—10BASE-T (standard) and USB (universally acceptable Ethernet standard).
- **Power supply**—85 to 250 VAC at 48 to 63 Hz.
- **Size**—220 × 168 × 64 mm.
- **Environment**
    - Temperature: −30 to +50 degrees Celcius
    - Relative humidity: 80%
    - Emissions: EN55022
    - Immunity: EN55024
    - Safety: EN60950

Figure 8-3 shows the finished product from Main.net (basic PLC modem).

Figure 8-3   *Modem Equivalent from Main.net*

Adaptations to PLC to Meet RBOC and ILEC Requirements

The key telcos that may be looking forward to deploying PLC solutions in the next few years are Regional Bell Operating Companies (RBOCs) and Incumbent Local Exchange Carriers (ILECs). The hierarchical nature of RBOC and ILEC networks has forced them to have complex requirements in terms of management systems. Therefore, some adaptations are needed to fit PLC solutions in RBOC networks (in particular). Figure 8-4 shows a conceptual diagram of the approximate adaptations needed to fit PLC technology into RBOC networks.

One of the main suggested changes is an interface between the RBOC network gateway (that is, either a router or a bridge) at point A in Figure 8-4. This would regulate the traffic flow into the first mile and also help provision and manage the first mile users efficiently.

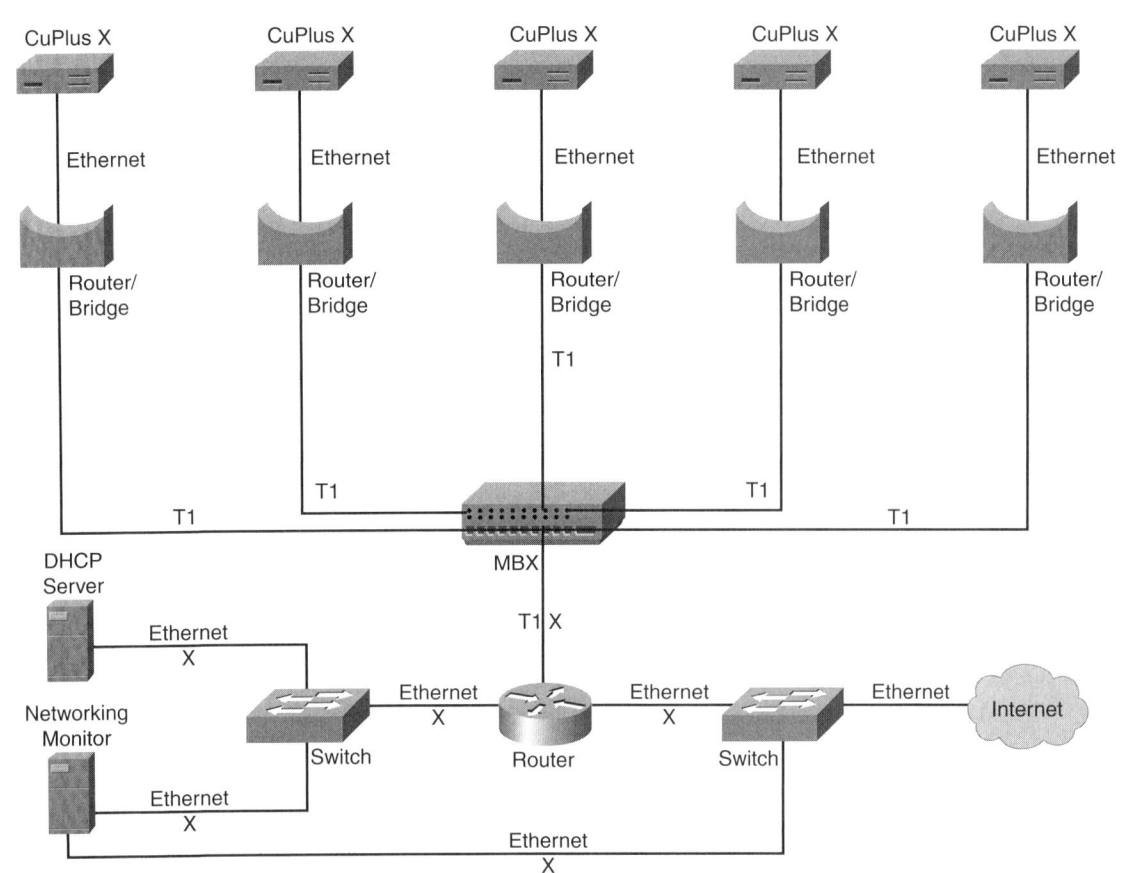

Figure 8-4 *Conceptual Realization of a PLC Network for Service Provider Implementation in the First/Last Mile (Figure Provided with Permission from Main.net)*

An Ideal First Mile Network: Coalescing PLC with EPON

In Chapter 2, "Passive Optical Networks in the First Mile," we discussed the benefits of PON and concluded that Ethernet over PON (EPON) was the simplest and best candidate for a high-performance broadband solution. However, one limitation inhibited the deployment of PON solutions worldwide: the high initial cost of deployment. Fiber to the home/office/end user or premises is an excellent solution that will, once and for all, solve the first mile access problem. In a world where economic necessities often dictate the way industries move, however, laying fiber to the home is not always easy to justify in terms of cost. That said, the cost is significantly reduced, and hence the problem greatly alleviated, when we have to lay fiber just to the curb. A curb here can be defined as the closest location from a home where the service provider has billing/management equipment already in place. This is usually within a distance of 300 to 500 feet from the end user. In this case, the ideal network is the one having two technologies: PON to the curb and PLC from the curb to the end user. This scenario would prove beneficial in a number of ways:

- Decreases the cost of deployment.
- Manageable solution making all parts of the network SNMP compliant.
- The cost per user is reduced substantially.
- Exemplifies the real meaning of broadband—gives each user a high-bit-rate solution.
- Is one solution that is able to bridge the digital divide.

Figure 8-5 shows a PLC-EPON network concept. Note that the EPON solution terminates at the curbside, and from there onward we have a PLC solution to the end user.

Figure 8-5 *PLC and EPON Together*

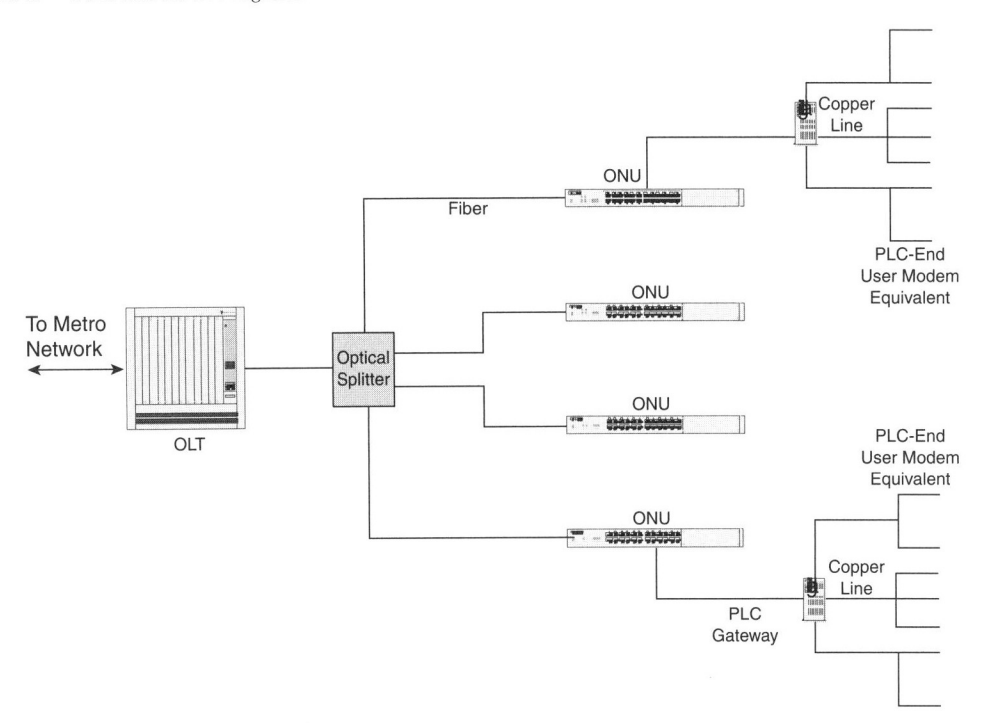

The advantage of a combined PON and PLC solution is that it acts as an excellent distribution network. The PON part exemplifies high bandwidth to the PLC network, whereas the PLC part means that the cost per user is sufficiently low (taking the copper wiring into consideration; because, as we know, this wiring already exists).

Comparison of PLC with Other Contemporary Solutions

Table 8-1 lists a detailed analysis of the PLC solution as compared to cable, wireless, DSL, and PON solutions. Note that for PLC we consider Main.net's solution in general. PLC provides some distinct advantages in terms of cost and feasibility. A disadvantage is that PLC doesn't provide a ubiquitous connection, one that has high bandwidth and is free of errors.

Table 8-1 *Comparison of Multiple First Mile Solutions (PLC Perspective)*

Features	PLC	Cable	ADSL	Wireless	PON
Plug and play	Main.net only	Yes	No	No	Yes
Enables home networking	Yes	No	No	Yes	Yes
Available in any geographical topology	Main.net only	Yes	No	No	Yes
Requires additional in-home wiring	No	Yes	Yes	No	Yes
Requires upgrade of existing infrastructure	No	Yes	Yes	Yes	No
Cost of first mile	Low	High	High	Low	High
Geographical coverage	Full	Limited	Limited	Full	Full
Target market	Home, SoHo	Home, SoHo	Business, SoHo	Business	Home and business

HomePlug 1.0 Specifications for PLC

Power lines were originally intended for electrical power distribution using the frequency range of 50 to 60 Hz. The HomePlug 1.0 standard addresses the use of this medium for high-speed broadband communication. An electrical distribution system is comprised of various materials with different impedance characteristics. In addition, as the consumer load pattern changes, the terminal impedance varies with frequency and time. Reliable data communication over an electrical power distribution system requires powerful forward error correction, automatic repeat request (ARQ), and adaptable MAC. These techniques ensure corruption-free communication to the end user.

The HomePlug 1.0 specification addresses the physical (PHY) layer, MAC layer, and the protocol specifics to overcome the challenges faced in PLC environments. The PHY defined by HomePlug 1.0 uses OFDM with a cyclic prefix (CP). OFDM splits available bandwidth into many small frequency bands called *subcarriers*. Each frequency band subcarrier can be modulated using various modulation formats. The PHY detects channel conditions using channel estimation techniques to acquire channels. It adapts by avoiding inadequate subcarriers and selecting an appropriate modulation method/coding rate for the acceptable subcarriers. Only the payload portion adapted to the channel conditions (preamble and frame control combination) are used as delimiters that start and end long frames. Physical carrier sense is performed by the PHY layer, and helps the MAC determine when the medium is busy (uses modified CSMA/CD). To avoid ISI, a CP of the last 172 samples from the inverse fast Fourier transform (IFFT) interval is prepended to form a sample OFDM symbol.

Connectors are designed to contact only one phase of the power-line system and the neutral line. The users can connect or disconnect the system whenever they want.

Signal Processing

Before OFDM symbols are formed in the analog front end (AFE), data is scrambled, Reed Solomon (RS) encoded, convolution encoded, and interleaved on the transmitter. The AFE consists of a constellation-mapping block, an IFFT block, a preamble block, a CP block, and a raised cosine (RC) block. The mapping block selects the type of modulation and the carriers to be used in the IFFT block, as specified by the tone map and tone mask. The IFFT block modulates the constellation points onto the carrier waveforms (in discrete time). The preamble block inserts the preamble. The CP block adds the CP, and RC shaping is used to reduce out-of-channel energy.

The PHY layer transmits four distinct entities: the preamble, frame control (FC), the payload, and priority resolution signals (PRSs). The FC and the preamble are always sent together, and form a delimiter. The receivers must detect and decode delimiters as well as the PRSs regardless of who is sending; hence they must use all subcarriers with the same modulation and encoding. The payload is adapted to the channel conditions by negotiation between the sender and receiver during channel estimation. Channel estimation determines which subcarriers to use and the type of modulation and forward error correction (FEC) rate to apply. Depending on channel conditions, various combinations of modulation types and FEC rates are available, enabling the sender to adapt to the channel to improve the data and error rates.

HomePlug 1.0 MAC

The MAC protocol defined by HomePlug is a modified CSMA/CA protocol with priority signaling. The protocol frame structure is shown in Figure 8-6. HomePlug devices communicate with each other without any centralized coordination. HomePlug 1.0 provides four priority classes, and priority resolution is achieved by asserting appropriate signals in the assigned time slots. Contention within the same priority class is resolved during the contention period using a priority-dependent backoff window.

Figure 8-6 *Format Structure for HomePlug 1.0*

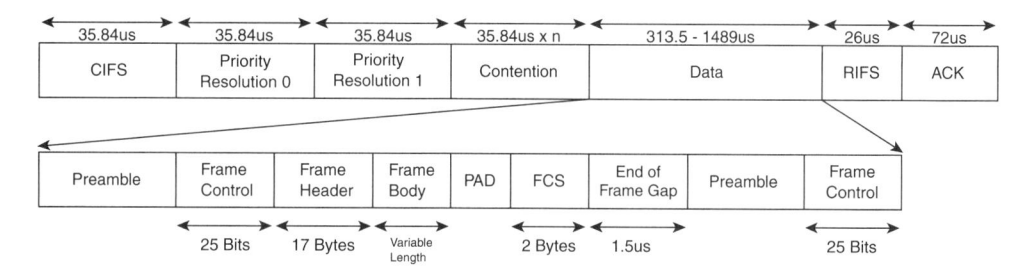

35.84us	35.84us	35.84us	35.84us x n	313.5 - 1489us	26us	72us
CIFS	Priority Resolution 0	Priority Resolution 1	Contention	Data	RIFS	ACK

Preamble	Frame Control	Frame Header	Frame Body	PAD	FCS	End of Frame Gap	Preamble	Frame Control
	25 Bits	17 Bytes	Variable Length		2 Bytes	1.5us		25 Bits

Carrier Sense and Collision Detection over Power Lines

The HomePlug 1.0 PHY layer reports physical carrier sense (PCS) by detecting preambles or priority slot assertions.

The MAC layer maintains virtual carrier sense (VCS) using the Length field of the Start of Frame (SOF) frame control, along with information on whether a response is expected or not (present in both the SOF and the End of Frame (EOF) frame control). Direct collision detection as used in Ethernet is unreliable due to attenuation, noise, and channel adaptation. Collisions can only be inferred from a lack of response after a frame is sent. This makes collisions very costly compared to CSMA/CD systems, so they must be avoided (by being less aggressive when the medium is busy, instead of transmitting as soon as the medium becomes idle [as in standard Ethernet]).

Market Analysis and Trends

As of now, the PLC market is a nascent market, best described as an emerging phenomenon. Among other predictions, a significant increase in the number of local bodies and communities that deploy PLC networks can be predicted. This is because service providers sometimes find it difficult from a right-of-way perspective to deal with utility companies, which are generally community owned/leased. PLC networking is gradually being accepted as a standard solution to the first mile problem. As the acceptance grows worldwide, the variety of available applications will broaden. Initially, PLC was used just for voice. As channel difficulties (noise) were overcome, the application of PLC was extended to data and later video. In addition, PLC can be used for monitoring and remote-sensing applications and in data acquisition.

As far as home networking goes, PLC represents a new technology. Because of its broadcast nature, a PLC network inside a house is like a LAN, making the PLC solution particularly more attractive. The possibility of putting native Ethernet over PLC is a key to the success of PLC. Because CSMA/CA can be performed over PLC, it is an excellent first mile technology.

Finally, PLC outperforms its main competitor (DSL) in terms of price and throughput. DSL solutions such as asymmetric digital subscriber line (ADSL) and Very-High-Data-Rate digital subscriber line (VDSL) are influenced by the distance bandwidth product limitation. PLC in contrast can regularly deliver high bit rates to the tune of 4 to 10 Mbps, substantially more than its competitors. Moreover, PLC gives the same performance (high-speed access) for both upstream and downstream communication.

Currently, the PLC market has several vendors that equally share the market. Several carriers have expressed interest in using PLC as a distribution technology in the first mile. Worldwide deployment of PLC has begun in a strong way. From the way PLC is picking up business, PLC will most likely oust DSL as the number one cable-based technology solution in the first mile. In the end, for all practical purposes, we will most likely find EPON, PLC, and WiFi as the three main components of broadband access in the first mile.

Hybrid Fiber Coaxial Plant as a Possible First Mile Access Solution

The rapid growth of cable technology for TV broadcasting was an important step in giving users a plethora of communication spectrums. Cable or coaxial cable technology supports wider bandwidth than the PSTN. Therefore, cable can be used for broadband communication in the last mile, particularly in the last inch. Cable modems were soon available on the market, and this form of broadband access became quite popular in areas where service providers provided broadband over coaxial cable. However, cable itself—that is, coaxial cable—has one limitation: It is prone to severe attenuation with distance. This means that an end-to-end cable solution is not possible. Present deployment of cable for broadband access is in the form of hybrid fiber coax (HFC) plants. In this setup, fiber connects terminate at a curb-site point of presence, and an HFC plant converts the broadband signals from the optical domain to multiple coaxial cables, which in turn service end users. Figure 8-7 shows an HFC plant with a converter box. The function of this box is as follows: It is connected to the central office (head end, essentially a video distribution network plus the gateway) by a fiber line. On the other side, it is connected to multiple homes through coaxial cable. The signal from the head end to the converter is fed through a fiber and hence is in optical format. This signal is demultiplexed and demodulated at the converter. The converter then distributes the converted electronic signal to the coaxial cable depending on the addressing format and so on. A single fiber can feed multiple coaxial cables. Each coaxial cable has 8 to 250 MHz of bandwidth available (whereas only 5 to 55 MHz is used for broadband). The cable comes in two grades: RG 59 and RG 60 (used for homes).

Figure 8-7 *HFC Plant*

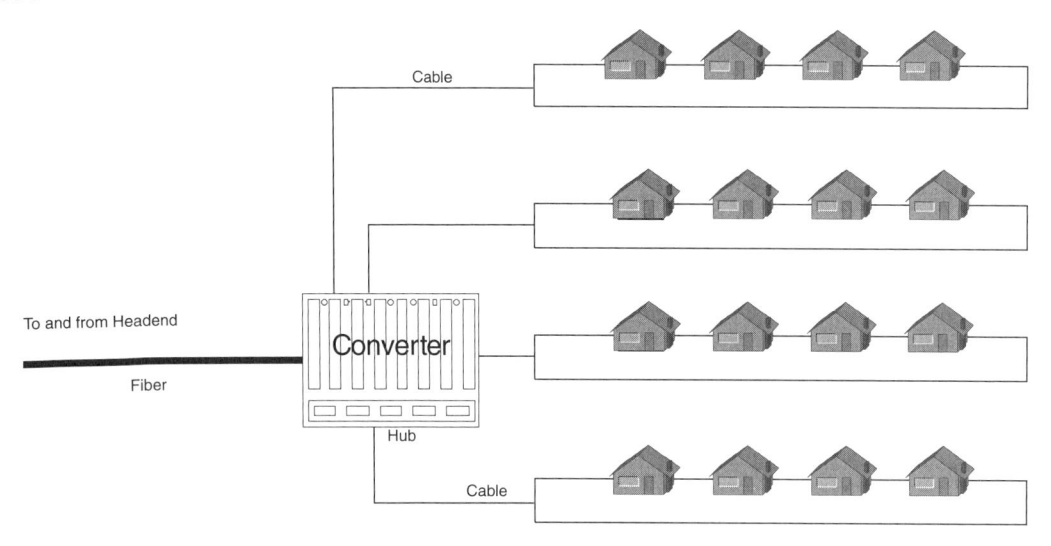

The attenuation of cable per 1000 feet is 60 dB at 100 MHz for thin Ethernet, 20 dB for thick Ethernet (RG 60).

The cable represents a shared medium for sharing the line access between multiple users. A CSMA scheme is deployed here. This results in low throughput, but overall access is much better than DSL or PLC. However, performance depends on the distance of the end user from the converter box and the number of end users at a given time trying to access the same channel.

HFC represents a mature technology and, therefore, is not the subject of interest for our study of first mile access networks.

Summary

In this chapter we examined PLC systems as a first mile access network solution. The chapter covered the PHY layer properties of the PLC network and also the various technologies needed to deploy a PLC network. We concluded this chapter with an analysis of PLC market trends and a quick look at an HFC plant as another alternative in the last mile.

Review Questions

1. What is the key technological limitation of PLC?
2. Which factors allow PLC despite the EMI from electrical lines?
3. Describe the interface requirements for PLC at the customer premises.

4 Describe OFDM modulation and state the advantages of OFDM over other schemes.

5 What are the key impairments for PLC in the last mile?

6 Discuss a PLC network layout from an RBOC perspective.

7 PLC is a more suited for which of the following and why?

  a High rises

  b Dispersed homes

  c Clustered communities

  d Campus environments

  e High-bandwidth applications

6 Describe the management challenge of a PLC network from a bandwidth-provisioning perspective.

7 What are the main features of SNMP that affect PLC networks?

8 Discuss HomePlug 1.0 and summarize your thoughts on a prospective new standard that uses PLC + PON as a complete last mile solution.

9 Discuss PLC market trends and explain the key business issues that may hinder PLC growth.

10 Explain why HFC has captured a large market share despite its average performance. Do you think that dual-use technologies such as HFC and DSL (same medium, two ways of implementation) will succeed in the first mile for broadband and ultrabroadband (bandwidth-killer applications such as video on demand and so on) communication? Compare this to a PON + WiFi solution and to a PON solution giving FTTH.

References

CENELEC, "Signalling on Low-Voltage Electrical Installations in the Frequency Range 3 kHz to 148.5 kHz—Part 1: General Requirements, Frequency Bands and Electromagnetic Disturbances"

Dostert, Klaus. *Power Line Communications*. Upper Saddle River, New Jersey: Prentice Hall PTR, 2001.

IEEE Communications Magazine April 2003

IEEE Transactions on Power systems

Courtesy of Jeff Norman from Main.net http://www.main.net

Courtesy of Jeff Norman from Main.net PLC Presentation

Lin, Y., R. Newman, and S. Katar. "A Comparative Study of Wireless and Powerline Networks," *IEEE Communications Magazine,* April 2003.

Main.net Communications: Bringing Broadband Internet to Every Electric Socket

Pavlidou, N., et al. "Power Line Communications: State of the Art and Future Trends," *IEEE Communications Magazine*, April 2003.

Onunga, J. and R. Donaldson. "Personal Computer Communications on Intrabuilding Powerline LANs using CSMA with Priority Acknowledgements," *IEEE Journal on Selected Areas in Communication*, Entire Special Issue 7.

www.cogent.com

www.plcforum.org

CHAPTER 9

First Mile Access Management and Business Model

First mile networks have the potential to directly and positively affect our daily lives. To realize this potential, however, we must justify these networks in terms of cost, confirm their technological viability, and integrate the technologies into the preexisting voice- and data-transfer world. In short, we want a business justification for such networks. This chapter introduces a new type of business model: the intelligent city. In such, a community, or otherwise socially connected group of individuals, builds a network and creates management tools to provide services efficiently. This chapter does not analyze the business case for these emergent networks. Instead, it covers these networks from a technological perspective, discussing the integration of multiple protocols in the same network.

Unified Management for Access Networks

The economic downturn and its effect on the telecommunications industry has prompted the accelerated development of this new concept of intelligent and publicly-owned networks. An intelligent city network represents a unique business model whose returns encompass social, economic, and technological aspects (those of a group of socially connected communities). This model enables multiple vendors to sell products and services to a larger customer base at a lower cost, and gives providers new opportunities to work in tandem directly with customers/communities. From a business perspective—namely, taking into account operational expenditure (OPEX) and capital expenditure (CAPEX) for building and operating such networks—there are several possibilities for enhanced service. These possibilities raise the primary issue in intelligent city networks: bandwidth management that supports provisioning in real time on a per-demand basis. Resolution of this issue is critical for the future success of such networks. This chapter addresses this issue and other concepts associated with a unified management scheme for such a network. Chapter 10, "Business Case for First Mile Access Networks," covers the finer points of designing intelligent cities while presenting the business case and the application (technology) case for intelligent city networks.

Intelligent city networks are also known as *community networks* or *municipality-owned networks*.

The Emergence of Alternative Models—Intelligent Cities

Recently, several cities and towns in North America have started to experiment with fiber-to-the-home (FTTH), fiber-to-the-curb (FTTC), and wireless-fidelity (WiFi) deployments, organized around an optical backbone, to build their *own* networks supporting high-speed Internet access and new services for their businesses and homes. The goal of such networks is to improve overall quality of life—including bringing new businesses into the city and creating more jobs. Many business and technological issues must be resolved before undertaking the building of such intelligent city networks.

Corporate financing of a fiber-optic network carries significant risks, especially from a return on investment (ROI) perspective, as evident from the recent demise of several service providers that failed to become profitable (even after spending billions of dollars over the past half-decade). Communities, municipalities, and local governments are therefore attempting to build community-owned metro-access optical networks based on a different/alternative (and more relaxed, from a business perspective) model. Their solution represents a significantly reduced effort, focusing on investment in infrastructure only, outsourcing management and active network deployment, and foregoing short-term ROI for longer-term socioeconomic community benefits.

To build an intelligent city network with fiber as its backbone—one that enables new services, the optimization of the network's bandwidth management is the key component to success. More than just an incremental change is needed to demonstrate the benefits of a truly high-bandwidth optical network to the curb or home. In other words, along with high bandwidth, bandwidth on demand (BoD) is also needed for home, for enterprise customers, and for communications service providers to make these kind of networks attractive, affordable, and practical. Application-specific service providers, for example, can use these networks to provide niche services without worrying about the investment in the physical layer. In the optical segment of the network, BoD can be obtained by real-time provisioning of wavelength and subwavelength (services), enabling high-speed real-time services to be effectively offered throughout the city. However, an optical MAN is often characterized by a diverse operational environment, consisting of various topologies and vendors that current optical bandwidth management systems fail to optimize. Traditional service provisioning, or circuit-switching systems, offer limited choices and rates, and are very slow. Even today, customers can wait days to obtain a new high-speed optical connection. To create a competitive environment in which providers can offer the community real-time, on-demand provisioning, along with the more traditional value-added services of network and protocol conversion, exchange services (including peering, aggregation, and protocol conversion, including SONET, ATM, IP, or Ethernet), traffic engineering and aggregation, bandwidth management, network security, service level agreements (SLAs), and private network connections with diverse network carriers, the emerging community providers are in acute need of new and better bandwidth-provisioning solutions. Currently, the metropolitan-area access relies on provisioning in which circuits are allocated, SLAs managed, and compliance enforced, all via a central point. Although the centralized operation yields a simplified management environment, it is generally agreed that a metro-access optical network cannot currently provide true optical connectivity, high efficiency, low delays and low data loss, real-time settlement for bandwidth trading, or BoD, or real-time provisioning and pricing using extant centralized bandwidth-management solutions. Under current economic conditions, better use of current bandwidth in terms of lower capital expenditures and operating costs, as well as improved bandwidth granularity, is essential for viable communication solutions. It is necessary, therefore, to develop solutions with added bandwidth-management intelligence that can support the emerging broadband applications while delivering the necessary bandwidth efficiency. Under such conditions, the need for advanced services at customer premises can justify the deployment of access optical networks that bring new telecommunication products, improve the quality of life, and open new markets, all of which can lay the foundation for accelerated regional development.

Intelligent City/Corporation Networks

A network infrastructure built by a representative authority for the deployment of broadband-based services represents a generic intelligent city/corporation network. An intelligent city network is a classic example of a centralized investment by the government/municipality for the betterment of the general lifestyle of the population per se. The intelligent city network becomes a platform for multiple services, allowing vendors to offer a plethora of these broadband services to engage enterprise and residential users on a voluminous basis due to the social diversity of the city population. Intelligent city networks are being developed with past optical experience in mind.

The recent downturn in the economy has led to a severe situation for telcos, particularly optical networking vendors, because of various reasons primarily low demand due to over-capacity in the core. This period of instability, referred to as "the photonic winter," has caused tremendous turmoil in the telco industry, leading to doubts regarding the business prospects of broadband communications. The business model of carriers taking on huge capital investments and creating abundant network infrastructure did not achieve the primary target of a fast ROI. The failure of this business model was partially due to a lack of foresight as well as the costing structure of the network infrastructure, and an incorrect assessment of the projected demands and revenue models of the carriers. An additional problem faced by carriers was the inability to provide the broad range of value-added services, many requiring efficient on-demand real-time bandwidth provisioning. The inability to support this type of bandwidth management were due to a number of reasons, primarily from the vast number of technologies that interacted with each other in a generic long-haul network. Cumulatively, the previously mentioned factors wrecked the all-optical vision for global optical networking.

Over the same period, metropolitan-area networking has seen some successes. Newer optical technologies such as WDM and optical cross-connects and so on were successfully applied in the metropolitan area and hence seemed to be a better investing reasons for good returns. First, the fiber represented a medium to carry significantly large amounts of data. This was seen as a paradigm shift from conventional network mediums such as copper and wireless. Second, the envisaged huge available bandwidth was seen as a safe revenue multiplier. Third, by keeping data in the optical domain, the network infrastructure was expected to be greatly simplified. Optical switching of bit streams was much easier and cheaper than in an electronic domain. This led to projecting lower capital expenditures and lower operating expenditures. By keeping data in the optical domain, network planning and deployment became simpler and more cost effective. These projections led the telco industry to believe that optical technologies were mature enough to cause a revolution. However, these projections were quite overoptimistic, too. Data streams were indeed easier to be switched in the optical domain, but the switching was purely between ports for end-to-end optical paths or lightpaths. Technological limitations prevented the deployment of packet switching in the optical domain. A trade-off between conventional lightpath switching and packet switching—the recently emerged paradigm on optical burst switching—was still a technology in its absolute infancy. Apart from light-trails (see " Light-trails: A Solution to IP Centric Communication in the Optical Domain"), no architecture could support such highly flexible schemes as expected from burst switching. Grooming traffic from multiple bit streams onto a homogenous high-capacity bit pipe or lightpath was not possible in the optical domain. Getting the desired granularity at an incumbent optical node meant time-division demultiplexing of the lightpath and segregating to create multiple slower rate streams. This could only be done in the electronic domain (often creating issues such as synchronization and so on to facilitate TDM and lead to a surge in the cost of provisioning such multiple bit rate pipes. These technological barriers demonstrated that no single technology solution will provide bandwidth at high granularity on demand, similar to electronic switching in the near future. On the other hand, purely electronic networking at high data rates does not provide a cost-effective solution, hence a mix of the two under a proper management suite offers the only cost-effective solution today. In the intelligent city network area, specifically, the metro access needs to be augmented by this last mile solution, often built around a wireless/copper/fiber network. The resulting integration of multiple technologies to suit the end-user requirements is highly challenging. Wireless networks built on top of the optical backbone primarily for voice and secondarily for data need to be provisioned keeping in mind the intricacies of both optical as well as wireless technologies, and the discrepancy in the speeds of optical and wireless transmission. Keep in mind, the entire wireless spectrum is a few gigahertz, whereas a single strand of optical fiber can cover several orders of magnitude more and yet have fiber reuse.

The emerging intelligent city network models drawn on this experience when building a new type of modular, centralized telco distribution framework. The intelligent city network model, in short, is a hierarchical implementation by a central authority to facilitate the exchange of information between residential and business users, and to create a uniform platform for service providers to invest in the most appropriate manner for optimized services that would be offered.

The three factors that most critically dictate the success or failure of an infrastructure of a project are invariably the social, economic, and technological implications involved. While balancing the effects of each of these factors, a central authority (government/municipality/corporation) has to decide how to best build a network infrastructure. The built network infrastructure called intelligent city network is further intended to create a level playing field for carriers and other telcos to provision their services in the best possible cost structure for the ultimate benefit of the community.

The intelligent city network, by technology classification, is a metropolitan-area access network on account of its proximity to the metro area and its multiservice capability. By provisioning multiple classes of service through a multivendor open system, an intelligent city network is intended to derive the best possible results in terms of economic, technological, and social implications, the justification for its deployment, as discussed next.

Nontechnological Challenges

An intelligent city network acts as a single unifying strata to reach a large population in the least expensive as well as most efficient manner. The intelligent city network showcases, by definition, an open system architecture for business and residential customers and provides for various access services. To develop a broad segmented network infrastructure to reach a diverse population profile with varied requirements can be facilitated by only a governing authority like municipality and so on that has the necessary and sufficient resources (both financial and otherwise). The central authority that commits to such an exercise benefits from the multiple service advantages stemming from community network deployment. Purely from a business-case perspective, the central authority creates an optical backbone for further leasing/renting of network resources to various specialized providers. Although slow to begin with, the ROI in an intelligent city network for both the leaser (central authority) as well as the lessee (the service provider) is expected on account of the broad structure of the network deployment strategy (discussed in technology section) as well as the multiservice platform. The intelligent city network business model also takes into account the fact that multiple service classes can exist to serve the various economic strata's of society to alleviate network bandwidth requirements from a long-term perspective. The intelligent city network can then be understood as a unified framework for carriers to provision their individual services and offer these services to specialized customers as well as consumers, consolidating the market share of the carriers, and assuring fixed minimum revenue to the providers. The business model also drives a need for carrier investment only at places where such investment is absolutely necessary, thus saving overall CAPEX. From a management perspective, the intelligent city network represents a unified management scheme with multiple technologies and services on the same high-bandwidth backbone. The services and technologies on the optical backbone can be managed from a wavelength or wave-service perspective, leading to resilient, scalable, and efficient networks with low operational costs (OPEX).

The single homogenous backbone in a geographic community created by the central authority represents a wide panorama of technologies coalesced into the backbone to exude user-confident services. From an intelligent city network perspective, the network backbone is an optical network, spread geographically across the community with various points of presence (POPs) to interact with users. The current level of technology for optical networks places this class of optical networks in the third generation optical networking.

From a historical perspective, the first generations of optical networks were legacy SONET networks, with tributary and distribution feeder lines. These predominantly catered to voice traffic and represented a synchronous TDM hierarchy. With the advent of data communication (primarily the Internet), voice services were outnumbered by data services, and pure SONET networks were inefficient (for reasons of cost as well as technology) for this kind of information transport. A multiservices platform paved the way to the second-generation optical networks, which were based on multichannel WDM systems. The benefit in efficiency of packetizing voice and data and maintaining a packet-oriented network made data networking on WDM a strong contender for second-generation optical networks. The packetized form of transport as well as the bursty profile of Internet (IP) traffic created a unique requirement for quality of service (QoS) for transport of information. QoS was then, more than ever before, considered a strong prerequisite for good service guarantee in packet-based communication modes. ATM and, much later, MPLS served as two generic protocols to ensure QoS to the traffic flows. These types of QoS protocols over optical backbone, through Gigabit Ethernet or Packet over SONET (POS), were still part of the second generation of optical networks. The third generation of optical networks is currently being consolidated as an amalgamation of new and old technologies to offer a class of service that is well defined. The data-transmission-oriented third generation of optical networks promotes a number of new technologies such as Gigabit and 10 Gigabit Ethernet, POS, MPLS, and RPRs (resilient packet rings) to create efficient transport over an intelligent optical layer. Despite these advances, traffic grooming and high-speed packet switching are still technologies in their infancy.

Intelligent Optical Layer

As opposed to past manifestations of optical networks, in which stagnated optical boxes had the primary function of data transport and providing elementary resiliency, the optical layer today serves as an intelligent means for transport of service-oriented data. WDM networking has now matured enough to provide smart features for seamless integration of the higher-layer services on the WDM physical layer. Among them, high-density, high-speed optical cross-connect architectures, highly granular optical multiplexing technology, and a smart provisioning layer are the key for the popularization of WDM networking. Among the biggest challenges in the third generation of optical networks is the ability to manage the vast number of different services according to network classification and user requirement. From an implementation perspective, metropolitan-area optical networks are the closest classification to intelligent city networks. Metro networks represent a class of networks that have multiservice platforms and have a sizable amount of internetwork as well as intranetwork traffic. Metro networks also have multiple providers sharing the backbone through wavelengths and wavelength (lambda) services creating a niche market for themselves. From a commercial perspective, metro networks represent a single high-growth area. Despite the recently witnessed downturn in the optical segment, metro networks still continue to experience steady growth, on account of their strong impact on users' needs.

Economic Implications

An intelligent city/corporation network represents a single broad-based infrastructure and a level playing field for multiple vendors to provide service to consumers. In contrast to a typical application-specific network, a city/county/corporation network has both consumers and customers as end users and both service providers and application-specific providers as the vendors. Moreover, it is built and managed by a central authority—the community leadership—which has the authority and the necessary regulatory laws to regulate the billing and costing structure of the network commodity, bandwidth. The community business model is a slow ROI model, with an emphasis on surety of returns rather than on a quick-return high-risk model. Despite this slow form of return, the business model is very stable, and bandwidth is well defined and well regulated. From the perspective of returns, in the present economic climate only a central authority such as the government or a local administration can take part in such an exercise on account of the very high investment initially needed in the network. The returns, although slow, are expected to be guaranteed in this environment, because bandwidth is now a secondary commodity, with applications being the primary requirement (the concept is also known as *content-delivery-based billing*). To draw an analogy, we consider the following: The network infrastructure can be envisioned as the "roads" in a remote area of a city suburb. Upon completion of this basic transport infrastructure, a section of the population would move to this part of the city on account of the ease of access (analogous to good bandwidth availability). The inhabitants can move only when there are houses and residential buildings on the periphery of the roads (access points). The access point creates revenue—realtors sell property, apartments, furniture (application services). As traffic increases, and vehicles and pedestrians use the roads, they pay tax to the government.

Business is conducted through this network of roads (analogous to a city network), and the government benefits through indirect taxation also. Over a period of time, the government is able to recover the entire investment through direct (to inhabitants and business owners) as well as indirect taxes. Moreover, by creating this infrastructure, a level playing field is formed for different business entities to sell their specialized services. The average lifestyle of the inhabitants is also bettered through such an exercise, showcasing the importance of building an infrastructure by a central authority. If a private entity were to build a township, the amount of resources it would need would be tremendous. Moreover, due to fear of competition and lack of trust, not many other companies would be willing to join such a private venture (analogous to service provider monopoly in the United States prior to 1984). In such a case, a model built by the central authority opens up the entire market to competition. The same principle can be applied to intelligent city networks also. Intelligent city/corporation networks from an economic perspective therefore showcase a single efficient model for revenue generation and creation of a bandwidth-on-demand (BoD) network for the end user.

Community Network Management

The third generation of optical networks (the first two were fiber and SONET) has found new application in intelligent city networks. One of the key issues in creating city networks and ensuring their overall success is the management of such networks. From the management perspective, city networks pose a multifaceted problem. Among the many issues that require special attention are multiservice provisioning platforms (MSPP), multivendor issues (interoperability), and data and voice bundling. Although far from complete, this list represents the key management bottleneck for provisioning services on a real-time basis in city networks.

The primary issue in managing a city network is the myriad technologies on the same network backbone. Data and voice, packetized and time-division multiplexed (TDM) technologies, coalesced and coexisting on the same platform, create several challenges for management and provisioning and billing. From a management perspective, it is desirable to provide efficient services as well as create a resilient network; whereas, from a provisioning perspective, it is desirable to make available third-generation network features such as BoD, a multiservice platform, and scalability. The wide range of network protocols that serve as transport mechanisms creates difficulties in provisioning such type of networks. Packet and TDM protocols when multiplexed together on separate wavelengths are quite easy to provision and manage, but issues arise when multiple technologies must be interconnected in an end-to-end scenario. The problem escalates when there is some additional QoS requirement for some of the end-to-end traffic flow. The right amount of grooming and the adequate QoS guarantees accorded are two critical issues in the provisioning of such services. In an optical access network, for example, an end-to-end user lightpath (optical circuit) may pass through several nodes, some oblivious of the data that flows through, whereas some, resorting to O-E-O conversion and creating protocol variations. These interconnecting nodes have the unique responsibility of ensuring seamless delivery of the data and creating a unique framework to facilitate good QoS-based communication.

Another issue in management is the equipment/element management system (EMS). Because of the slowdown in the economy due to overinvestment and the absolute resolve by network planners to use every available piece of network equipment for deployment, a typical metro-access or first mile network may have multigenerational network equipment. On average, a broadband network may consist of equipment from at least two of the three known generations of networking. Managing such networks can become extremely difficult, because each generation creates its own functionality and hence promotes an entirely different set of network parameters. For example, ATM and Frame Relay represent the previous decade of networking, whereas MPLS and its much more optical cousin GMPLS represents a newer generation of networking. The former were (virtual) circuit based, whereas the latter were purely packet based. Hence such networks must be based on industry standards. For most known protocols, standardization is a key to their global acceptance. In other words, as long as network equipment conforms to a standard, the possibility of good internetworking (and hence provisioning) is high. Despite adherence to standardization process, networks today may still not be manageable due to protocol nonconformity or overhead wastage. For example, the overhead wastages in ATM and SONET are different, creating a mismatch for transport of data first over ATM and then over SONET. On the other hand, by using a statistical TDM scheme such as Gigabit

Ethernet or 10 Gigabit Ethernet, the bandwidth-use issue gets partially solved. It is the opinion of the research community that GigE and its faster-mode 10 GigE will be the most effective methods of data communication in the next decade or so.

The multilayered approach to internetworking by itself creates a serious management problem. To provision services on an end-to-end basis, the network has to be so managed such that the selected services through multiple technologies can be optimized to yield the best results. This can be done using a generalized framework for management at the optical layer. When IP routing was transformed into high-speed gigabit-capable routers, there was a need to have a smart provisioning scheme for the router ports, to map ingress and egress ports in an efficient way. This scheme, called *Multiprotocol Label Switching (MPLS)*, is a method to create forward equivalence classes (FEC) for different traffic schemes and switching packets (provision) in routers by purely looking at Layer 2 scheme instead of decoding the entire Layer 3 header. This results in shorter routing cycles, as well as accorded good QoS when desired. This scheme, when extended to a larger spectrum from just switching to the optical layer as well as the multiple technologies is now being standardized as a generalized framework called Generalized Multiprotocol Label Switching (GMPLS).

Historically, optical network management itself has evolved in synchronous and asynchronous methods: TDM and LAN technologies. The TDM technology management—that is, management of SONET networks—was done by TL-1 commands. On account of the difficulty in deploying TL-1 commands universally, TL-1 was proposed to be altered with a more prolific method-CMIP or common management information protocol. CMIP is a machine-to-machine language, intended to be vendor neutral. CMIP has been popular in SONET networks, but it did not spread beyond SONET technologies. For packet-based technologies such as routing, switching, and so on, SNMP became the accepted standard. With the integration of packet and TDM technologies, there also emerged a need to integrate the management schemes of the two. In principle CMIP and SNMP were incompatible. CMIP uses a connection-oriented model of operation, whereas SNMP, by virtue of being packet mode, is connectionless (UDP). Moreover, CMIP executes management strategies and actions through a handshake protocol, thus creating a chronological hierarchy for execution. On the other hand, in SNMP, being entirely connectionless, there is no need for a hierarchy to establish communication between clients and servers. With the need for maximum utilization of the network, and the wide spread of data traffic through the Internet, the integration of the two technologies became necessary. SONET networks originally designed to carry voice circuits now were transformed for packet-based communication to maximize their bandwidth utilization. Likewise, established packet networks could carry packet-based voice (VoIP and so forth) and hence avoid the need for expensive additional deployment of SONET gear. This created an additional bandwidth management problem. Another fundamental difference between packet and TDM networks is that of availability and reliability. TDM networks such as SONET are very reliable. They have excellent resiliency. By having carrier class reliability (50 ms restoration and so on), SONET networks are good for voice communication. Packet-based networks, in contrast, operate on the principle of availability (best effort) Resiliency in packet-based networks is not fully automated as in TDM networks. The times required for locating and correcting failures may vary. Therefore, such networks may not ideally be suited for voice traffic. To ensure voice communication in packet networks, it is then important to have QoS of the packet streams. Newer technologies such as differentiated services (DiffServ) and integrated services (IntServ) allow such. Provisioning such packet-based network with the hierarchy of the protocols is a management challenge.

The evolutionary model in networks today clearly emphasizes a scheme that begins and ends with Ethernet-like technologies at the ingress and egress of a network periphery. SONET/SDH networking may now only exist in the core of such networks creating all optical pipes on WDM channels. The end-to-end flow in a network needs to be provisioned across these multiple segments of Ethernet, SONET, and back on Ethernet through a WDM-compatible core. Another issue here is that of routing and wavelength assignment in the WDM core. Having a fixed number of wavelengths can be a bottleneck—moreso, if we consider the limitations in tunable laser technology and the bottleneck of finding the right transponder cards to provision wavelengths on demand. Opto-electronic conversion-based transponder cards are typically not bit-rate independent, thus creating a management optimization problem for traffic grooming. All this means the necessity to increase the equipment stock at network elements.

Mobility Management

In a last mile network, wireless is an important revenue-generating segment because of its commodity and mobility characteristics. The wireless network runs on the optical backbone for transport. Many of the wireless services are voice based. This creates an intricate end-to-end model. The wireless traffic has to be sent seamlessly across the optical backbone. The management strategies in the optical backbone have to be now extended to incorporate this extra delay-sensitive wireless traffic. Wireless traffic by itself may be geographically very diverse, and provisioning in the backbone might have to be done at very fast speeds. The constant changes in wireless traffic create management problems for the entire network. On average, wireless circuits for voice and so on do not have a long duration. Access points for wireless are generally quite well defined, but the traffic variation is unclear. Allotting bandwidth to such access points on a real-time basis is a challenge. Wireless data networks based on principles such as the 802.11b standards need a separate provisioning procedure. Moreover, as pointed out earlier, a serious bandwidth mismatch exists between the optical layer and the wireless technologies. Scheduling and routing wireless traffic through interconnects of these two layers is a bandwidth management issue.

Because most of the wireless traffic is mobile oriented, bandwidth requirements at access points change dramatically. Therefore, network bandwidth optimization is not always possible. A vision of an ultra-broad GMPLS-like layer can be viewed as the best possible alternative to manage such unified networks. Assigning bandwidth on a real-time basis to end-to-end traffic flows spread across geographically as well as protocol-varied (heterogeneous) networks is a classic problem. Optimizing such networks for real-time bandwidth means creating a management platform that has good interaction with the various technologies as well as the protocols involved. Each network part may have a management system well tailored to meets its specification, while still able to interact with other networks. This type of micro- and macro-level management creates an appropriate platform for optimization of the entire network as well as the subnetworks in the access and the core. The management plane must be aware of the various resources in the network to create the optimal virtual connection topology for facilitating good end-to-end communication. After the management plane constructs the virtual topology, it must provision traffic paths. Figure 9-1 shows the mobility management problem as a seen as part of an end-to-end solution in a network.

Figure 9-1 *Mobility Management*

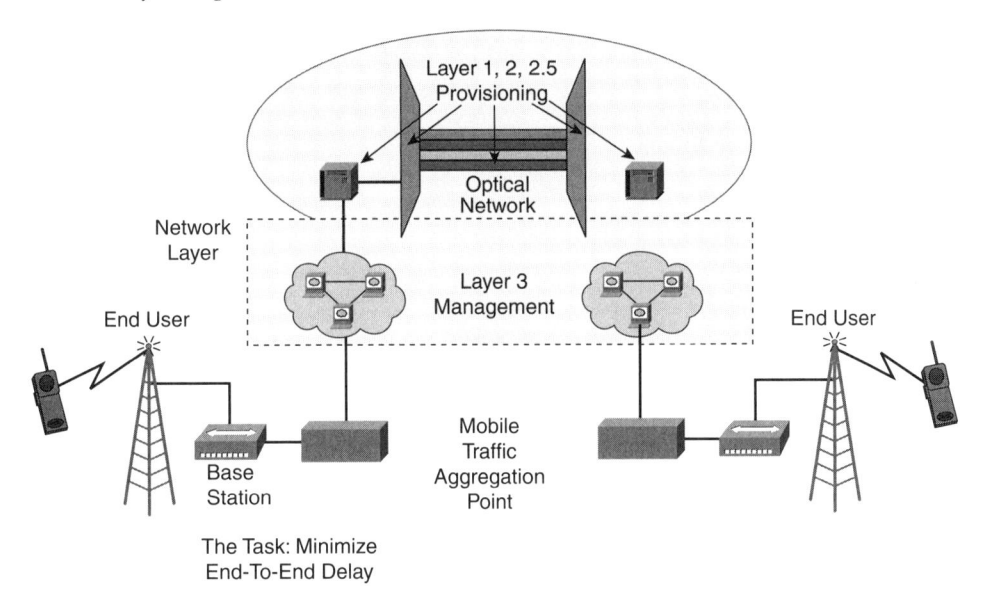

Potential of Effective Bandwidth Management

The automated scheme of managing networks by discovering the resources, creating a connection topology, and then managing or provisioning traffic can yield significant benefits in intelligent city networks. This scheme results in significant advantages, including the subscriber is billed, most appropriately, only for the specific services that are used (that is, content-delivery-based billing). In contrast, earlier schemes created billing anomalies due to multiple management planes and lack of centralized computing of billing the end users. Second, the provisioning time required for networks with a nonunified management schemes can be very high. As Figure 9-2 shows, the provisioning time decreases with efficient management schemes. A unified management scheme also allows carriers to better utilize the network infrastructure. The management plane has full knowledge of the network infrastructure and can efficiently use each network resource to create the best connection topology. This can result in a significant cost saving. For instance, in "BITCA: Bifurcated Interconnection of Traffic and

Channel Assignment," we see a mean 30 percent cost savings in optical equipment for provisioning lightpaths by having a unified control algorithm for routing and wavelength assignment in an interconnected ring network. A unified management strategy also avoids duplication of network resources, thus eliminating undue redundancy. Redundant inventory creates undue surges in capital expenditure. This can be avoided by having a unified management strategy. Another effect of an efficient unified management scheme is the facilitation of growth of the network. A strong management scheme allows for modular and incremental growth. A unified scheme also fine-tunes the network, to best place network resources to create homogeneity of resource utilization throughout the network.

Figure 9-2 *Provisioning Times for Services Today (From "The Need for Flexible Bandwidth in the Internet Backbone")*

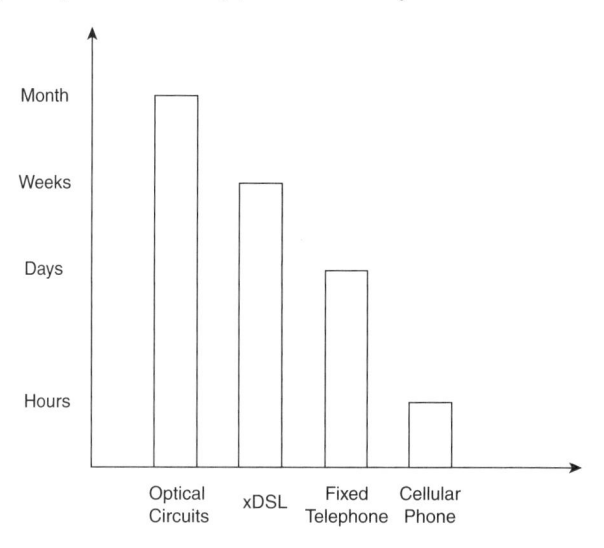

A good management scheme should provide a resilient network with low protection times. The management scheme should be able to provision the network in the least required time, under different loads (variations). Despite a diverse range of protocols, the management plane has to interact efficiently with each and every protocol. Apart from the economic objectives of saving costs and optimizing network resources, the management scheme has to minimize the blocking probability of connection-oriented services. This, in other words, means that the call-admission probability has to be maximized to ensure maximum traffic in the network. It is often not easy to dynamically cater to such network "churns." Churns are variations in traffic schemes.

Despite an efficient management scheme, dynamic network provisioning may be suboptimal, if the network was initially not well planned. Although outside the scope of this book, network planning and capacity planning serve as a necessary framework for network optimization.

Signaling is yet another important aspect of a good management scheme. While reserving bandwidth, efficient and bipolar signaling creates fast provisioning of the resources. Efficient signaling helps secure BoD and is a key tool for a good management policy. See Figure 9-3.

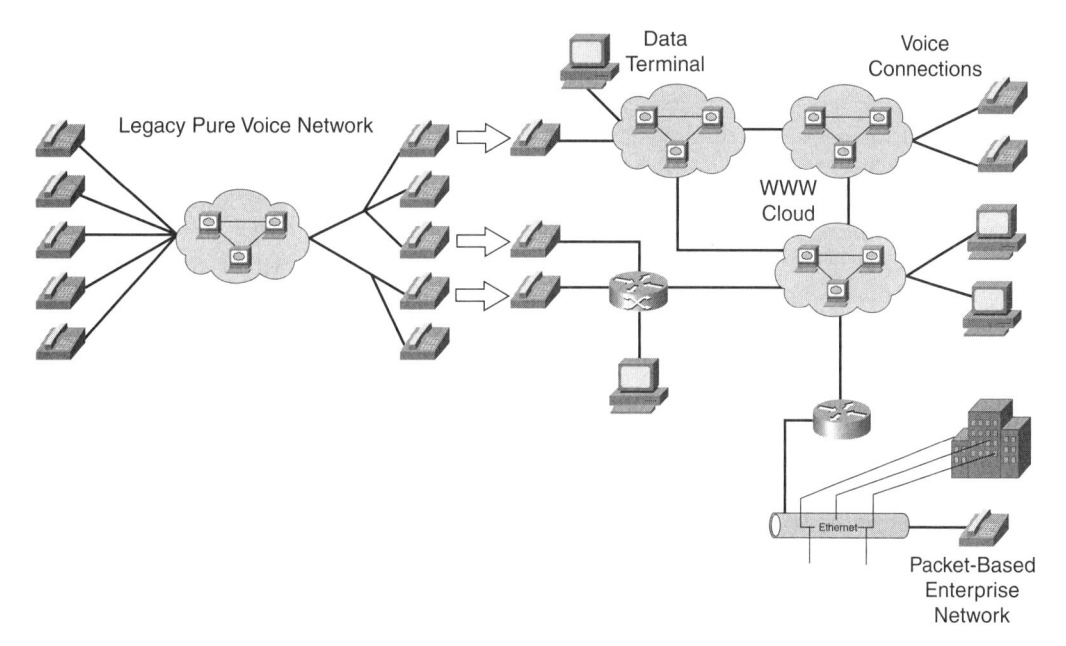

Figure 9-3 *From Circuits to Packets—from Voice to Data*

Directions for Network Management

The importance of last mile networks is in creating an efficient business model, which further can be best optimized by a good management scheme. The design of a unified management scheme is complicated by the diverse range of protocols and technologies that interact in producing the network hierarchy. A good management scheme results in cost savings for capital as well as operational expenditure. We summarized the needs and properties of a good management scheme necessary for operating a first mile/last mile/intelligent city network, with low OPEX and CAPEX and providing a suitable platform for fast provisioning. The following sections outline the protocol interaction and the qualitative requirements of the last mile access solution. Note that hardware interoperability is quite simple and achievable because most hardware is standardized and, therefore, matching is easy. In contrast, software or protocol interoperability is difficult to maintain due to the proprietary and distributed nature of software coding. Vendors differentiate from the competition by having a multitude of software features that are often not compatible with one other. There needs to be a middleware that allows multiple protocols to interact. This middleware is the key platform for software interoperability. However, practical implementation of middleware is difficult because of the complexity of the interacting software programs and variations in product features.

QoS in the First Mile

Networks worldwide can be characterized primarily by two parameters: the quantity of service and the quality of service. The quantity of service is the net throughput (goodput) offered by the network, whereas the quality of service (QoS) is a set of parameters that characterize class of the throughput that is offered. In internetworks today, the QoS of a traffic stream often determines the revenue-earning capability of the network (for example video, real-time applications, and so forth) Internet Protocol (IP) by itself is a best-effort service paradigm, which creates several anomalies in network communication. Among these are the delay associated in a stream of packets in a flow, the basic minimum throughput of a flow over a period of time (session life), the jitter associated with a flow, and end-to-end sequencing of the packet hierarchy. Delay is the most fundamental QoS parameter for optimization, because of the prominence of delay-sensitive applications in networks today (for example, video file transfer and, of course, voice). The minimum service guarantee dictates the minimum throughput that can be achieved for a particular flow irrespective of other flows or parameters in the network. For "live" applications such as VoIP or videoconferencing, packet sequencing is an important attribute for providing consistent service. The orderly delivery of packets in real-time applications creates a streamlined flow of the application. No flow can be delivered to the egress node without some delay. The delay experienced by multiple packets may vary. The variation in this delay, known as *jitter*, is critical to some real-time applications for good service. QoS in networks today is an important consideration for providing revenue-bearing services in an optimized manner. In the year 2000, despite the Internet data boom, the ratio of revenue from voice to data was 9:1. Although this is expected to fall back and stabilize finally, with data revenue overtaking voice and other services, the main proponent for that would be the packetized transmission characteristics such as VoIP and video transfer over packets using a QoS mechanism such as, for instance, MPLS.

Legacy TDM networks were based on circuit-switched identities where the bandwidth for a flow was predetermined. QoS in terms of bandwidth allocation was static and nonvariable. The delay for data flow in a circuit was quite fixed and the only variable aspect was the buffer occupancy of the data which created the flow. SONET/SDH links had little to offer in terms of QoS to end users, as they showcased a broadband technology for raw unintelligent bit transfer. With the migration of networks from circuit to packet switching, QoS became a more important aspect for study, because the basic communication unit was a single multibyte packet that could traverse a network at a speed and in routes that were not predetermined. This created anomalies in the offered throughput, delay, jitter, and so on for consecutive packets. To reduce this anomaly, packets were accorded levels or QoS parameters that helped network elements distinguish consecutive packets in a flow from each other on the basis of their requirement and take curative action to facilitate packets reaching their intended destination under the constraints as given. One way to accord QoS in small-scale networks (with a limited number of users, typical of first mile access) was to distribute the delay tolerance and delay variation across the logical graph connection in the most optimized way. However, as the number of nodes and traffic intensity grows, this technique is quite cumbersome and ineffective. It is cumbersome because of management issues—shifting the delay tolerance around the network consistently leads to the same effect of congestion and lesser throughput. For instance, we can consider the shifting of delay in a first mile network as the solution of the far-near problem in EPON networks.

From first mile access network's perspective, QoS is an important factor for design and implementation considerations. Apart from the economical aspect, QoS is important in right administration of resources to provide a unique solution to end users. QoS is also important from the aspect of designing networks. We show in the next section with the help of a toolkit aid how QoS enhances network performance and the generic rules for traffic engineering.

Toolkit for QoS and Traffic Engineering

Traffic engineering (TE) deals with the performance of a network in supporting a network's customers and their respective QoS needs. Two aspects of TE in first mile access networks detail with measurement of traffic and control of the traffic. The latter operation ensures that the network has resources to provide for the requisite QoS. In a single class, the key traffic-oriented performance objectives include minimizing traffic loss, minimizing delay, maximizing throughput, and enforcement of service level agreements (SLAs). Resource-oriented performance objectives deal with the network resources such as communication links, routers, and servers that contribute to the realization of traffic-oriented objectives. The management of all users' traffic to prevent congestion is an important aspect in the QoS picture. Congestion translates to reduced overall throughput and substantial increase in end-to-end delay. Most networks provide transmission rules for their users, including agreements on how much traffic can be sent to the network before traffic flow is regulated (flow controlled). *Flow control* is an essential ingredient to prevent congestion in a first mile access network. This can also be considered as a policy for admission into the network to a particular data flow. Congestion can be described by two cases: First, there may be insufficient network resources; second, there may be sufficient resources to support the users' QoS requirement, but the traffic streams are not mapped properly into the available network resources. Therefore, some parts of the network become underutilized, whereas other parts become overutilized. The first issue can be solved by building first mile networks with more overall bandwidth (for example, PON). Congestion-control techniques such as window control operations, are also possible methods to alleviate the issue. The second issue, that of insufficient resource allocation, can be addressed through good traffic-engineering solutions. The most feasible and optimum solution is load balancing of the network—that is, providing the best routes to a given set of traffic demands keeping in mind network capacity and churn (future projected dynamic increase in traffic).

Service classes can be used to organize traffic flows in a network and are organized according to the following operations:

- Timing between sender and receiver
- Bit rate (variable or constant)
- Connectionless or connection-oriented sessions between sender and receiver
- Sequencing of user payload

- Accounting for user traffic
- Segmentation and reassembly of user protocol data units (PDUs)

Table 9-1 shows classes of ATM traffic.

Table 9-1 *Service Classes of Traffic (ATM and Frame Relay in First Mile Access Networks)*

Class	Performance
Class A	Constant bit rate (CBR), TDM-based, connection oriented; requires timing and minimal flow control; prevents average loss
Class B	Variable bit rate (VBR), statistical TDM based (caters bursty traffic sources); connection oriented; requires timing; minimal flow control required; prevents data loss
Class C	Variable bit rate (VBR), statistical TDM based (caters bursty sources); does not require timing; extensive flow control; does not permit loss of data; connection oriented
Class D	VBR statistical TDM approach; does not require timing; permits flow control no loss permitted; connectionless scheme

Using MPLS to Queue Traffic

Queues represent the single most fundamental access units for traffic shaping and QoS demarcation in a network. Queues are manageable and scalable entities that dictate the nature of traffic shaping in a network. Routers, switches, and other network equipment are built on the following four primary kinds of queues, which shape the traffic and allocate QoS priorities to the network:

- **First in first out**—Transmission of packets is based on their order of arrival. For example, MPLS uses this method for a given FEC.
- **Weighted fair queuing (WFQ)**—The available bandwidth across queues of traffic is divided based on weights. Given its weight, each traffic class is treated fairly. This approach is often used when the overall traffic is a mixture of multiple classes. For example, Class A traffic is accorded a heavier weight than, say, Class D traffic. WFQ is well suited to manage MPLS flows. According weights is a network parametric operation depending on QoS and protocol dissemination.
- **Custom queuing (CQ)**—Bandwidth is allotted proportionally for each traffic class. It guarantees some level of service to all traffic classes.
- **Priority queuing (PQ)**—All packets belonging to a higher-priority class are transmitted before any lower-priority class. Therefore, some traffic is transmitted at the expense of other traffic.

There are other type of advance queueing techniques such as Round-Robin (RR) and Modified Round-Robin Queues and so forth.

Failure of OSPF and BGP for Providing QoS

In the current form of Internet, route dissemination as well as set up is done by conventional protocols such as OSPF and BGP. These protocols are not designed to provide QoS parameters to a traffic flow. The route establishment has nothing to do with bandwidth available or burstiness of the flow or so forth. Logical connectivity determines the network routing.

Constraint-Based Routing

Constraint-based routing (CBR; analogous to QoS routing) is designed to provide a route through a typical access network based on the user's QoS needs. It is demand driven and is aware of the traffic trunk attributes and the attributes of network resources. Each router (network element) automatically computes an explicit route for each traffic trunk based on requirements of the trunk's attributes subject to the constraints of network resources and the administrative policies of the network. ATM and Frame Relay use CBR quite effectively. CBR can be extended to OSPF and IS-IS operations also.

Attributes of CBR

The attributes of CBR are as follows:

- **Peak rate**—Defines the maximum rate at which traffic should be sent into the network. The peak rate is useful for purpose of resource allocation. If resource allocation within a domain depends on peak rate value, it should be enforced at all the ingress nodes. Typically, if we consider peak rate value for each stream, the network would be generally underutilized.

- **Committed rate**—Defines the rate that the domain commits to be available to a set of paths

- **Excess burst size**—The excess burst size (EBS) may be used at the edge of a network domain for the purpose of traffic conditioning. The excess burst size may be used to measure the extent by which the traffic sent on a constraint route exceeds the committed rate. The possible traffic conditioning actions such as passing, marking, or dropping are specific to the MPLS domain.

- **Weight**—Determines the constraint based route's relative share of the possible excess bandwidth above its committed rate.

Differentiated Services (DiffServ)

The primary idea behind DiffServ is to (1) classify traffic at the boundary of the network and (2) regulate (shape) this traffic at these ingress boundaries. The classification algorithm entails the assignment of the traffic to behavioral aggregates. Behavioral aggregates are a collection of packets with common requirements (namely QoS). In this way, we have a small number of classifications for packets, which helps in determining flow types and, finally, according resources. The identified traffic is assigned a value: a DiffServ code point. In IP, this may be the 6-bit TOS (type of service) field (as in IPv4) or the traffic-class octet (as in IPv6).

Following the classification of the packets at the boundary of the network, they are forwarded through the network based on the DiffServ code point (for example, TOS). The forwarding is performed on a per-hop basis, in other words the DiffServ node alone decides how the forwarding is to be carried out. This concept is called per-hop behavior (PHB). At each node, the DiffServ code point is used to select the PHB, which in turn determines the scheduling treatment for the packet and also the drop probability for the packet.

DiffServ uses a DiffServ domain, a collection of networks operating under the same administration (for example, an ISP). The DiffServ domain consists of contiguous set of nodes that are compliant to the DiffServ protocol and agree to a common set of service provisioning policies. After a packet is injected into the network (subsequent to the ingress router), the internal nodes forward packets based on DiffServ code points.

Three types of per-hop behavior are defined in DiffServ specifications:

- DiffServ defines a default PHB in which there is no special treatment accorded to the packet.
- DiffServ also defines expedited forwarding (EF), by which packets undergo low delay and are subjected to low loss. Typically, the service of such packets is done in a way that these packets spend the minimum time in the queue.
- DiffServ also defines assured forwarding (AF). By this method, it is possible to provide different levels of forwarding assurances to packets. For instance, WFQ is a good method for such a queue-management scheme.

Integrated Services (IntServ)

The term IntServ refers to an overall QoS architecture that guarantees end-to-end QoS for applications that need it. Such guarantees ensure that applications that have a minimum bandwidth requirement and that have a maximum bound on end-to-end delay have their needs met. In IntServ, service classes have been defined to meet the needs of the end users for various application types. The service group established the IntServ tenets also defined how RSVP (Reservation Protocol) could be used as a signaling protocol for ensuring QoS for end-to-end streams. Note that there is distinction between IntServ and RSVP, and that RSVP is just one of the several protocols that can be used for signaling purpose by IntServ. The IntServ architecture includes a number of components in addition to a signaling protocol. An application can specify the type of traffic it would send to the network, and the characteristics of the traffic as well as the QoS it expects from the network. The traffic specification is known as *TSpec*, and the request for a certain QoS is known as *RSpec*. IntServ requires network elements such as routers and switches to perform a variety of functions, including the following:

- **Policing**—Verifying that the traffic conforms to its TSpec and taking action such as dropping packets if it does not
- **Admission control**—Checking to see whether there are enough resources to meet a request for QoS and denying if requests are not available
- **Classification**—Recognizing those packets that need some degree of QoS
- **Queueing and scheduling**—Making decisions about when packets are transmitted and which packets are dropped, consistent with the QoS requests that have been granted

IntServ has two defined service classes: guaranteed service (GS) and controlled load. GS is intended to serve the needs of applications that require a firm guarantee of bandwidth/delay. To obtain a hard bound, the application provides a TSpec that precisely binds the traffic it will inject, including such parameters as peak rate, a maximum packet size, burst size, and a token bucket rate. The burst size and token bucket rate together comprise a token bucket specification, which is the standard way to represent the bandwidth characteristics of an application that generates data at a variable rate. The most important parameter in guaranteed-service RSpec is the service rate, which describes the amount of bandwidth to be allocated to this flow. It has been shown that by knowing this parameter, plus those in the TSpec, the maximum delay that could possibly be experienced by a packet can be calculated. Further, an application can control its delay bound by increasing the service rate in its RSpec. Guaranteed service comes at a cost: It requires that every flow using GS be queued separately, and this results in less-than-optimal network utilization.

Controlled load overcomes these drawbacks by giving up on hard mathematically provable delay bounds. Controlled load ensures that an application receives service comparable to what it would receive if it were loaded on an unloaded network of adequate capacity for just that application alone.

Reservation Protocol

RSVP is intended to provide guaranteed performance by reserving the bandwidth at each network element that participates in supporting the flow of traffic. RSVP is a backward reservation protocol, in the sense that it reserves the resources from the egress to the ingress nodes. RSVP requires that receivers of the traffic to request QoS flow. The receiver of the application must determine QoS profile and then pass the required information in terms of an acknowledgement of the demanded resources or a negation of the demand back to the ingress host.

802.1Q VLAN

The IEEE 802.1Q standard defines an architecture for virtual bridged LANs, the services provided in virtual bridged LANs and the protocols and algorithms involved in the provisioning of those services. No QoS mechanisms are defined in this standard, but an important requirement for providing QoS is included in this standard (for instance, the ability to regenerate user priority of received frames using priority information contained in the frame and the user priority regeneration table for the reception port). The advantage of this standard is that multiple MAC addresses are "tagged" in a way to provide a VLAN environment. This is particularly important and useful when we look at first mile networks from the perspective of home office and small enterprise networking. In a Layer 2 domain, the QoS mechanism is based on the value of 802.1p bits (CoS), which are 3 bits in the 802.1Q header.

QoS in ATM

For ATM, let us first consider the peak cell rate reference model: The ATM forum specifications provide a reference model to describe the peak cell rate (PCR). For conformance definitions, generic cell rate algorithm (GCRA) is used for ATM. Similar to constraint routing CR-LDP (Constraint Routing–Label Distribution Protocol) parameters and attributes are listed here:

- **Peak data rate**—The maximum rate at which traffic can be sent to the constraint routing–label switched path (CR-LSP). It is defined with a token bucket with parameters: peak data rate and peak burst size.
- **Committed data rate**—The rate that the MPLS domain commits for the CR-LSP. It is defined with parameters committed data rate and committed burst rate.

Implementation of Different Network Architectures

The previously described toolkit provides a glimpse through the different approaches in providing QoS in first mile networks typically in metropolitan environments and networks in general based on the requirement and the resources present in building these networks. These techniques may be efficiently applied to produce the desired results for flow streams that need QoS. Two generic and one futuristic architecture scheme for QoS can be assimilated: architecture and framework for voice-oriented communication, architecture and framework for data-oriented communication, and future-oriented frameworks for adaptable QoS.

Architecture and framework for voice-oriented communication: Voice traffic can be understood by its inherent requirement for low latency and packet synchronization. Quantized and sampled voice packets need to be sent through a network with bounds on the round-trip delay time leading to total transmission delay time. SONET/SDH networks seem to be the most convenient form of transport for voice oriented traffic. In the first mile, this is replaced by DS 0, T1, DS3 and finally OC-3 going ahead to OC-48 and eventually OC-192. Most importantly, voice traffic is typically not bursty; therefore the requirement of bandwidth for voice traffic is quite constant. SONET/SDH and TDM networks are good exponents for this kind of traffic requirement. SONET/SDH networks have a very uniform data rate as well as resiliency is quite constant (delay and jitter). Data traffic, on the other hand, is bursty and, therefore, the interpacket arrival times can be variable. Data traffic can therefore require multiple bandwidth levels that can change (subject to time and requirement). For data traffic, packet-oriented QoS schemes have good performance.

ATM is the first such scheme that can be used. ATM is connection oriented, so somewhat creates efficiency issues (apart from high overhead; 5 out of 53 bytes are overhead). However, ATM has good performance for transport of data traffic subject to QoS issues. IntServ and DiffServ are two well-defined approaches to packet-oriented QoS schemes. IntServ, as mentioned before, uses multiple signaling protocols between network elements to reserve the necessary resources. DiffServ creates an easier method for differentiating and, therefore, switching/routing traffic flows based on the QoS requirements. DiffServ also is fast in computation and efficient for a large network. MPLS-based system is also a good paradigm for high-speed links, especially to provide good QoS. By distributing labels that act for fast switching as well as for distinguishing flows and flow types, MPLS represents a unified protocol for generic data QoS requirements. MPLS, when implemented over various physical hierarchies—namely, the optical layer—creates GMPLS. In summary, we can observe that the migration of networks to accord QoS from ATM to MPLS has been steady and efficient. MPLS can do most of what ATM can, at a lower cost, although scalability in an MPLS network is somewhat debatable. Attempts to provide lower-priced QoS technologies in networks today have focused on management and provisioning issues, as exemplified by MPLS as a key QoS technology for networks of the future.

Directions for Access Networking

QoS guarantees in a network are important considerations that affect the revenue of the network. From a service provider point of view, QoS creates not just revenue but also optimizes the network from an application perspective.

Comparison of ATM Solutions with MPLS for Core Networks

Data-link technologies have matured greatly since the inception of the Aloha protocol about three decades ago. For efficient and resilient communication, many implementations of link-layer and higher-layer technologies have been carried out. Each technology represents a new paradigm toward the resolution of some successful parameter to facilitate network communication. In this regard, Ethernet represents the first multivendor effort for effective link-layer communication in LANs and recently extended to WANs. The surge of Internet traffic has made IP a prominent protocol for transport applications. Because of its best-effort characteristic and variable-size length, however, IP finds providing QoS and fast switching requirements too stringent in its present form. Data-link protocols, such as Ethernet are comparatively good substitutes for Layer 2 switching. ATM evolved into a protocol that provided QoS to traffic flows. IP over ATM creates a high rate and effective service scheme, for core networks. The massive overhead requirement in ATM networks and the stringent requirement of ATM interfaces created the MPLS paradigm. MPLS is a smart alternative for providing QoS as well as creating fast switching for IP routes. MPLS uses labels that are tagged to IP packets before the address creating a Layer 2.5 protocol. These labels map routes (flows) to different network resources such as ports and enhance the switching speed of network elements.

ATM in Core Networks

ATM works by identifying flows into two categories, a broad virtual path circuit (VPC) and a narrow virtual connection circuit (VCC). By deploying this flow-based scheme in the network, ATM creates connections between source destination pairs. The ATM layer defines transmission of data in fixed-size cells and also defines the use of logical connections. The use of ATM creates a need for an adaptation layer to support information transfer protocols not based on ATM. By creating VPC and VCCs, the time required for processing and connection setup is substantially reduced. Moreover, this results in a simplified network architecture and network functions can be grouped on virtual circuits and virtual paths. Due to VP/VC establishment scheme, the network has to aggregate fewer entities and thereby performance improves. Performance benefits in ATM are generally expected in speed and QoS. Fixed cell size makes it easier for intermediate ATM switches to switch cells between flows as compared to variable-length packets. By allocating resources to a particular VC and creating a virtual channel connection (VCC), QoS can be achieved. The VP/VC scheme also results in reduced processing and short connection setup time. After a path has been set up along a route, adding a channel is very easy. This does result in some degree of scalability to the flow.

Some of the characteristics of VPs/VCs in ATM networks can be summarized. A VCC user is provided with QoS as specified by parameters of cell loss ratio, end-to-end latency, and cell sequence integrity. In addition to dynamic VC setup, permanent VC (PVC) and semipermanent VC (SVC) also exist. Traffic parameters can be negotiated between a user and a network for each VCC. The input of cells to VCC is monitored by the network to ensure that the negotiated parameters are not violated. The control signaling in ATM networks for setup and so on is essentially on separate channels. It could be using user-to-user VCs or user-to-network VCs or just by meta-signaling (by keeping a permanent signaling channel).

The ATM forum has defined the following service categories:

Real-time services

- Constant bit rate (CBR)
- Real-time variable bit rate (rt-VBR)

Non-real-time services

- Non-real-time variable bit rate (nrt-VBR)
- Available bit rate (ABR)
- Unspecified bit rate (UBR)

CBR is the simplest service to define. It dedicates a VC/VP to a source destination pair, creating fixed-bit-rate access. Real-time VBR is intended for time-sensitive applications—that is, those requiring tightly constrained delay and delay variations. The bit rate may vary with time, and this is the principle difference as compared to CBR. rt-VBR allows the network more flexibility, and hence the network is able to statistically multiplex more connections over the same dedicated capacity and yet provide the required service to each connection. For bursty applications that do not have a tight constraint on delay, non-real-time services are the preferred approach. For some non-real-time applications, it is possible to characterize the expected traffic flow so that the network can provide a substantially improved QoS in the areas of loss and delay. Such applications can use the nrt-VBR service. For all other traffic that needs best-effort service, UBR is the best alternative. To improve the service to bursty sources, which would otherwise use UBR, ABR service has been defined. An application using ABR specifies a peak cell rate, a minimum cell rate. ABR mechanism uses explicit feedback to sources to ensure that capacity is fairly allocated. Any capacity not used for ABR is then available for UBR traffic.

The ATM adaptation layer (AAL) has the following services defined for effective communication: handling of transmission errors; segmentation and reassembly to enable larger blocks of data to be carried in the information fields of ATM cells; handling of lost and misinserted cell conditions; flow control and timing control. The AAL layer is organized in two sublayers, the segmentation and reassembly (SAR) sublayer and the convergence sublayer (CS). The SAR sublayer is responsible for packaging information received from the CS into cells for transmission and unpacking the information at the receiving end. The CS provides the functions needed to support specific applications using AAL. AAL offers five basic types of service.

Multiprotocol Label Switching

Multiprotocol Label Switching (MPLS) as defined by Rosenberg in an IETF draft and has major features such as label assignments, swapping, merging, and aggregation and results in faster switching of traffic flows and QoS guarantees in a network domain. MPLS can be used with a variety of Layer 2 networks (namely, Ethernet and so on). Being compatible with OSPF and BGP, MPLS finds integration into the present Internet quite easy. By virtue of its capability as a protocol, MPLS supports source routing (explicit routing). MPLS is a Layer 2.5 protocol that circumvents the need for IP header processing at intermediate nodes, by adding a label to distinguish flows based on routes and ports. The labels are distributed through the network using variants of label distribution protocol (LDP). Labels map an FEC to a particular port of a label-switched router (LSR). The LSP can often have packets that have multiple MPLS labels just ahead of the Layer 3 header. LSRs do not need to process higher-layer address of packets, but can forward packets to the correct port based on a binding between the FEC and the label. Basic label assignment to an FEC may be through independent control (incumbent LSR assigns label independently by using local routing protocol) or through ordered control (incumbent LSR assigns label based on some scheme such as ingress-to-egress assignment). An MPLS header consists of a label (20 bits), a stacking bit (for multiple label stacking), and TTL. Through this format, without looking at the IP-level address, a packet can be and QoS provided.

Label stacking: Labels are local entities; they are bound to an FEC by an LSR. Therefore, along an LSP a route may be interpreted by various LSRs as per their localized label distribution and convenience. By stacking labels, one on top of the other, the route can be simplified, and stacking allows designated LSRs to exchange information with each other and act as border nodes to a large domain of networks and other LSRs. Pushing of labels is analogous to adding labels to create a stack, while popping strips the topmost label of the stack. An LSR may merge labels to create a single label for various packets going to the same interface (port). Two fundamental routing philosophies for LSPs are hop-by-hop routing and explicit routing. In the former, each LSR chooses the next hop for a given FEC independent of other constraints. Whereas in the latter, the ingress LSR chooses the entire route and hence the intermediate LSRs that are to be on the chosen LSP.

Comparison of ATM and MPLS: ATM Perspective

The inception of ATM solutions was to create QoS-based flows in core networks and wipe out the best-effort nature of IP traffic. From a comparison perspective, ATM is more reliable than MPLS for voice flows. By according service parameter such as CBR to voice flow, voice circuits can be easily set up, creating dedicated service with low latency and good sequencing. The constant available bit rate provides a dedicated channel for voice communication. MPLS, on the other hand, can create QoS by indicating so in the label itself. However, MPLS is still not guaranteed due to its questionable performance in a highly congested network.

After setup, VPC and VCCs are easily scalable and do not need frequent signaling. In case of PVC, signaling is almost negligible compared to an MPLS network. Due to the fixed nature of the ATM cell structure, the end-to-end delay is quite constant. MPLS, on the other hand, can have different routes (through hop-by-hop routing) and have variable overhead (due to label stacking). This leads to several issues in creating variable delay.

From a management perspective, ATM management is quite simple and highly scalable. MPLS management is also simple, but complexity increases with scale. Label distribution and allied functions such as label withdrawal, label request, and so on can be a management disaster as the number of LSRs in a network grow.

As a summary of the comparison, ATM can be considered a very high-end protocol for QoS requirements, whereas MPLS can be a lower-cost (much) alternative and can do most of what ATM can.

MPLS Versus ATM: MPLS Perspective

MPLS was sought to be a low-cost, scalable, manageable solution to surmount the delay in Layer 3 routing as well as provide QoS to traffic flows. Its direct comparison to ATM is obvious and there exist some strategic benefits in deployment of MPLS technology over ATM. In ATM, we require an ATM adaptation layer for delivery to higher layers as well as acceptance of higher-layer packets. In contrast, no such layer (and hence protocol processing) is required in MPLS formulation. This greatly reduces the processing at nodes. Moreover, labels are local entities and therefore their processing is straightforward at LSRs.

MPLS can do some degree of routing by itself: hop by hop and explicit routing. Further explicit routing can be loosely or strictly explicit depending on the LSP path definition at the ingress LSR. The routing is with cognizance of a routing protocol such as OSPF. Explicit routing can be particularly valuable for traffic with QoS requirements. Unlike ATM, in MPLS we do not need an entirely different channel for overhead signaling.

Although fewer bits are available for the label, labels can be stacked to create arbitrarily complex MPLS label stacks. This makes MPLS addressing and trunking vastly more flexible than that of ATM, because there is no need to impose an arbitrary boundary between VP and VC switching

Finally, as the name suggests, MPLS is invariant to other protocols. It can take a wide panorama of protocols after the label. After the label is stripped at the egress LSR, the PDU may be any protocol. No special processing (such as AAL in ATM) is required for encapsulation of the label on the PDUs.

ATM and MPLS in Conjunction

Overlay model: N routers can configure into a logical fully meshed topology, this though attractive creates significant overhead for mapping the N(N – 1) possible source destination pairs in the network. The idea here is to keep the IP routers invisible to the ATM layer. MPLS labels are mapped to VPIs and VCIs to merge with ATM layer. For mapping ATM and MPLS: The 16 bits of VCI and 8 bits of VPI in the ATM cell are replaced by a single label field of 20 bits, packed into a 32-bit label header.

MPLS packets can be 1500 bytes long, plus any labels added to the packet. (Note that this requires the use of "baby jumbo packets" if Ethernet is used as the transport for MPLS.) This compares well with the 48-byte cell of ATM and reduces encapsulation overheads, particularly in the case of small packets: for example, it allows a minimum-length TCP packet to reside in a single MPLS packet, rather than two cells as in ATM.

ATM and MPLS Implementation Costs

From a deployment perspective, ATM has a higher CAPEX than MPLS solutions. First, ATM equipment is hardware driven; therefore, there is an additional cost of ATM interfaces. In contrast, MPLS solutions are primarily software driven. Routers can be upgraded to incorporate a label switching philosophy by software alone. Whereas ATM solutions are good for trimming operational expenditure. The net requirement for provisioning and maintaining ATM is quite low. ATM is not easily scalable but provides guarantee of service. MPLS is highly scalable, but as the size increases management of both MPLS and ATM networks can be rigorous. Keeping track of the labels, FECs, and bindings can create severe memory-processor convolutions (and therefore delays) in network performance. In summary, ATM is an expensive protocol to deploy and unless dictated by economics not worth to pursue, whereas MPLS is comparatively simple protocol, with relatively low deployment costs and can accommodate both voice and data easily (and thus provides an excellent path for converged networks). It provides overall service better than best-effort IP service but is limited in scalability by the network heuristics. Figure 9-4 shows a protocol overviews for QoS in data networks.

Figure 9-4 *Conceptual Overview of Protocols for QoS in First Mile Access Networks*

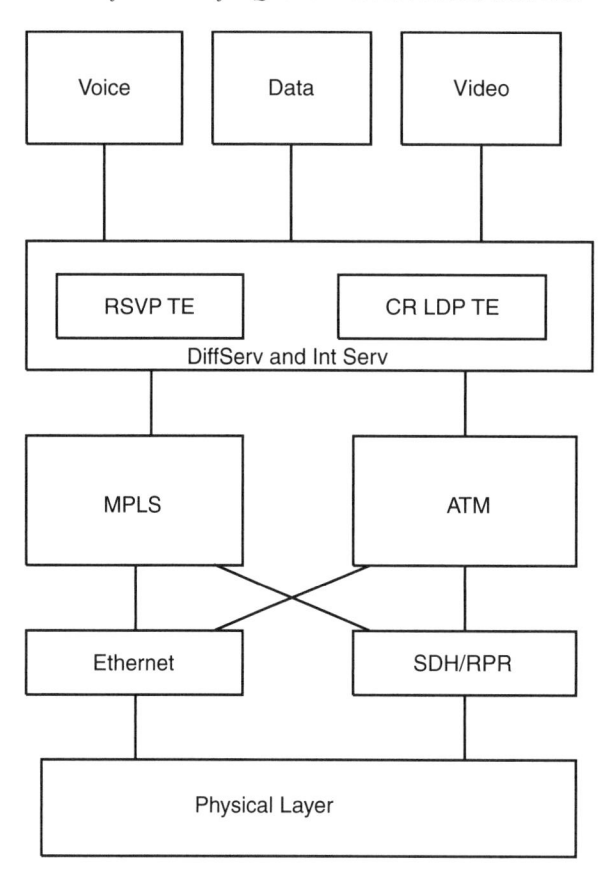

Summary

This chapter covered multiple management technologies and approaches to managing first mile networks. This chapter also showcased the intelligent city approach, both from a technological as well as from a business point of view. The chapter then focused on multiple protocols that are needed to provide end-to-end service in access networks. The management planes and the evolution of technology over the past three decades was discussed. This chapter outlined the problems seen in management of first mile networks and the final solution needed to provide a unified management scheme (that of having a generalized platform to encompass multiple technologies and multiple product portfolios and suites to enable a standard management plane).

Review Questions

1. What is an intelligent city network and how is it different from a service provider network model?
2. Describe the management challenges in the first mile.
3. What are the key challenges for managing end-to-end flows in a mobile plus optical network?
4. State the key points to consider for managing a cellular network overlaid over a SONET backbone.
5. Define and describe the multiple ATM classification for QoS.
6. Which technology is more scalable and why: ATM or MPLS?
7. Describe and differentiate between DiffServ and INtServ. How can IntServ be applied to first mile networks?
8. Describe the RSVP scheme.
9. Describe the gradual change in worldwide networks from circuits to packets from a SONET to GigE perspective. From an ATM to MPLS perspective.
10. Describe VLANs and discuss how QoS is achieved in Layer 2 VLAN networks.

References

Awduche MPLS TE Optical IETF Draft.

Black, Ulysses. *MPLS and Label Switching Networks*, Second Edition. Upper Saddle River, New Jersey: Prentice-Hall, 2002.

Chlamtac, I. http://www.cnuce.pi.cnr.it/Networking2002/conference/chlamytac.pdf. Networking 2002: Pisa, Italy.

Gumaste, Ashwin and I. Chlamtac. "Light-Trails: A Solution to IP Centric Communication in the Optical Domain." Proc of HPSR, IEEE Workshop on High Performance Switching and Routing. Torino, Italy: June 2003.

Gumaste, Aswhin, et. al. "BITCA: Bifurcated Interconnection of Traffic and Channel Assignment." Proc of OFC 2002: Anaheim, CA.

Keshav, S. "An Engineering Approach to Computer Networking."

IETF MPLS Draft Rosenberg

Stallings, William. *High-Speed Networks: TCP/IP and ATM Design Principles*. Upper Saddle River, New Jersey: Prentice-Hall, 1998.

Xiao, G., J. Jue and I. Chlamtac, "Lightpath Establishment in WDM Metropolitan Area Networks", SPIE/Kluwer *Optical Networks Magazine*, special issue on Metropolitan Area Networks, 2003.

CHAPTER 10

Business Case for First Mile Access Networks

The previous chapters showcased some of the key technologies that make up the first mile access network for broadband communications. This chapter illustrates the merit of these technologies through a technology case and a business case. This chapter also discusses the salient features of the first mile business case—the most promising part of the first mile access overall solution.

The technology solutions discussed in the previous chapters were each primarily developed to solve a specific problem. These were evolutionary solutions rather than revolutionary ones. When data networking first became prominent, for instance, there were three clear solutions the networking community tried to adapt. One was in trying to get maximum leverage from the existing telephone lines, which resulted in frequency division sharing of the Public Switched Telephone Network (PSTN). This meant that the copper line could be partitioned into frequency domains, one for voice and another (larger one) for data, thereby leading to popularization of Digital Subscriber Line (DSL) technology. The second solution was the use of cable technology. Although not technically very effective for full duplex communication, cable-modem-based broadband solutions are gradually becoming more and more popular. The third technology that evolved was wireless data networking and the finalization of the 802.11b/g standards. As we note here, these technologies are basically evolutionary in nature and, hence, don't really represent a paradigm shift in the way networks worldwide behave. A fourth technology, one that existed for a long time but was not deployed widely, was passive optical networking (PON). PON, based primarily on passive optical components, was first seen in the late 1980s. However, cost and technology limitations prevented widespread deployment. The maturation of the technology and lowering costs made PON a reality. Eventually the fiber to the home (FTTH) and fiber to the premises concepts became more widespread and adopted by the networking community. This led to more communities worldwide beginning to deploy fiber to the curb or home.

While these technical developments were happening, the telecommunications sector suffered a tremendous setback because of a global recession that markedly affected this particular sector. What was needed to pull out and away from the conventional telecommunications business was a paradigm shift in generic telecommunications business model itself. This shift in the business model propelled the appearance of a new and alternative business model, called the community network concept. The ideas implicit and explicit in the community network (or intelligent city) helped develop a business case for first mile access networks. From Chapter 1, "Introduction to First Mile Access Technologies," it is apparent that there is a strong business case for first mile access networks. This case is further justified in Chapters 2 through 7, which outline of each technology component of the first mile. However, there still remains substantial technical and business aspects that need to be understood to complete the picture of a first mile access solution, particularly from business point of view. We still need to cover, in terms of technology, the integration of individual technologies into the complete picture of the first mile solution. The management aspect in this chapter is covered through a community network model. Furthermore, we look at the business aspect of first mile access networks and how first mile access systems define the market. The idea here is to define the business case for first mile access networks.

The Application Business Case

Before we describe the multiple business approaches for providing a solution to the first mile access problem, we must first understand the applications that are growing in demand which create the need for broadband networking. While multiple applications have emerged that seem trivial to describe, considering their non-technical nature, there is still a need from a networking perspective to identify the bandwidth hungry requirements of such applications. It is these applications that are dictating the shift in first mile networks from dial-up networking to more broadband networks. These descriptions enable you to understand the requirement that these high-bandwidth applications put on the underlying network. Recent years have brought about an exponential increase in high-bandwidth applications. Now we look at some of the factors that affect the first mile access area.

Security

Post 9/11 security needs have played an important role in the planning and designing of networks. Backbones have to be very secure and run applications that are efficient and have highly relational and distributed processing. The aspect of security that is created due to the amalgamation of information processing with variable bandwidth requirements places an additional requirement on the backbone serving the network. The need for high-resolution processing and data mining among dispersed and fundamentally large data-storage centers requires rapid network provisioning with high security.

In the first mile access for residential users, an important aspect of high-speed networking is remote surveillance of assets. Houses and other community assets need to be kept under constant observation. This can be done remotely from a central control room or in remote locations, such as an office. Images and video files are sent from the location to the observation post. Having circuit-based connections between remote locations and observation posts is the best option but is seldom deployed because of financial and technological constraints. It is not very tempting to keep a circuit live constantly because the problems are usually rare. Sending video files and images over a packetized system is a convenient and low-cost solution. To deploy video and file transfer through the first mile network by using a packet-based communication system, for instance, the system has to accord a high level of QoS. For example, images have to be sequential in delivery; if they are not, assume a disaster scenario like this: An individual observes his home and his children in the house through a video transmission system. The images are delivered out of synchronization. He observes his kid crying first, falling from a couch, and then laughing. With a correct delivery sequence, he should have observed his child laughing first, falling down from the couch next, and then crying. Of course, networking technology has matured to a level where such types of message delivery are not permissible at the IP level itself. This example still highlights the basic need for QoS and, hence, a high-bandwidth solution for first mile access networks.

Healthcare

Current trends indicate a global increase in health awareness that has fueled a need for multiple opinions which allow doctors to drill to the core of health-related problems, all requiring substantial bandwidth. Current networks are unable to meet the growing challenge of transporting large image files (such as MRI scans) on a real-time basis over connections that typically do not always need the peak bandwidth levels. Also, geographically dispersed and individually specialized bodies need to investigate the same files and images and then have multiway conferences to build a consensus for further medical action. This leads to added pressure on the network, requiring multicasting and excellent response from the underlying network. Moreover, the voluminous nature of the healthcare business makes it such that the network has to have a large number of points of presence (POPs); thus, it needs be economical to deploy.

Networking Economy

Alongside the heavy investment seen in the e-commerce industry over the past few years, there has been a substantial investment and interest in the telecommunications industry with a particular aim to achieve the *vision*—that of making "broadband to the end user" available as a commodity solution. However, much of the investment faded due to the immaturity of the technology and the absence of a solution that delivered the necessary requirements to create a strong network business model that fosters growth. To bring back investor confidence in a bleak industry defined solutions are needed, and first mile access is the most effective.

First mile access networks, as mentioned before, are a solution that are economically viable and are almost certain to pave the way for a strong recovery in the telecommunications industry. From the point of view of broadband carriers or system vendors, the first mile represents a strong growth-oriented sector. The advantage first mile networks have compared to core networking, is that they provide a way to make networking consumer-focused, thus generating business by the volume—quantity as opposed to just quality.

Data-Centric Applications and QoS-Demanding Applications

The tremendous surge in Internet traffic over the past decade has led to the exponential growth of the Internet, dwarfing the traffic generated through voice-based systems. Despite several implementations and evolving generations of networking technology, no pragmatic solution allows data-centric architectures that readily permit a tractable model for IP-centric communication. Those that do allow IP-centric communication are often too expensive due to their reliance on an expensive optical and opto-electronic approach, highly inefficient provisioning of the optical bandwidth, expensive and inefficient control of the network, and an inability to provide QoS requirements and support SLAs for bandwidth-sensitive applications. Consumers are demanding, but are not able to obtain, various degrees of granularity to support high quality applications, such as video on demand and so forth, at a price that is affordable and with technology that can be seamlessly upgradeable.

Based on the above four-pronged generic need for bandwidth-killer applications, we can now summarize the generic application portfolio that creates the business case for first mile access networks.

Product and Services Portfolio for the First Mile Business Case

In the first mile access area, we have showcased multiple technologies that are combined together in the same market space. The business motivator for these technologies is a set of applications (like healthcare and telemedicine) whose demands are peculiar and for whose sustenance we need the basic first mile access solution. Let us now see some of these services that make up the first mile.

Voice

Traditionally, voice communication was the first revenue bearing service that drove the communication industry. Voice communication requires a bidirectional duplex channel with approximately 64 kbps of bandwidth. The primary technical issue with voice is that the end-to-end delay must be kept below a certain level (latency requirement). Voice communication is a service provider's biggest revenue generator. The number of telephone lines across the world has increased tremendously and has given rise to the PSTN network, on whose framework we have built a few first mile broadband solutions (for example, DSL).

Voice systems are traditionally circuit switched. The idea is to connect with the help of switches—leading to circuits that are end to end. There are some problems with this approach. First, circuit-based voice systems are not efficient because their use of the channel bandwidth is poor. This can easily be understood when we consider a voice communication: Numerous times in a duplex conversation neither party is talking. Second, for circuit-based voice communication, we have to ensure end-to-end path availability. Sometimes this can be a problem. By using electronic circuit switching, the approach to end-to-end reservation is fairly easy; however, it is not cost efficient. Consider a system where we have to use ATM tunneling and then SONET for transport. Both SONET and ATM platforms have very high costs associated with them. Moreover, for the bandwidth they offer, they do not do much! As the World Wide Web moves toward an all-IP core, SONET/SDH and ATM terminals are not just expensive but also don't do justice to the vast potentials of gigabit-switched routers. Therefore, there is a need to move away from traditional circuits for voice communication and migrate to packet-oriented communication.

Voice over packets is an interesting concept in the first mile. Among the many advantages seen, the primary is the ability to lower system costs. Packetizing voice and placing voice in packets means selecting a packet-based protocol as a carriage for voice packets. Voice streams are sampled (at discrete intervals faster than the Nyquist rate) and then quantized (i.e., rounded off to the nearest integer) into digital codes. These digital codes generate binary bit streams, a method popularly known as *pulse code modulation* (PCM). The choice of protocol for transport of course is IP.

Voice over IP (VoIP) is a very important paradigm that has been proposed for packet-based communication. Instead of hard-wired circuit switches, VoIP packets can be switched in what is called *soft switches*. These are software-based modules that look at the header of a VoIP packet and then route it to the respective port using fast and intelligent software. The advent of soft switches seemed to provide the necessary flexibility in first mile networks, especially while differentiating traffic flows from voice to data and so on. However, as the size of the soft switch grew, it became evident that the soft switch could not provide the necessary scalability needed for such high port counts. In other words, as the number of input and output connection points grew, the switching demands became more severe, leading to longer switching times. The soft switch worked on a software module that was typically serial in nature. This meant that for the switch to look at a particular packet, the packet needed to be queued before it could be processed. Voice communication, being delay sensitive, suffered extensively in such a system. Recently, some approaches have been proposed to solve this issue. Among these one of the fundamental differences from the original soft-switch approach is to use a parallel processing algorithm either in software or hardware (by hard-coding inside a programmable array) format.

At Layer 2, VoIP packets are encoded into an Ethernet carriage, and the simplicity and universal acceptance of Ethernet enables excellent communication. In the future (and in some ongoing implementations), VoIP flows can be switched at Layer 2 itself using MPLS to allocate a label to the Ethernet frame, one that depicts the QoS issue and the port diversity (which port to switch to, resulting in good mapping). This makes VoIP technology an extremely interesting, affordable, and ideal fit in the first mile access where bursty data traffic and VoIP are all together as Ethernet frames but yet are differentiated in their service needs. It is our estimation that until there is a severe change in the business model that governs worldwide networks and we adopt a billing policy to bill data traffic also, voice communication will be the key revenue generating service in the years to come.

Data Services

The recent surge of data traffic worldwide has been propelled by the information technology (IT) and Internet revolution. Just as money in the hand is equivalent to possessing a tool, being able to access quality information can be considered a strong asset. The importance of data networking and the move toward e-commerce has automated the way we live and has given a new meaning to broadband communication by being able to connect a desktop at home to the enormous World Wide Web. To the end user, information on the web is available at the click of a button, and this means that there is a great interest in exploring this data and analyzing it to satisfy and suit one's needs.

Data networking is a non-QoS-based service, but the various ways in which data networking is being carried out these days enables the end users to download multiple kinds of data. This means that data networking has led to the use of multimedia communication; hence, interactive multimedia is growing in popularity. Finally, the growth and the subsequent stabilization of e-commerce has led to a lot of data exchange that directly affects businesses. Online banking, online purchasing, data warehousing, data mining, and so on are all different aspects of the e-commerce industry. Among others the premier issue that affects the e-commerce industry through first mile networks is to make these networks as secure as possible.

Secure Sockets Layer (SSL) is an encryption methodology that makes sessions for e-commerce promulgation secure and safe from hazards like leaks and hacking. SSL is an open non-proprietary protocol for managing security of message transmission on the Internet. It uses the public and private key encryption system from RSA (Reed-Solomon algorithm) for security and also includes the use of digital certificates for authentication purposes. Some adaptations are always needed at central offices (COs) and local servers for the successful carrying out of such sensitive services. The first mile area is no exception to these modifications.

Video

Video communication over a network is the ultimate real-time application in communication. There are several applications and approaches to video communication that are proposed. Among the many applications for video communication are the following:

- Videoconferencing
- Video distribution
- Video programming
- Telemedicine and second opinions
- Visualizations and distributed computing

Videoconferencing

Video communication on a commercial level, was first proposed to promote the concept of the home office and to transform the conference room from an inhouse central facility to a global distributed setup. Videoconferencing is a key communication tool for making business happen over distances. Apart from the direct advantages in saving travel costs and being at the leisure of one's own office space, videoconferencing adds an extra human angle to real-time communication. It has been observed that e-mail communication can often be misleading—words can be minced and misconstrued. This can lead to poor interpersonal communication—the key to effective business communication. Also, voice communication too can be quite misleading at times. Words spoken are often not taken in the right spirit, jolting business efforts. Videoconferencing is one business proposition that puts an end to these anomalies.

From the first mile access area networks, videoconferencing is broadband communication in the true sense. Voice and images have to be transmitted and received through a network, and both need multiple QoS guarantees. Among which, delay and frame synchronization are the key. A point-to-point videoconferencing system often means setting up a circuit between the two points and dedicating a bandwidth requirement to this circuit. However, the real challenge comes when there are multiple users involved. This means that each user broadcasts into a multicast tree. This can be a nearly insurmountable challenge. Each user must be well synchronized. If not, when he or she returns a response to a particular action by another user—all of them go in complete disarray.

Videoconferencing servers in the first mile are subject to two design issues: first, as mentioned, in according the right synchronization and scheduling and, second, in managing the bandwidth that is allocated to the videoconferencing channel. Finally, this bandwidth management has to take into consideration the multiple services in a way to maximize network performance, not a task that is always possible. Traditional videoconferencing was carried out using ISDN dedicated lines. However, these lines are expensive to deploy and difficult to provision dynamically, leading to leaving them static.

Video Distribution and Programming

A simpler case of videoconferencing is video distribution. Videoconferencing is a two-way duplex phenomenon, whereas video distribution is a unidirectional scheme where a video signal is broadcast to a multicast tree. However, video distribution differs from cable distribution where the viewer can tune to any of the channels. In video distribution, first the quality of the picture is digitally encoded and, thus, vastly superior to cable. Second, the viewer has the right to choose a program and have interactive ability. Video distribution has some degree of "duplexity," in the sense that management information (the user can tune to any movie) is fed on duplex channels.

Telemedicine

The use of IT in the field of medicine is quite natural. The distributed nature of inhabitants and the further random geographic distribution of healthcare professionals make telemedicine an attractive approach. Considering especially the fact that smaller pockets of population may not have geographic access to a particular specialty of surgeons. Images are sent to specialists and to superior investigation centers to form concrete opinions. Most of these images (for instance, MRIs and CT scans, are large file sizes and in addition need to be delivered effectively (good QoS). Telemedicine places an uneven challenge on the first mile network. This challenge is to make large bandwidth pipes dynamically available for short durations.

Visualization

Apart from the need for telemedicine, the distributed nature of the medical field sometimes means that there exists a need for opinions on images and reports from multiple specialty centers. This is called *visualization*. In other words, visualization means distributed access of information and transfer of the information intelligently. Cancer research and its subsequent treatment, geoscience monitoring, and security systems are all applications belonging to the concept of visualization. First mile access networks need particular fine-tuning (in terms of QoS) for such high-bandwidth, fast-provisioning approaches.

The Technology Business Case

Up till now in this chapter, we have discussed a series of applications for the first mile access area. Now in this section, we provide an integrated approach for creating the basic business case in a first mile network. We show how multiple technologies can be laid together, side by side, in a way to maximize network performance and create an effective business case.

The first mile access area can be divided into two subareas: (1) to provide bandwidth to the curb and (2) to distribute bandwidth from the curb to the end users. The reader must note here that end users can be customers (such as in small and medium-sized enterprises or consumers (such as in homes and localities). In fact, a research community calls the area from the curb site to the end user as the last inch solution—different from the last mile solution. When we consider providing bandwidth to the curb, we have to consider both these areas from the perspective of

- The volume of bandwidth needed
- The types of services at each end point on the curb required

The first point basically is a summation of the end users' requirement (peak need), whereas the second point defines the QoS guarantees that must be satisfied. A further comment can be added here that the first mile network, as of today in most areas, is the PSTN network. Wherever possible, the PSTN network is used. However, where it is not possible (for example, when the number of end users is high or if the QoS guarantees cannot be met through the coaxial PSTN), it becomes important to upgrade the PSTN to a more broadband-conducive network. There are multiple ways to do it, and we discuss a most heuristic (i.e., non optimal—though quite near that—solution based on previous data sets and iterative algorithms) method to do so.

A first mile network can be viewed as a collection of multiple end users who fall into one of many categories. The number of categories varies, but as a rule of the thumb, typically a dozen categories are sufficient to describe the types of end users in the first mile. Further, each end user is interested in a number of services that he or she needs to subscribe to. Again, the number of these services from the product and services portfolio is not more than 10. So if there are N end users (usually N is in the range of hundreds of thousands), we can note that each service to each end user has two components c,f where $N_i(c,f)$ describes a logical connection from a CO to user N_I, such that c is the bandwidth (by volume of the connection) and f is the service being offered on this connection. Again, note here that $N_i(c,f)$ is just a logical connection to user i and this is not a physical connection.

Figure 10-1 shows that the basic topology is that of a star with a single source (CO) and multiple endpoints. A logical connection can be specified shown in Figure 10-2. If we compare this to a PSTN scheme, the granularity provided to the end user is that of maximum shielded twisted-pair (STP), which may not be able to satisfy the bandwidth guarantees needed at all times. In a PSTN scheme (as shown in Figure 10-2), there is universal connectivity, and the volume of bandwidth provided to the end user is uniform (in the multiples of voice lines or DS0). This means that although we have universal connectivity, the quality of the connections is limited to what DS0 can do.

Figure 10-1 *Distributed First Mile Network*

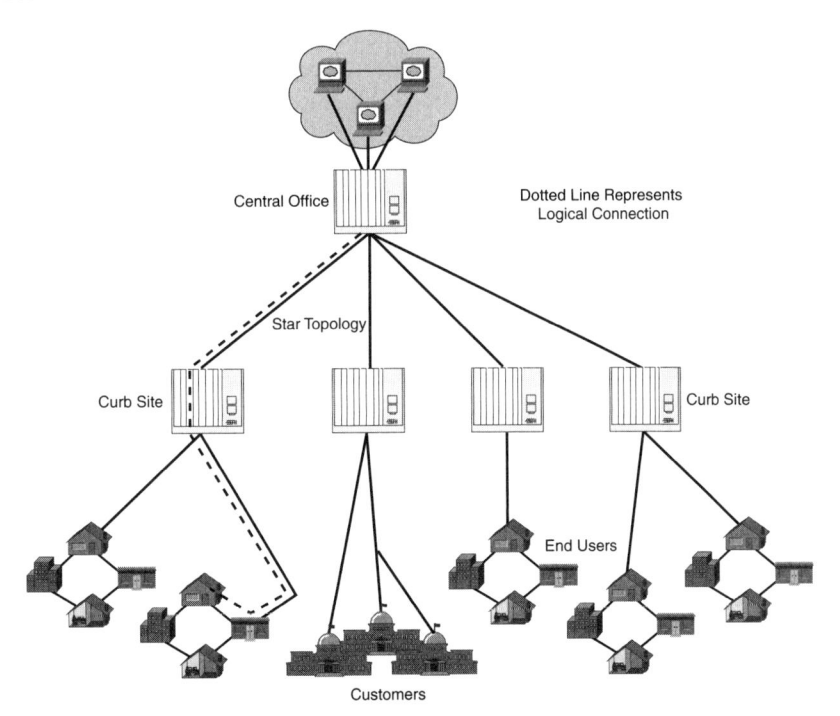

Figure 10-2 *Current Schematic for PSTN*

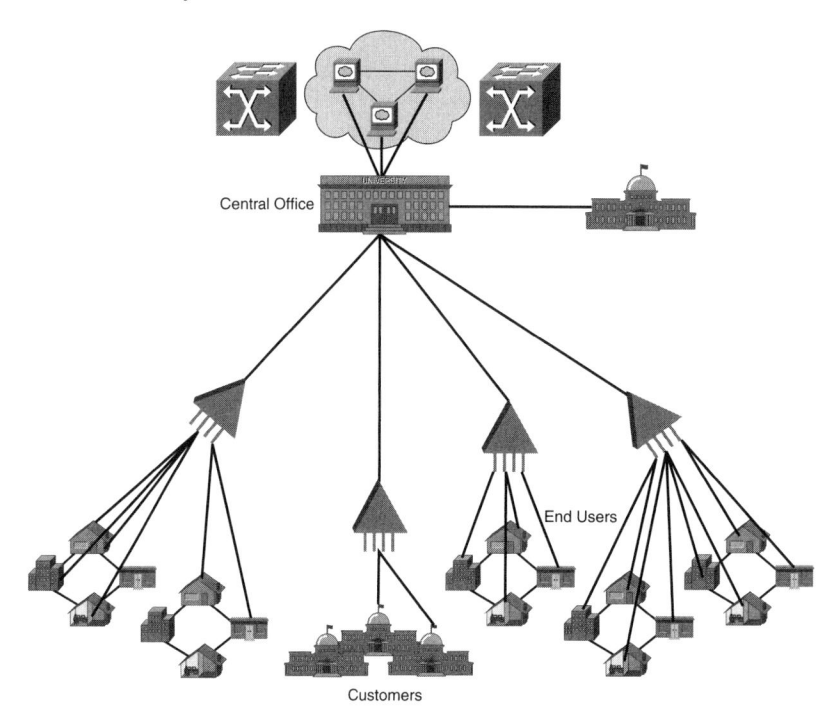

The initial broadband first mile access solution was an adaptation to the PSTN. DSL and later xDSL (ADSL and VDSL) were innovations that formed an evolutionary (rather than revolutionary) approach to broadband networking in the first mile. However, this approach was not enough. DSL lacked in multiple technological areas and, further, DSL resulted in a business model that was not very lucrative in terms of revenue because of high maintenance and scalability costs. ROI was low (for the provider), and due to management issues, consumers were not satisfied with DSL connectivity. The technological case for first mile, hence, was clearly not to be based just on the PSTN alone.

As DSL was maturing there was a parallel thrust in passive optics as it gradually became a feasible concept. PON was now more affordable than before due to a mature technology that went inside the creation of PON components. However, despite this sharp fall in PON pricing, PON was not sufficient to promote the concept of fiber to the end user (FTTU). Hence. it was realized that a hybrid solution was necessary: fiber to the curb (FTTC) and PSTN or some other communication methodology to the end user. One should note that by using a complete 100-percent PSTN scheme, the potential of PSTN is not fully utilized—primarily because for short distances (for example, within the last inch solution) PSTN can deliver high data rates. Hence, the best approach would be the one we mentioned—that is, providing fiber to the curb and then using PSTN or use some other scheme like WiFi, fiber (multimode), PLC, or FSO to the end user.

Note here that after fiber is laid to the curb, this fiber line can offer service to a number of end users. Also note that each end user may have different needs, and we can use different last inch technologies to solve these needs. If the end user is an office with multiple connection requirements, for example, a single fiber solution should suffice. Thus, we can have a fiber solution right up to the end user. On the other hand, if the user needs some sort of intrahouse mobile connectivity, we can use 802.11b from the curb site to the end user. Of course, the wireless 802.11b solution works under the constraint that the end user is in the vicinity of the curb site. But then this is a last inch problem and not too difficult to solve.

In the wireless solution for the last inch, the end user has the liberty to move within the range covered by the 802.11b/g base station antenna. Lately, there has been commercial activity where a user seamlessly can move from the range offered by one 802.11b antenna and into an area that is under the influence of another antenna. This kind of networking is called *ad hoc*, or simply *roaming*, and is described in Chapter 4, "Data Wireless Communication." Furthermore, there is the extreme case where an end user can move from one 802.11b/g antenna serviced area to another area that is under the influence of a normal cellular network. She can then use the GPRS (General Packet Radio System) to get broadband connectivity, for example. This means a handoff is needed between 802.11b and GPRS. This technology is expected to gain momentum in the next few years.

Finally, the end user may have needs that are stringent and, in that case, may use the PSTN from the curb site onwards. In that case, she may just use the copper network (STP). The implementations then can be either over DSL or using the new concept of long-reach Ethernet (EoVDSL).

The one thing that deserves mention here is that in a hybrid network it is important to have protocol transparency or, rather, have seamless communication from the CO to the end user via multiple network classes. It is the experience of network planners and capacity analysts that networks which have a wide range of protocols and which need multiple protocol transformation are expensive to operate and extremely hard to manage because of protocol interoperability issues. Hence, it is our strong recommendation that for the first mile area where economics is the key driver (not that it isn't in other places), it is beneficial to have a single protocol over multiple technologies in the first mile. We strongly believe that each available technology can fit and solve a set of problems, provide some granularity in bandwidth, and adhere to a range of protocols. Thus, it is important to choose the set of technologies that are closest to each other in terms of the protocol transparency they offer. The technologies in a network may vary and should be deployed, keeping in mind what is needed. For now, there are two conventional approaches to protocol uniformity in the first mile:

- The Ethernet solution
- The time-division multiplexing (TDM) solution

The Ethernet Solution

With several million Ethernet ports deployed throughout the world, Ethernet is one of the most native and widely accepted LAN protocols in the world. It has been around for three decades now and is regarded as extremely easy to use and commercially is very popular. The migration of Ethernet from LANs to WANs by the creation and standardization of Gigabit Ethernet and subsequently 10 Gigabit Ethernet resulted in Ethernet ports being found across the entire network. From a worldwide network (Internet) perspective this meant that Ethernet could now be found in the access, the metro, or the core (long haul) area with equal likelihood. The ubiquity of Ethernet is extremely beneficial for protocol dissemination. This also guarantees uniformity. Furthermore, it alleviates the problems that are faced in conventional distributed systems such as management and provisioning.

In the Ethernet approach to the first mile, we envision a PON solution (namely, Ethernet over PON to the curb). The network uses a star topology. This results in an OLT-ONU solution from the CO to the curb site. The transmission of information within the first mile, is of a bursty form—that is, a particular user may desire bandwidth, in spurts or bursts. Even dedicated voice and video channels can be sampled and made to fill up bursts or TDM slots. Ethernet then serves to carry this payload in the first mile. The generic bursty nature (heavy tailed message distribution) of data traffic, further strengthens our need for an Ethernet-like, non connection-oriented scheme in the first mile. From the curb site to the end user, we have three ways of transmission (also shown in Figure 10-3):

- Wireless using 802.11b, which again transmits Ethernet frames
- Optical using the fiber and low-cost optics such as multimode fibers—again using Ethernet technology
- Copper using Ethernet over VDSL (EoVDSL) or also called long-reach Ethernet in the nomenclature given by Cisco.

Figure 10-3 shows a three-pronged (wireless/optical/copper) approach to provisioning bandwidth on demand to the end user by dividing the access area into 2 components: the last mile and the last inch.

The TDM Solution

Since the advent of digital communication and the need to maximize the channel efficiency to facilitate users to use the same channel, TDM has been proposed as a tangible way to timeshare the channel bandwidth amongst these multiple users. TDM is a well-defined legacy technology that is adaptable because of several decades of acceptance by the communication industry. In TDM, users are allocated timeslots, and a particular user transmits only when his or her timeslot arrives. If there are N users and each timeslot is R seconds long, then for any user, the time difference between any two consecutive slots in which he or she transmits is going to be NR seconds. In principle, however, TDM technology is expensive to deploy on account of certain stringencies exhibited in deployment. All the users that timeshare a given channel, need to be synchronized with one another to be able to correctly time their transmissions in the timeslots that are allocated to them. The problem in TDM is the need for synchronization. Electronic synchronization is necessary and expensive at high speeds. At lower speeds, however, the cost is low. The second disadvantage with TDM is that it is not friendly to bursty traffic. TDM slots (allocated to different users that share the channel) are of fixed size. This often leads to severe utilization issues. This can be understood from the fact that a particular user may have no data to send when its slot arrives, while at the same time another user running some bursty application may have data to send but its slots arrive with fixed periodicity—hence, hampering the utilization of the system.

Figure 10-3 *A Community Network*

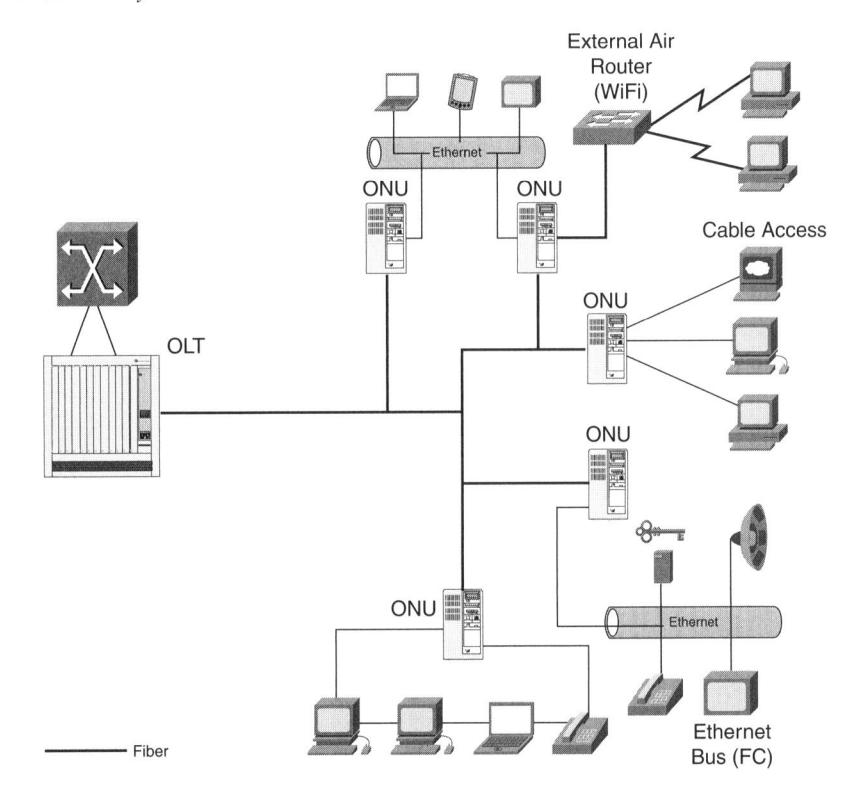

In addition to simplicity, some of TDM's advantages are universal acceptability and good performance. Also, TDM is a fair protocol—all users have equal access—subject to conditions dictated under SLAs. From the first mile point of view, we can have a TDM end-to-end solution with BPON or GPON to the curb and then use a variant of DSL technology from the curb to the end user. This is realized using ATM technology with its QoS guarantees, for which it was specially developed to provide. Provisioning and so forth is easy as long as the bandwidth needs are not very dynamic; when dynamic, they are almost impossible to provision. Figure 10-4 shows last mile BPON/GPON solution.

Figure 10-4 *BPON/GPON Solution for the Last Mile*

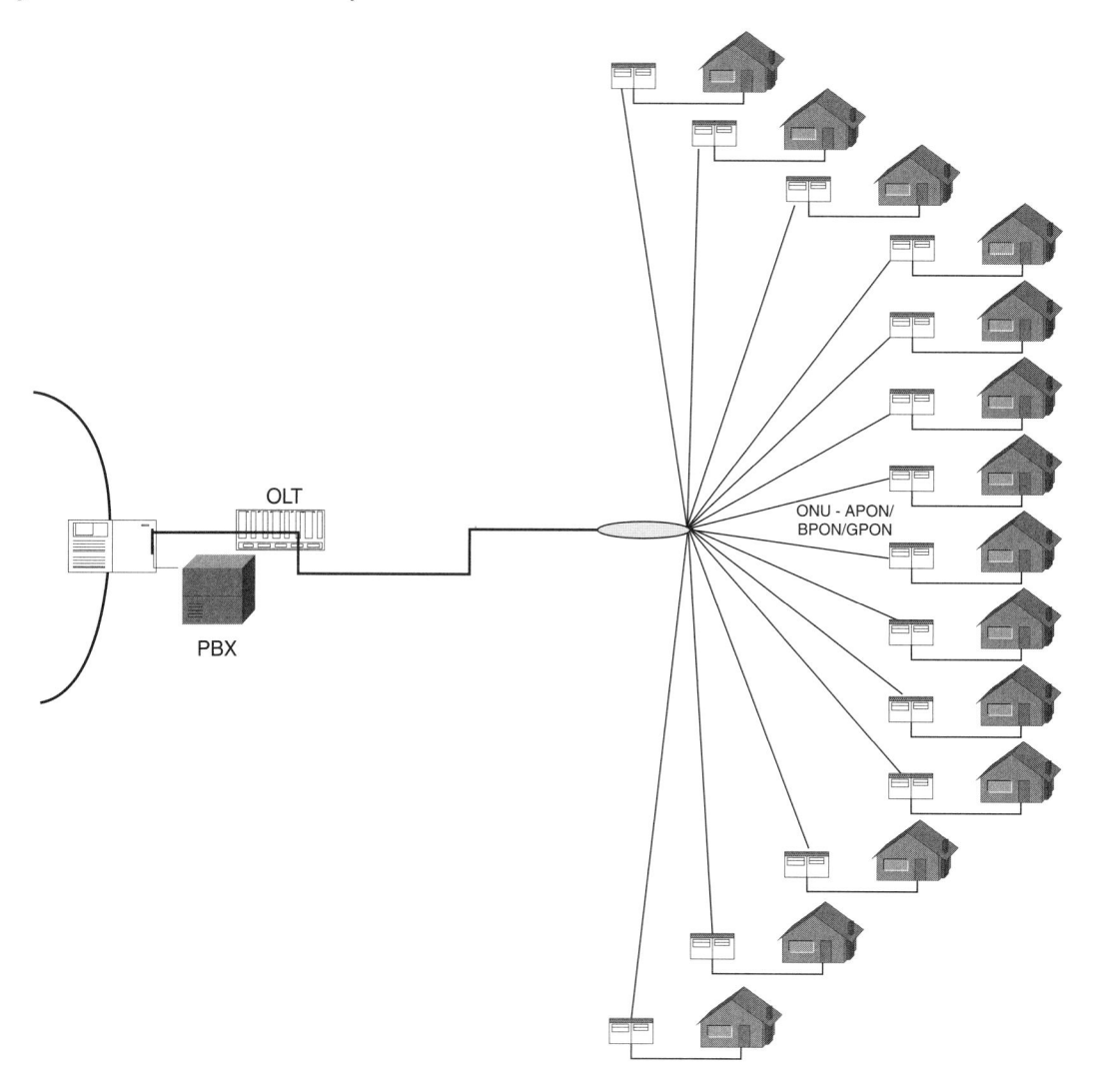

Business Case: Characterizing a First Mile Access Network

Having discussed the technological aspect of first mile networks, we are now in a position to build its business case.

Among the issues we need to consider to build the business case are

- **System architecture in the first mile network that is related to the specific product/service that is sold to the end user.** The architecture basically is necessitated from the needs and the requirements of the end user. The architecture then translates into a final solution for the first mile access network.

- **Definition of the product/service components and user interfaces.** $N_i(c,f)$ can be used to describe a logical connection. Here i means the ith user, c the amount (granularity) of bandwidth required by this i^{th} user, and f the QoS desired. This gives an idea of the interface needs for each end user.

- **Definition of the QoS parameters.** Multiple QoS parameters have been studied in Chapter 9, " First Mile Access Management and Business Model." The most important ones are delay and synchronization. In addition, some other nontrivial QoS parameters are peak bandwidth, committed rate, and so on.

- **Specification of the support system functions necessary for providing the service:**

  — Network management functions—Management of these distributed resources is a serious challenge in today's first mile networks. Auto provisioning and maintenance are important functions that need special attention.

  — Help desk functions—Technical support and customer care are important after sales features to be considered while operating a first mile access network.

  — Billing functions—The distributed nature of the demands means that there is a strong need for a good billing paraphernalia. Billing systems have to be accurate and fast. In addition, remote billing functions are also helpful.

  — Some of the other functions that need special attention while designing operation support system (OSS) are security, performance, fault service, business, accounting, and configuration management.

Case Study: Distributed and Unified Technology Solutions in a Community Network

We have so far studied multiple technology solutions and are now in a position to evaluate these through an example of a first mile access network business case. Before we actually build the business case, let us first examine two business models carefully and then choose the most appropriate model for our consideration. The first business model is what we will term the Conventional Distributed Model (CDM), while the second model, can be called the Cumulative

Comprehensive Model (CCM). The CDM can be applied by an ILEC also, so it is not specific to a community or intelligent city network but more of a first mile access network model case.

Conventional Distributed Model

Consider a network infrastructure with multiple service types being offered to the end users. For example, DSL, direct PON (fiber to the enterprise/fiber to the premise), WiFi, PLC, and FSO. Now each of these services is generally based on components that are vendor specific as well as standardized to suit only a certain set of requirements and functions (depending on the technology). So when we have a network where multiple users have multiple service needs and there exists a distributed deployment of all these multiple technologies, we are actually talking about our first model, CDM. In this, a service provider lays the infrastructure and invests a substantial sum of money in this process. The money it invests in building this network is called capital expenditure (CAPEX). The service provider generally has one to two technologies to offer. It tries to keep its portfolio for multiple services (riding on these different technologies) but only uses a minimum number of technologies for provisioning these services. The key here is that each new technology means a new round of CAPEX, and with the revenue/volume ratio being the only consideration for the service provider, it makes more sense to have minimum number of services and cater to as many users and services as possible, even if it means neglecting certain pockets of users or certain classes of services. Consider a scenario where we have the two service providers A and B providing service to a town. Assume provider A specializes and offers DSL, WiFi, and PON services (hence, respective granularities). Also assume provider B offers services using PLC, DSL, and FSO. In all likelihood, the mention of PLC means provider B has graduated from a utility company to a broadband provider. The end user has the option to choose the provider, and the end user then chooses either A or B depending on the service offerings by A and B and also what A and B will charge the end user. Among the many reasons to make the choice, some of the points the end user considers are as follows:

- Availability of service in the geographic area of the user
- Provisioning issues
- Credibility of the provider (no bankruptcy)
- Additional services offered (bundled services)
- Incremental service offerings (future upgrades)
- Customer support
- Maintenance
- Cost of service
- Backward compatibility with existing infrastructure (especially for ILEC)

The provider has to invest large sums of capital in laying the network and installing the equipment to provision services to the end user. It is a near impossible task for providers to accurately forecast what equipment to place where and determine the exact granularity of the equipment. This means there are several iterations of trial and success before the most immaculate configuration for a particular POP is found. Though this may seem to be a crude optimization procedure, in practice it requires a quite minimal number of iterations and a service provider's expertise comes handy in predicting of the configuration. Another way that service providers can find the correct configuration for a particular POP is to create a mathematical model of what is called the Classical Terminal Assignment problem, the solution of which leads to an optimal number and the location of the terminals (components) that are placed throughout the network. The end user pays the service provider a monthly fee (or as decided in the SLA) and the service provider uses this money to do the following:

- Recover CAPEX
- Recover operational expenditure (OPEX) incurred during day-to-day operations—for instance, overheads, salaries, and so on—for the service provider
- Generate profit

The high costs of laying the equipment and the network layout means that the ROI cycle for service providers is generally of large duration. Coupled with this is competition in terms of both rivals as well as emerging technologies and service requirement trends. This eventually leads to situations where, for a given geographic area, multiple service providers build their own networks that overlap. This overlap means that most of the networks operate at fractional of their potential utilization. The low network utilization has a direct negative effect on the incoming revenue. This often means that the actual ROI cycle is much longer than the ROI cycle initially calculated because the network utilization is often lower than what was originally projected. The effect: long-term and often unrecoverable debts. The distributed model also has a technological disadvantage: that of not being able to provide incremental services. Incremental services simply mean the ability to increment service requirements from a lesser to a higher value seamlessly by according the correct scalability feature to network equipment. For example, if provider B wants to graduate from DSL to a service that is faster (for example, PON), it needs to refurbish its entire network with fiber. This is a solution that is not always economically viable. Hence, the disadvantage of the distributed approach is that providers risk losses and potential bankruptcy.

From the consumer point of view, if a consumer is stuck with a particular provider who is offering service set X and the consumer wants to have service set Y, there is no in-service upgrade method available. In-service upgrade is particularly important if we consider that the service provider has to conform to the 99.999 percent reliability and availability requirements. The consumer either has to change the provider (generally by paying fines and so on) or wait until the provider is able to provide this new service (which generally happens very slowly). All in all, it's a loosing proposition for both the consumer and provider. We require a solution that makes a justified business case for first mile broadband solutions. The next section describes one such case.

Cumulative Comprehensive Model

The CCM is a paradigm shift from the CDM. The CCM is a more recent model (post 1996), proposed to respond to the sagging telecommunication industry. The CCM model is unique in nature and is based on the approach of community networks or intelligent cities. In the CCM model, a community organization, such as a county, township, corporation, or municipality, forms the organizing body for the network. This organizing body creates investment to build the access network.

After adequately planning and then building the network, the organizing body invites service providers to provision services in the network. The service providers in return are able to tap revenue for the volumetric nature of services that they can now offer to the entire community's inhabitants and that at a substantially lower investment. The organizing body charges its "inhabitants" (end users, customers, enterprises, and consumers) based on the service they subscribe to. The consumers now have access to any service they want (chosen freely from application-specific providers). The organizing body being inherently a nonprofit organization has a relaxed payback period for the loans it has taken in building the network.

The CCM is a more relaxed model than the CDM and, hence, proves to be more successful for both the providers (ROI cycles are shorter) as well as the inhabitants. As a hidden benefit, the organizing body—be it a municipality or a community—gains access to its inhabitants. This can foster a strong hierarchical relationship between the organizing body and its citizens. This strong bond helps to bridge the digital divide by bringing together those who cannot pay the high costs of IT and those who can. This also facilitates e-governance and e-democracy: two unidirectional ways for the government to reach out to the people. Also, because the price for provisioning additional incremental services are vastly lower than the CDM approach, this model is extremely investor friendly both technologically as well as economically due to its overall network flexibility.

As can be seen, the CCM approach is definitely a new alternative business model, and in this book, we choose to compare it with the traditional CDM scheme. The CCM model as described is easily seen as the preferred approach in building access networks. However, to really justify this approach, we will now build the business case for the two models and then quantitatively determine the benefits of the two schemes from a high level perspective. As a disclaimer, we would like to add that there could be multiple in-detail and in-depth analyses that lead to more intriguing results from different perspectives; however, research into these is not within the scope of this book. This work basically is a generic approach to building the two models and comparing them to validate the argument in the preceding section.

Before we compare the two methods, we first need to show approaches to designing the network infrastructure and show how the network can be laid as well as what the thought processes are involved in laying the first mile solution.

Planning the First Mile Access Network: CCM Perspective

We have so far discussed the principles behind the two approaches that can be used to formulate a business case for first mile access networks: CDM and CCM. CCM, as we have seen, is a more novel business concept and guarantees low duration ROI cycles that ultimately result in strong growth. However, while making a business case for the CCM approach we need to first quantify the network. A typical CCM network like a community network either can be a greenfield development or can be a takeover by a community of a faltering provider network. The basic underdevelopment in the first mile means that the earlier case has a higher probability of happening compared to the latter case. This implies that the community or organizing body needs to carefully plan the access network. There are several proven approaches to network design and capacity planning but most of these are centered around the commercial aspect to facilitate service providers generating revenue. These approaches invariably build networks only to cater the higher stratas of bandwidth consumers, leaving lower end consumers to the mercy of alternate choices. However, in city networks, we must note that the prime objective of the organizing body is not just to make revenue but is to ensure a betterment of the lifestyles of the inhabitants through the offerings of broadband services to the entire community.

The best-case scenario in community networks is when a community network can claim that since its inception and deployment, there has been an increase in the overall GDP of the community thanks to the ability of this network to bridge the digital divide. This can happen only when the aim is to ensure good broadband services to most of the community. On the other hand, the service provider's chief objective is to maximize returns. One more differentiator in the design of first mile access networks, communities, and service providers is that service providers seldom lay a network to cater to satellite townships and small residential sectors that do not justify the network layout commercially. An organizing body for a first mile, on the contrary, tries its best to reach out to each and every inhabitant and offer the best possible services.

Apart from the reach of the CCM scheme in being able to facilitate services to the entire community, the basic CCM model for the organizing body itself makes good business sense. On the community owned backbone, multiple service providers can provision their service, without having to invest heavily in terms of equipment or fiber. Their investment is now core-centric and that's where there the main business interest lies (SLAs between enterprises and so on). From the service provider's perspective, by getting a readymade, fully connected (to end users) network it now has access to a lot more customers than by laying its own small network. Also, this development allows for competition among multiple service providers—one that is healthy and rules out monopoly, which has the drawback of stagnating both technology and the economy. The final benefit of all these is the research and development effort involved in churning out new technologies that are incremental and affordable by the community.

The above discussion makes it quite evident that the CCM approach needs a quantifying method for building and planning the network. This shall now be described.

Consider a township/municipality of U users. Let there be C logical connections that need to be deployed in the town. We don't know C, but that is one of the objectives of this exercise. Let the geographic area be represented by a coordinate system (x,y,z). Let a particular point be described as say T(x,y,z) in this network. Let us first define our objectives in laying the network:

- We want to find K_1: the number of POPs (aggregation) in the first mile network and find K, the set of geographic locations mapping K_1.
- Find NOC_1 (the number of network operating centers) and their corresponding geographic locations given by the set corresponding to {NOC}.

- To find the first two bullets, we need to find N, the total number of logical connections in the network with their characteristics in terms of services and flows (bandwidth).
- Let S be the set comprising the services that the network offers (voice is S_1, video is S_2, and so on, and these are a subset of S).

Solution to Designing the CCM Scheme for Access Networks

Let us first assume a point of reference T(0,0,0) as the frame for our network. This reference frame may be a CO (central office) or the primary NOC among the many NOCs in the network.

1. Do a census to find the number of users in the area represented by T. Note, that T gives us the location of each user. For example T(22,2,17)=9 means that at location corresponding to T(22,2,17) there are 9 users. Let the total number of users be I, and we can represent a particular user i_I.

2. Do a survey in the geographic locality given by T(x,y,z) such that for each user i find $N_i(c,f)$, where c denotes the user requirement of bandwidth and f the service profile set.

   a. For example, f for a particular user, in the term $N_i(c,f)$, may be the services set corresponding to say two video distributions lines, one surveillance line, two broadband connections, and three VoIP connections. Thus, f basically gives an idea of what service guarantees are needed to the user denoted by i.

   b. Likewise, c, in the term $N_i(c,f)$, for a particular user denotes the cumulative bandwidth desired by user i. Here, it is necessary that there be some future load projection but note that when a broadband network is being designed generally we do take into account some of the futuristic demands and key provision for these while laying the network.

3. If we have to do a detailed account for calculating projections, then one way of doing so is as follows:

   a. Calculate $f(c_i)$ that is the bandwidth of ith user.

   b. Create a table which has information of $f(c_i)$ for each i and name this table f_i.

   c. Create table $f'(c_i)$ where f' signifies the projected demand for i^{th} user after two years.

   d. Similarly create table $f''(c_i)$ where f'' signifies the projected demand after five years.

   e. Finally you can average these for medium term projections or do a trend analysis for incremental services.

Planning the First Mile Access Network: CCM Perspective   253

Figure 10-5 *A Township Network*

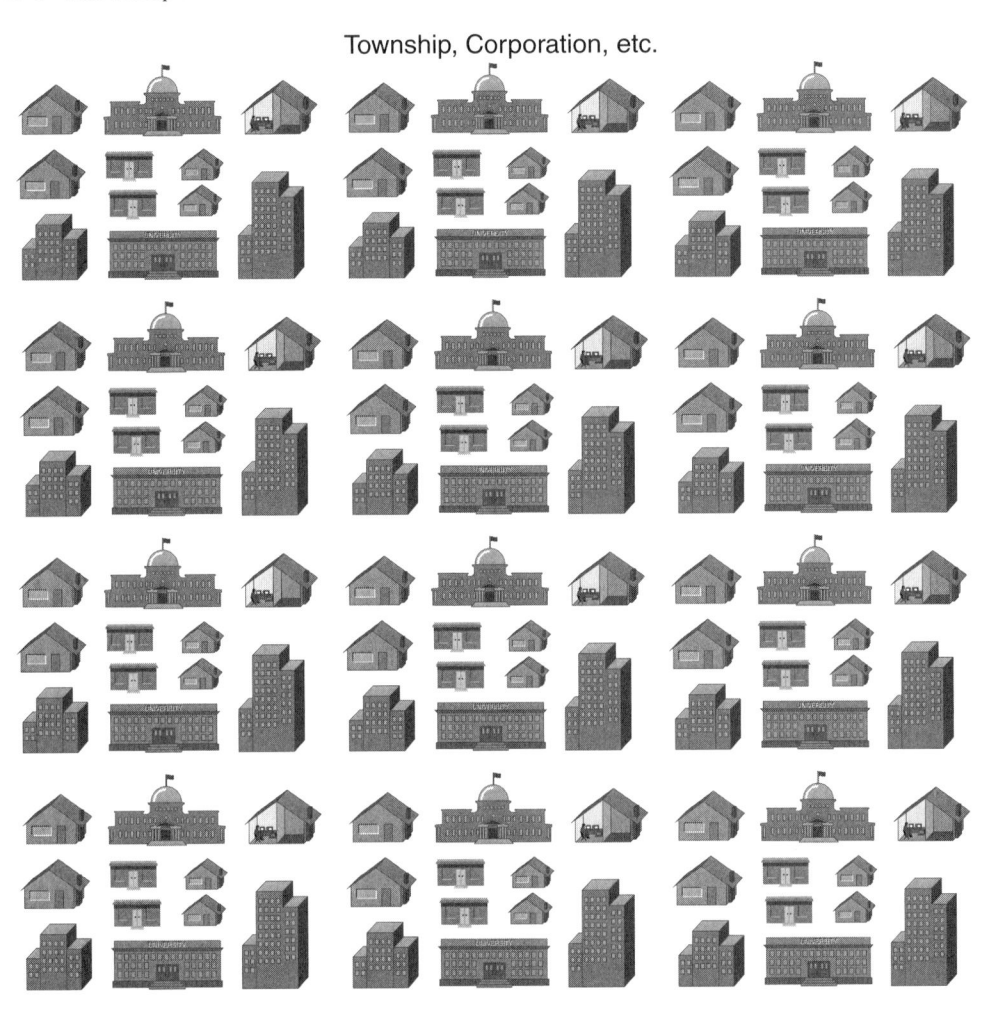

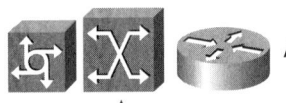

Township, Corporation, etc.

Service Provider's Central Office—
Also Organization Body Central Office
by Default

254 Chapter 10: Business Case for First Mile Access Networks

Figure 10-6 *Networking Planning*

Township, Corporation, etc.

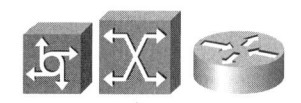

Service Provider's Central Office—
Also Organization Body Central Office
by Default

Shaded Area
Represents a POP
Coverage (Functionality
Dependent on Distance
and Usage)

Planning the First Mile Access Network: CCM Perspective 255

Figure 10-7 *Next Phase*

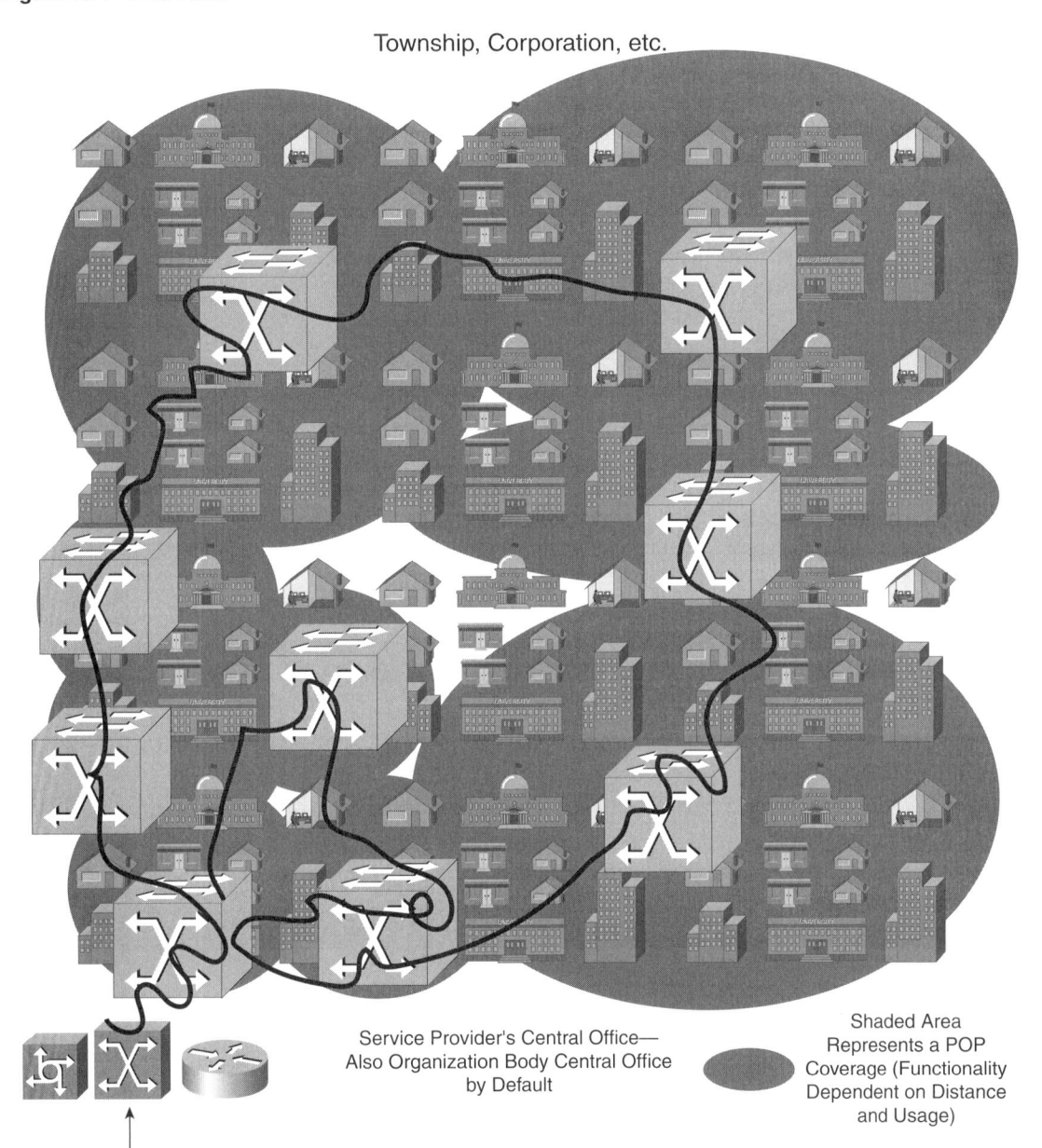

Township, Corporation, etc.

Service Provider's Central Office—
Also Organization Body Central Office
by Default

Shaded Area
Represents a POP
Coverage (Functionality
Dependent on Distance
and Usage)

Figure 10-8 *Final First Mile Network*

4 Finding POPs

   a Through a section of iterations we can break the geographical area into various POPs. The number of POPs is K_1, which is a variable.

   b Calculate possible POP sites and POP functionalities.

   c Create a functionality map that draws color filled circles with the POP being the center of the circle.

   d This exercise gives a functionality map with overlapping circles.

   e Remove circles that are overlapping to create a map with minimum overlap but no area uncolored. Figure 10-6 shows the optimum configuration of the township with POPs

   f A circle is now colored red if and only if the POP cannot satisfy the demand of the users in the circle. Note here that a circle here may not specifically be round but depends on the z coordinate. If the area is flat, then the circle is round, or else this "circle" is now shaped in a way to best describe the "reach" of the POP (distance wise) and the cumulative bandwidth reach. So in principle it does not remain a circle any more!

   g If a circle in the functionality graph is red in color repeat the procedure for the entire township in steps "a" through "f" to just the red circle. One way is to break up the circle into smaller constituents, either by having additional bandwidth or functionality allocated to the POPs, which are basically smaller areas within the larger circle. In conventional networking this means placing more "cards" at a POP site and populating network elements adequately.

   h Note: mathematically each circle can be described as

   $$\pi_k(B) \le \sum_{\theta=0}^{2\pi} \sum_{r=0}^{d} B(\theta, r)$$

   where d is the lengthwise maximum reach of the POP

   B indicates the bandwidth requirement of user at $B(\theta, r)$ and that is given by $N_i(c)$- neglecting f of course.

5 Let us now calculate the revenue model.

   a Let S={$x_1, x_2, ..., x_n$} be the services that are provided.

b Calculate:

$$A_1 = \sum_{i=1}^{i_{max}} F|f_i/x_1|$$

$$A_2 = \sum_{i=1}^{i_{max}} F|f_i/x_2|$$

and so on for each service type.

A_s indicates the total revenue throughout the network for service S.

c The total revenue generated through the network using distributed service pricing is

$$Y = r_1 A_1 + r_2 A_2 + \ldots\ldots r_s A_s$$

where r_S is the rate of the Sth service.

d Now the total number of POPs is NOC_i. Hence the total cost in the network is given by the following:

$$C = cost(NOC) x NOC_i + Network-layout_{total-fiber+copper} + Management + labor + admin$$

Similarly we can calculate the revenue per time cycle t as

$$Revenue = Y*t - C$$

Note here that we have postulated one typical solution for designing a network. There are and will be many possible approaches, each giving a set of results. The approach that most suits a network depends on the quantity that needs to be maximized, which in most cases is revenue. Likewise in a bearish economy, the most suitable approach is the one where network CAPEX and OPEX need to be minimized. Algorithms can be written for such maximization and minimization problems and are a subject beyond the scope of this book. Furthermore, one very apparent approach is the formulation of a linear program (LP) that works to maximize a particular quantity of our interest, subject to the conditions given. Such a linear program can be solved using any commercial linear program solver.

Case Study Distributed Pricing Versus Cumulative Pricing in a First Mile Network

We have illustrated in the previous section the merits of a self-sustained effort by an organizing body in a first mile area network. We can now build a financial business case for a first mile network. Because this exercise may seem trivial to begin with, we can compare two methods of business in the first mile, a cumulative pricing scheme and a service-based pricing scheme. We shall describe the two schemes briefly and then evaluate both of the schemes in a first mile network:

- **Cumulative pricing scheme**—Based on the approach in the previous section, we can see the qualitative benefits that arise when an organizing body creates a network infrastructure taking into account primarily the interests of the inhabitants (CCM). The organizing body offers a collection of services at discrete flat rates. For example, it may offer two voice lines, a videoconferencing line, and two lines for high-speed broadband access all coalesced for a flat rate of $50 per month.

- **Service-based pricing**—In contrast to the earlier pricing scheme, a service provider may offer end users services for which it would charge a rate. This means that two voice lines could come for $20 a piece and one videoconferencing line for say $25 and so on. This means that the user has to pay for every service she subscribes to independently. This scheme generates more revenue but also means that there are more providers. Thus, the infrastructure costs are also high, meaning less profits (or rather in today's telecommunication scenario, more losses).

In the example to follow, we evaluate both of the schemes through the pricing structure of an actual network.

Consider a community size of 60,000 users. Half the users need broadband Internet access (DSL or higher). Seventy-five percent of the users need telephone connections (assume a developed country). Let us assume 10 percent need videoconferencing and other video-based broadband services. Finally, assume 30 percent need home office equipment and connectivity. In Table 10-1, we show the income matrix (of a service provider) for a service-based pricing. The first column shows the services, the second one shows the number of customers, the third one the charge the provider levies for that service per customer, and the fourth the speed (connectivity needed per customer). Then we have total revenue that is obtained through the service followed lastly by the total bandwidth consumption for the service.

Table 10-1 *Revenue Matrix for Service-Based Pricing*

Distributed Approach	Number of Users	Cost per User (Charged)	Data Speed Provided (Mb)	Total Price	Total Bandwidth
Internet	30,000	10	0.256	300,000	7680
VoIP	45,000	20	0.128	900,000	5760
Video	6000	20	3	120,000	18,000
Teleconferencing	3000	15	2	45,000	6000
Home office	18,000	25	3	450,000	54,000
	102,000		Total	1,815,000	91,440

In Table 10-2 we show the pricing model for cumulative pricing scheme. We assume an average number of connections. We calculate this number (i.e., the number of connections) by the adding the total bandwidth for all the users (for each connection) and dividing by the attrition rate, which gives us the probability that a connection is used. The worth of this exercise translates directly into the reliability and availability of the network connections. For a 60,000-person community, we get the number of connections as 71,400. This is a conservative number and will be higher once actual demographics are considered, but it suites our concept demonstration here effectively. In the table, we show in column 2 the bandwidth we offer per user, the cost that we charge is shown in column 3, and the "effective revenue" is shown in the penultimate column. The last column shows how much bandwidth is needed by this network for the given number of connections.

Table 10-2 *Cumulative Pricing Scheme*

Users	Speed (Mbps)	Cost per User ($/mo)	Price	Bandwidth (Mb)
71,400	5	45	3,213,000	35,7000
71,400	5	55	3,927,000	35,7000
71,400	5	65	4,641,000	35,7000
71,400	5	75	5,355,000	35,7000
71,400	5	85	6,069,000	35,7000

Note that in Table 10-2, we assume a flat service of 5 Mbps, of which a QoS portion can be allotted as desired. Note that the 5 Mbps is more than the cumulative rate offered by all the services in Table 10-1. Second, we note pricing schemes, from $45 per month per user to $85 per month per user.

In Table 10-3, we show the expenditure matrix for the distributed services model. The first row describes the cost of the bandwidth the community pays per month to the service provider(s).

The second row gives the cost of bandwidth for the entire fiscal year. The third row gives the cost of equipment. Note here that the cost for equipment is higher than the cumulative approach.

Table 10-3 *Distributed Service Pricing Costs*

Total cost of bandwidth (community pays to the carrier)/per month	91.44 x 5000	457,200
Total cost of bandwidth (community pays to the carrier)/per year		5,486,400
Distributed cost of equipment per month	1240 x 91.44	113,385.6
Distributed cost of equipment per year		1,360,627.2
Fiber cost per year (inclusive of management)	350M/9	3,888,888.889
Total cost		**11,306,501.69**

In Table 10-4, we describe the costs incurred while setting up a network through an organizing body like community. Note here that the cost of equipment also reduces (per Gbps is the calculating mean). This is because in a cumulative scenario a single multiservice provisioning platform can provide multiple services and, hence, save costs. Row 1 describes the total cost that the organizing body (community) pays to the service provider per month. Row 2 describes this cost for the whole year. Row 4 illustrates the cost of equipment on a per year reducing balance method. Likewise, row 6 shows the cost of laying the network on a yearly basis of repayment. Row 7 shows the management layer costs for example NOCs and so on.

Table 10-4 *Costs Incurred in Distributed Service Pricing*

1	Total cost of bandwidth (community pays to the carrier) per month)		1,125,000
2	Total cost of bandwidth (community pays to the carrier) per year)		13,500,000
3			
4	Distributed cost of equipment per year	900 x 225 x 12	2,268,000
5			
6	Fiber and copper		3,888,888.889
7	Management layer		65,000
8			
9	**Total cost**		**20,846,888.89**

Finally, in Table 10-5 we show the comparison of the income part. We see that in the distributed pricing approach the profit percent is 97 percent, whereas in the cumulative approach the profit is always higher yet the cost to the end user is lower.

Table 10-5 *Income Through Both Distributed and Cumulative Pricing Approaches in the First Mile Network*

	Income (from Consumers)	Surplus per Year	% Profit
Distributed approach	21,780,000	10,473,498.31	92.63252772
Cumulative approach @ 45	38,556,000	17,709,111.11	156.6276785
Cumulative approach @ 55	47,124,000	26,277,111.11	126.048118
Cumulative approach @ 65	55,692,000	34,845,111.11	167.1477758
Cumulative approach @ 75	64,260,000	43,413,111.11	208.2474337
Cumulative approach @ 85	72,828,000	51,981,111.11	249.3470915

In fact what is not shown in Table 10-5 is the result of the simulation that shows some additional numbers such as the following:

- In the cumulative pricing model built around the community network concept, we see that the end users pay less, yet they get more bandwidth (and hence other services). In fact if we consider an intermediate scheme like the end user paying a flat $65 per month, we see that the end user on an average gets around 5 Mbps of bandwidth. On the other hand, in the cumulative approach the end user gets on average a 1.2 Mbps peak rate.
- From the service provider's point of view, it does not make good business sense to make large CAPEX and yet reach only a section of the population, resulting in limited revenue.
- Hence, from the service provider perspective, it has lower revenues. So the distributive pricing model doesn't work for them. This is one of the fundamental reasons behind the telecom bubble burst of 2001.

Market Analysis for CCM Method in the Access Area

Potential target customers:

- For the first mile, all small and medium-sized enterprises, households, communities, corporations, municipalities, and other forms of bandwidth-consuming bodies are potential targets.
- ILEC/IXC's focus on ITU standards and need products to address their needs.
- NSPs, community networks, local cities, and utility companies focus on low cost and service delivery speed (EFM standards).
- Public utility commissions (PUC).

Competitive analysis of the first mile:

- Positioning for each market is critical and determines overall success. Positioning for first mile issues can be based on technology advantage, pricing, quality, service, and implementation methods.
- Products can be differentiated by having management functions, and so on.
- In addition, service differentiation is a key competitive tool that is very effective in the first mile.

Strength-Weaknesses-Opportunities and Threats (SWOT) analysis

- Strength: Low cost, in service upgrade, technology, and incremental service offerings.
- Weakness: Lack of standardization and cost issues in infrastructure, multivendor interoperability.
- Opportunities: Business by volume (the entire community/city).
- Threat: Public policy is not completely predictable.

Marketing strategy insights for first mile:

- Product positioning strategy:
    - Differentiate the product based on performance
    - Differentiate to different market group
    - Metro access
    - Metro super access
    - Last/first mile
    - Last inch
- Price strategy
    - This follows an inverted U curve with revenue in the Y-axis and price in the X-axis. Initially revenue increases as price increases, but after a certain threshold, any further increase of price only lowers the revenue (customers go to competitors and the provider has achieved its maximum market cap).

Sales and promotional strategy: This entails smart market penetration leading to market capture and further consolidation.

Financial-economical analysis for first mile:

- Financial plan of the development of the product
- Forecast of the CAPEX expenditures
- Forecast of the operational costs
- Revenue forecast
- Financial summary

Summary

In this chapter, we have showcased the business case analysis of first mile networks. The chapter illustrates a method to build a business case for first mile networks. We show how to build the access network, taking into account the network economics as well as technology factors.

The chapter treats four levels of business case analysis for the first mile. In the first level, we study the applications of the first mile—the bandwidth-killer applications that are the key motivators of the first mile network. In the second level, we discuss the products and services portfolio in the first mile—that is, how these applications can be facilitated, as well as how the first mile network can sell bandwidth-products to its consumers. In the third level, we discuss the implementation of the network taking into account the first two levels of the business case (namely, applications and products) and building the network design from there onwards. In the last level, we study the financial business case and validate the motivation and intuitive correctness for the first mile. This level also gives an idea on billing, and we discuss one popular billing mechanism and propose an alternative billing scheme that is reliable and pragmatic, yielding a more efficient business case for first mile advance access networks.

Review Questions

1. What are the key products that will need bandwidth-killer applications in the next three to five years?
2. Describe the CCM model in detail and discuss some of the pitfalls it may have in the future.
3. Why is a strong legal framework absolutely necessary to safeguard the CCM model?
4. Research the optimization problem for assigning the minimum number of POPs to a given network. Formulate a linear program for the assignment based on cost minimization and maximize the utilization of cards at each POP site. Make suitable assumptions.
5. What is the advantage of billing based on services in a first mile?
6. How is the method in Question 5 not sustainable in the future?
7. Describe the biggest threat to CCM scheme.
8. By extrapolating, numerically verify the results of CCM and CDM as well as billing schemes for large number of end users (say, 700,000 in a city).
9. For Question 8, make suitable assumptions and plot the inverted U-shaped curve for services against pricing to show how this inverted U shifts with number of users.

10 Describe the interoperability issues in a last mile/last inch network for the following:

PON and WiFi

PON and PLC

PON and xDSL

CWDM and long-reach Ethernet

References

Chlamtac, I. and Ashwin Gumaste, "Bandwidth Management in Community Network." IWDC 2002 Keynote address: Calcutta, India.

CHAPTER 11

The Complete First Mile Network: Integration, Interfacing, and Management

So far, this book we have focused on the multiple technologies that may contribute to the first mile access space. As you can understand, the last mile area is a cluster of multiple technologies, standards, and approaches, each designed to solve some particular problem, but none so well designed to solve the entire first mile access issue of providing broadband-capable access to end users. Therefore, it becomes imperative for us to discuss and detail the manner in which these multiple technologies fit the first mile access space and discuss the levels of abstraction that detail their interaction with each other. Here we digress to study the ways each technology is designed and how these fit together, inclusive of the effort needed to integrate these technologies. This chapter also examines design examples of last mile networks based on the principles of technologies, requirements, and economics.

Interfacing Requirements in the First Mile Access Space

The vast number of different technologies in the first mile means that there is a substantial amount of interfacing between multiple technologies. Interoperability between these technologies is crucial to ensure an end-to-end solution. This chapter covers different approaches that help build the first mile network. This chapter should be viewed as a possible vision for the future, because most of the proposals in this chapter are not yet formalized; they are just guidelines for engineers to design the next generation of first mile networks.

The preceding chapters discussed multiple implementation philosophies in the first mile, among which the most prominent were passive optical networking (PON), wireless (WiFi), digital subscriber line (DSL), power-line communications (PLC), and long-reach Ethernet (LRE). The first mile network can be further classified into two broad categories:

- Network to the curb from the central office (first mile)
- Network from the curb to the end user (last inch solution)

There are technology solutions that are predominantly deployed for the first category only, whereas there are solutions that are designed specifically for the last inch. For providing solutions to the curb site, the primary technologies used are PON, DSL, coarse wavelength-division multiplexing (CWDM), and wireless MANs. On the other hand, the solutions for the last inch are broader, with fiber-based multimode, WiFi, PLC, free space optics (FSO), and DSL among many other possibilities and combinations. Figure 11-1 shows a first mile network with the clear distinction of the network to the curb and from the curb to the end user. This makes it imperative that two or more technologies intermingle with one another.

Figure 11-1 *First/Last Mile Integrated Network*

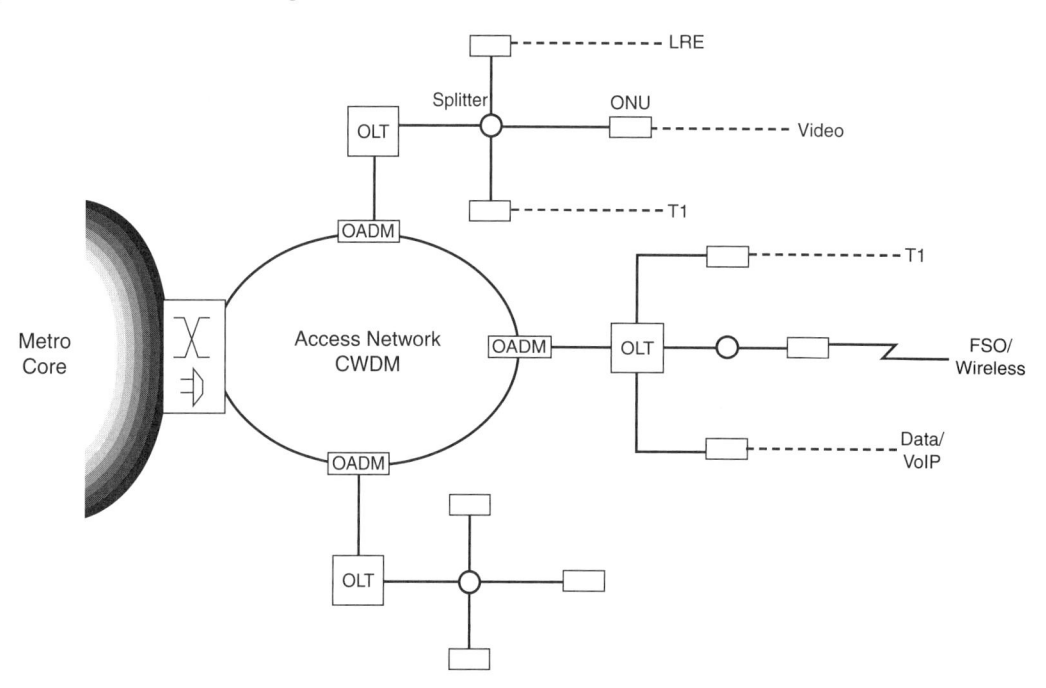

This chapter discusses the interfacing needs of two given technologies from the broad set of technologies shown.

CWDM and PON

PON networks can be easily laid over CWDM access networks, as shown in Figure 11-2. CWDM networks can support up to eight (16 for some vendors) wavelengths and can be used to scale the first mile PON solutions for high-bandwidth applications. CWDM is transparent to various PON technologies such as Ethernet over PON (EPON), broadband PON (BPON), gigabit PON (GPON), and Gigabit Ethernet PON (GE-PON); therefore, migration in the PON networks does not affect the access networks. CWDM network elements by themselves are passive and protocol transparent, and hence do not require any management or provisioning. A wavelength is dedicated for each optical line terminal unit (OLT) and can support up 2.5 Gbps worth of bandwidth (1 Gbps if Gigabit Ethernet interfaces are used). Protected OLT designs are also possible by reusing the same wavelength in both east and west direction and using redundant OLTs.

Figure 11-2 *PON Networks Can Be Easily Laid over CWDM Access Networks*

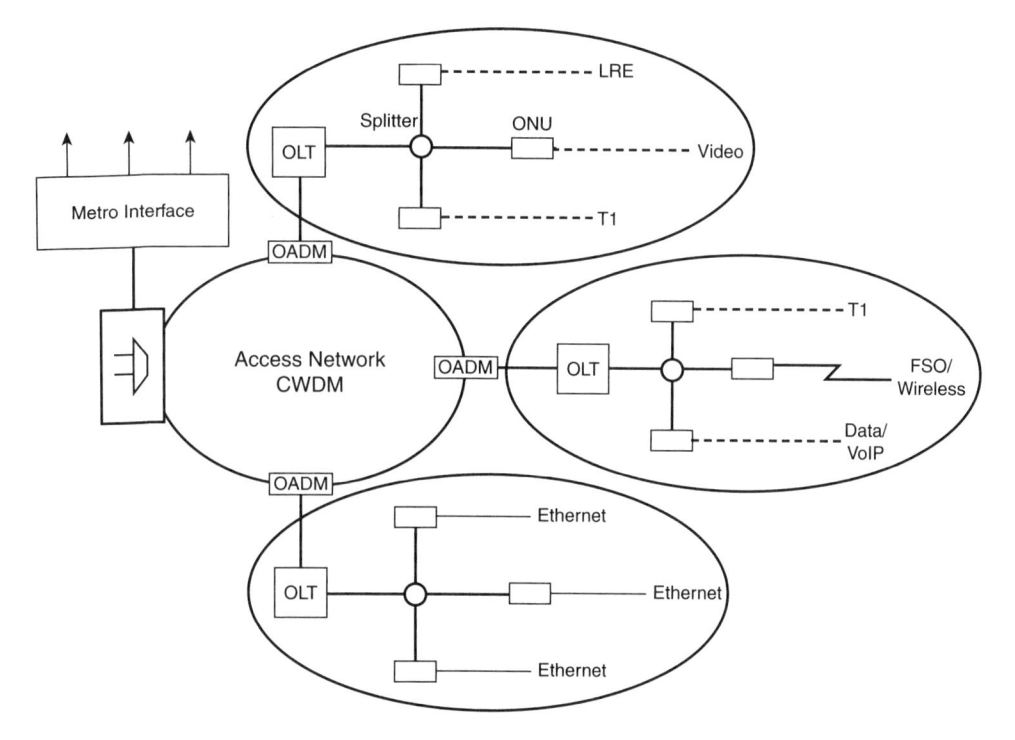

PLC and LRE

PLC is a smart way to provide broadband access using high-speed copper lines originally meant for electrical power transfer. LRE, on the other hand, is used to provide point-to-point Ethernet services between two ends of a last inch solution. Typically, an LRE solution and a PLC solution go hand in hand. Consider, for example, a high-rise building that is to be given broadband access. Without laying new network media, the best way to do so is to use the existing infrastructure. In this scenario, DSL is one way to meet the demands. However, DSL suffers from speed limitations. Therefore, PLC and LRE are clubbed together. From the base of the building to each floor, we set up an LRE network. The LRE network is limited primarily by distance and secondarily by the product of distance and bandwidth. Therefore, an all-LRE solution is not a viable option to reach to each cubicle in the office or every room in a house. So at each floor, the LRE interface is connected to a PLC interface. The PLC interface is able to provide bandwidth to each end user (on workstations or elsewhere) on each floor. As you can understand, there is a need for a strong matching interface between the LRE and the PLC endpoints. Because the interface should support Ethernet frame transfers, a switch is the best alternative. On the other hand, we can also use Ethernet interfaces for connecting LRE to PLC.

EPON and WiFi

EPON is the technology of choice for broadband access to the curb. PON means a low-cost network, and Ethernet signifies universal protocol acceptance and availability. However, having said that, the main issue with PON is that PON is comparatively more expensive than DSL (although, at the same time, PON provides much more capacity than DSL). This is a motivation to distribute the massive capacity of the PON network at the curb site among multiple end users. There are numerous ways to do so, among which WiFi represents one possible method. The seamless integration for EPON and WiFi is possible because both adhere to the Ethernet protocol. Therefore, an interface between EPON and WiFi is not limited by protocol. However, it is limited by transmission characteristics and the variable-rate nature of WiFi (backoff and so on). As we have discussed previously in this book, WiFi transmission can be rolled back in case a channel becomes noisy. This means that the EPON optical network unit (ONU) must also be told of the rollback. It is the responsibility of the interface to inform the EPON ONU and thereby the OLT of the rollback. Apart from this, the other important function that the interface needs to carry out is traffic engineering for video transmission over WiFi. This is a nascent field for research and is expected to grow along with the improvements in technology.

EPON and ADSL

The choice of method for broadband is EPON, which guarantees high-speed access to the curb site. From the curb site to the end user, there are many ways to transmit. One of the low-cost approaches is to use the existing Public Switched Telephone Network (PSTN). The method uses one of the variants of DSL over the telephone network. This is a relatively popular method on account of the low cost and ease of deployment. The variant of DSL most suited for this approach is asymmetric DSL (ADSL), whereby the upstream and downstream rates are different, and are so to adhere to the access characteristic of the end users. An EPON system gives broadband access to a curb site through an ONU. From the ONU, the high-bandwidth data needs to be coupled onto the public network leading to the ADSL. Note that ADSL is a time-division multiplexing (TDM) protocol, whereas EPON means Ethernet, which is statistical TDM. This means that the interface needs a fair amount of intelligence to do traffic shaping in addition to the task of tunneling virtual connections and ensuring quality of service (QoS) to the end user.

BPON and PLC

At the time of this writing, there is heavy debate about which variant of PON will be commercially acceptable. Although there may be many contradictory statements from carriers and providers, we maintain our vendor neutrality and do not talk about which technology is more commercially viable. Of course, we agree that the Ethernet version is more effective in terms of both an economic and technological solution in the first mile access, and we symbolically present EPON as a good choice. However, we also consider BPON, the other TDM version of PON, which has grown from conventional models of ATM and has now digressed to provide gigabit-capable bandwidth with its new version—namely, GPON. Coming back to BPON, one of the interfacing needs of BPON is when BPON terminates at a curb site through an ONU. This ONU then feeds to a PLC network. Note that a PLC network is quite indifferent to the protocol, and TDM is also welcome. However, for a PLC network, considering its passive access medium, the best protocol is Ethernet. Therefore, the interface between the PLC and the BPON has to cater to the medium requirements of PLC. A PLC network is restricted by bandwidth anomalies and losses, which are random, and there is generally a need for retransmission at the interface.

Similarly, on the upstream side a scheduler is required for successful operation. Here we have to make a successful transition from Ethernet to TDM. Again this means scheduling and QoS issues.

BPON and DSL

BPON and DSL are two technologies that can be seamlessly integrated. Both are TDM based. BPON feeds directly to a DSL network through a DSL access multiplexer (DSLAM) architecture as described in Chapter 7, "DSL Technologies." Here the advantage is of a simplified interface (and therefore management). Because both technologies are TDM based, the management becomes very simple. TDM management itself is expensive, but the uniformity in the solution is a distinct advantage.

X to FSO

In a business complex, we have ONU in some buildings, but the other buildings do not have fiber connections. However, the other buildings are obviously static and, therefore, an FSO solution is possible for these. The interface requirements are fairly less stringent, apart from the fact that there may be a need for retransmission if the channel quality fades, a phenomenon that occurs randomly.

Multimode Fiber

The future of first mile networking is extremely bright. Technologically, we can forecast a direct movement from copper to fiber and wireless. Where single-mode fiber (SMF) deployment is expensive, there may be increased deployment of multimode fiber (MMF). MMF, by principle, allows the transmission of multiple modes of light. Modes are solutions of Maxwell's equation for transmission in an optical fiber. Multiple modes means more attenuation per kilometer. However, the relatively shorter distances seen in the first mile are good motivators for the deployment of MMF. MMF is cheaper than SMF. It is more tolerant to mechanical damage; therefore, fiber-laying strategies are much more relaxed in this case. MMF can easily be graduated from existing LANs to future first mile networks.

Management of Multiple Technologies in the Network

Due to overwhelming usage of multiple technologies and the wide array of services offered in the last mile, service provisioning and management is very important for successful implementation of these solutions. Seamless provisioning systems that interface with all the previously described technologies are essential. Management systems should be able to seamlessly support initial customer contact, turn up of the services, customer service, and billing systems (all of which should be transparent to the customer). The most important issue facing broadband service providers today is their lack of customer service and support. The management system designs typically involve an integrator to integrate multiple network elements into a common console that can be used for reporting, resource allocation, provisioning, and customer support.

Figure 11-3 shows a diagram of a typical management system. The management systems should also support service multiplexing to support multiple services on the same port.

Figure 11-3 *First Mile Network Management*

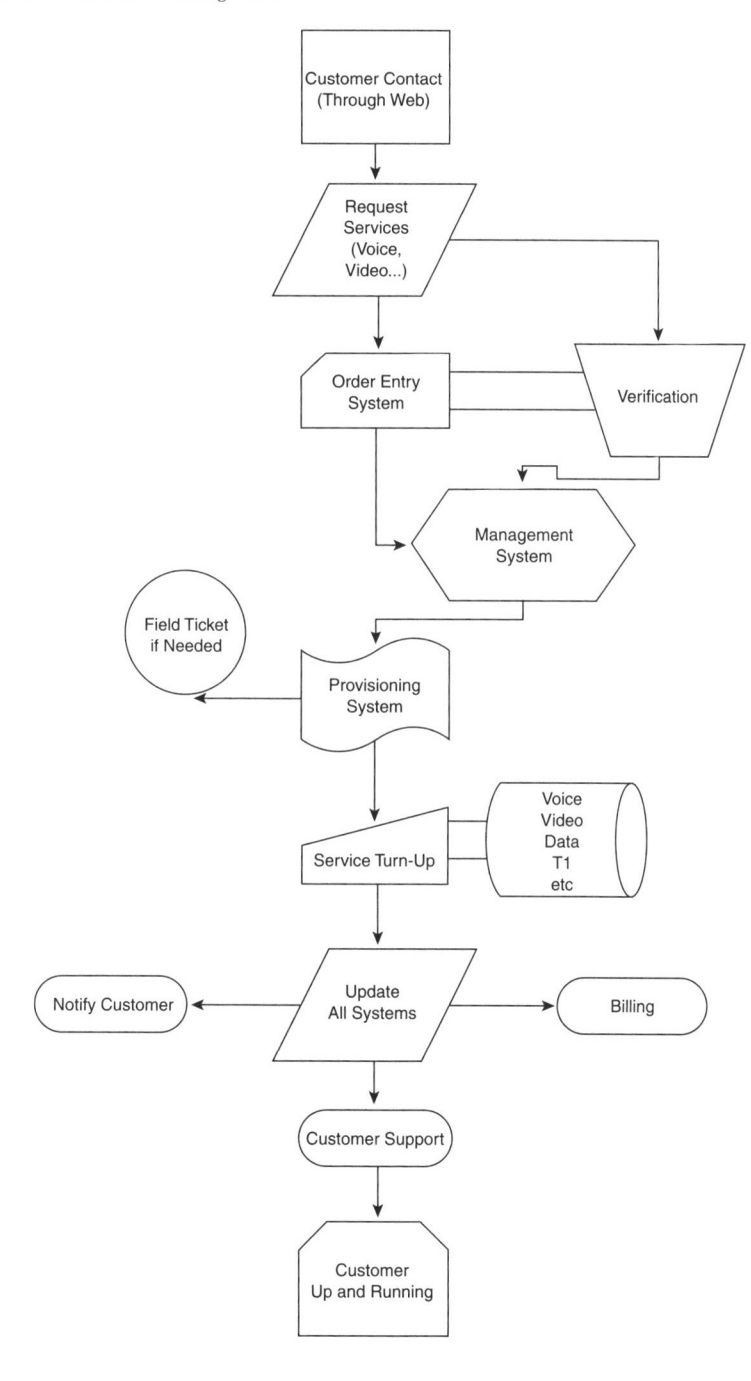

Billing systems are typically based on flat rates or are usage based. It can be very challenging to set up the billing system based on usage, and this system is not recommended. The usage-based systems are considered in packet-based networks, but experience has proven the difficulty of keeping track of usage (making such systems not really worth the effort). The billing systems should be able to track the multiple services (voice, video, data) over the same interfaces if necessary. The idea should be to provide a billing system that bills end users based on content delivery.

Billing Issues

A first mile network is a primary revenue generator for service providers, because the first mile network directly interacts with the consumers. Because there is a plethora of technologies in the first mile, there are multiple solutions that need to be considered for actual billing. The multiple technology suite means that the cost per bit is not fixed and depends on factors such as QoS, delay, the network user location, and so on. For example, a user who is mobile needs to be charged more than a user who is not. Likewise, a user who is surfing the Internet should not pay the same as a user who is watching a movie on demand. This kind of service-based differentiation is the key to billing the first mile network hierarchy. Billing strategies depend on the management philosophy of the network. Generally, it is a rule of thumb that remote management of the network leads to excellent billing as well. Issues arise when the network is stratified into multiple technology layers. There is no plausible way to find out whether a data packet used a route for which the bill was higher than another packet that was routed through a best-effort scheme. This often leads to underbilling, a phenomenon that causes service providers to lose revenue. If the network is made to obey highly stringent rules, it may overbill. That is, customers end up paying for more than they get. This is generally seen as a bad business practice. A good way to bill is to deploy a good management system and divide the network into domains that control the flow of network data in service-specific ways.

Future Applications: First Mile Networks

First mile networks are poised to be the gateway to the world of broadband. These networks are the future communication gateways to the new world of high security, good health care, and an excellent quality of living. These future networks will be the platforms for providing high-speed applications that will then put the onus on the network infrastructure. Among these, video applications are extremely important but are difficult to implement due to their stringent QoS needs. Teleconferencing is a good substitute for physical traveling and also helps overcome the communication misunderstandings inherent in voice conversations or text messaging and e-mails. The use of first mile networks will not be restricted to a certain section of society but will be promoted to an entire audience that broadly includes the complete citizenry, with the chief objective to bridge the digital divide. The aim is to provide individuals a knowledge base through information technology that they can use to deal with issues that affect every aspect of our lives.

A number of issues related to the first mile, such as regulatory restructuring and other legalities, need a great deal of attention from legislators and lawmakers the world over. Without due support and public interest, it will be difficult to widely lay first mile access networks.

In the future, broadband networking will be a commodity that will be accessible to all. Developed and developing nations will use this technology common denominator to foster strong economic fundamentals and ultimately foster good growth. Broadband networking will change the way we live our lives. To ensure a successful implementation of broadband networks, it is imperative that we build extremely innovative and economically viable first mile access networks.

Summary

In this chapter we covered the interfacing requirements for first mile networks. The chapter discussed multiple popular combinations, approaches, and ways to interface first mile technology. The chapter also touched on billing management functionality required in the first mile. The chapter concluded with a focus on the future of first mile access networks from both a technological and business perspective.

Review Questions

1. Describe the main drawbacks to interfacing the following:

   PLC and LRE

   PON and WiFi

   EPON and DSL

   BPON and WiFi

2. What are the advantages of EPON and LRE as compared to EPON and ADSL?

3. What is the key to a successful billing management system?

4. Research and describe multiple OSS functions that need to be addressed in the first mile.

5. What role do applications have in the future growth of first mile networks?

APPENDIX

Units of Optical Power Measurement: Decibel

The power level in optical fiber communications is too wide ranged to express on a linear scale. A logarithmic scale known as decibel (dB) is used to express power in optical communications. The wide range of power values makes the decibel a convenient unit to express the power levels that are associated with an optical system. The gain of an amplifier or attenuation in fiber is expressed in decibels. The decibel does not give a magnitude of power, but it is a ratio of two powers. See the following equation:

$$(\text{Loss or gain}) \text{ dB} = 10\log_{10}\left(\frac{P_2}{P_1}\right)$$

Example:

Calculate the gain of the amplifier in dB, when 1 watt is applied to the input and 2 watts is measured as the output:

$\text{dB} = 10\log_{10} 2/1 = 3 \text{ dB}$

Measured output is 2W

Gain of this amplifier is 3 dB.

dBm is the power level related to 1 mW.

$$\text{dBm} = 10\log_{10}\frac{power(mW)}{1mW}$$

INDEX

Numerics

2B1Q, 166
802.11b standard
    key operations, 107–108
    performance, 108–109
802.11g standard, 109
802.16 standard, 113
    adaptive modulation schemes, 116
    line-of-sight radio propagation, 131
    MAC CPS, 124
        MAC addressing, 124–126
    modulation schemes, 121–122
    network design, 131–132
    overlapping network solutions, 131–132
    privacy sublayer
        key management protocol, 123
        packet data encryption, 123
        SAs, 123–124
    reference model, 114
        IEs, 119
        MAC CPS, 115
        MAC layer, 115
        physical layer, 115–119
        privacy sublayer, 115, 122–123
        service-specific CS, 115
    scheduling services, 126–127
        GPC mode, 127–128
        GPSS mode, 127–128
        requests, 127
    typical network architecture, 114
    wireless MAN-SC air interface, 113

A

AAL (ATM adaptation layer), 225
absorption loss, 153
access networks, 7, 10
    bottlenecks, causes of, 27
    CCM approach
        market analysis, 262–263
        planning phase, 252, 257–258
    comparing, 22–23
    MANs, 10
    wired, 10
        cable networks, 17–19
        DSL, 11–12
        fiber, 12–17
        PSTN, 19–20

wireless, 20
    fixed wireless, 21
    free space optics, 22
    infrared waves, 22
    WiFi, 22
acquisition techniques, 152
active scanning, WLANs, 108
ad hoc networks, 101–103, 243
adaptive modulation schemes (802.16 standard), 116
addition of holes, 140
ADSL (asymmetrical DSL), 11, 170–171
    interoperability with EPONs, 270–271
    RADSL, 172–173
    splitterless, 171–172
ADSL lite, 172
aerosol absorption coefficient, 153
air as wireless medium, 87–88
amplitude modulation, 9
antennae
    EIRP, 96–97
    half-power beamwidth, 96
    smart antennas, 97
        major lobe, 98
        multipath phase problem, 99
        transmission strategies, 98
antenna theory, 95–96
APD (Avalanche photodetector), 145
APONs, 36
APs (access points)
    ad hoc mode, 103
    infrastructure mode, 102
architectures
    implementing, 222
    of CWDM networks, 72
ARPANET, 3
ARQ (automatic repeat request), 187
association (WLANs), 108
asymmetrical DSL services. *See* ADSL
asynchronous impulsive noise, 186
ATM (Asynchronous Transfer Mode), 223
    implementation costs, 227–228
    in core networks, 224–225
    PONs, 16
    versus MPLS in core networks, 223, 226–227
ATM forum, 224
attenuation, 14, 87
    of PLC channels, 185–186
attenuation coefficient, 153
attributes of CBR, 219
ATU-C (ADSL terminal unit-central office), 170
ATU-R (ADSL terminal unit-remote), 170

auto-tracking, 152
Avalanche breakdown, 146
Avalanche multiplication, 145
Avalanche photodiodes, 145–146

B

backoff, 128
bandgap energy, 140
bandwidth management
    attenuation, 14
    impact on first mile access, 5–6
        resource provisioning, 6–7
    intelligent city networks, 213–214
    last inch requirements, 240
    maximizing with WDM, 59
    on access networks, 7, 10
        cable networks, 17–19
        DSL, 11–12
        fiber, 12–17
        MANs, 10
        PSTNs, 19–20
        wired networks, 10
        wireless, 20–22
    protection bandwidth, 68
bandwidth on demand, 4
BE (Best-Effort) service, 127
beacon frames, 108
beam spreading, 154
BER (bit error rate), 62, 148–150
best-effort service, 216
BFSK (Binary Frequency Shift Keying), 93
billing, 274
BLSR (bidirectional line-switched ring), 69
BoD (bandwidth on demand), 204
bottlenecks
    causes of in access networks, 27
    commercial solutions to, 27–28
BPONs (broadband PONs), 16
    advantages of, 37
    cells, 36
    CTS, 36
    data rates, 36
    interoperability with DSL and PLC, 271
BPSK (Binary Phase Shift Keying), 103
BRI (bit-rate interface), 166
broadband services, 7
    DSL, 161–162
        ADSL, 170–172
        asymmetrical services, 164–165

        CAP, 176
        DMT, 175–176
        DSLAMs, 163
        HDSL, 166–167
        HDSL2, 169–170
        IDSL, 165–166
        incremental services, 179
        limitations of, 162
        poor customer service, 177
        RADSL, 172–173
        service provisioning, 178
        SDSL, 168
        symmetrical services, 164
        typical xDSL network, 176
        USP, 162
        VDSL, 173–174
BS (base station), backoff, 128
BSS (basic service set), 102
business case for first mile networks, 247. *See also*
  technology case
    CCM business model, 249–250
        first mile planning process, 250–252,
            257–258
        pricing versus CDM model, 259–262
    CDM business model, 248–249
        Classical Terminal Assignment
            problem, 249
        incremental services, 249
    data services, 238
    intelligent cities, 203–207
        backbone, 207–208
        bandwidth management, 213–214
        BoD, 204
        economic implications, 209
        managing, 210, 212–214
        optical layer, 208
    video services, 238
        telemedicine, 240
        video distribution, 239
        videoconferencing, 239
        visualization, 240
    voice services, 236–237

C

cable networks, 17–19
calculating
    BER, 148–150
    EIRP, 97
    Mie scattering, 153

optical power of optical splitter output, 32
power budget, 152–153
SNR, 150–151
total loss, 152–153
CAP (Carrierless Amplitude/Phase) modulation, 176
CAPEX (capital expenditure), 248
case studies
    CCM business model, 249–250
        first mile planning process, 250–252, 257–258
        pricing versus CDM model, 259–262
    CDM business model, 248–249
        Classical Terminal Assignment problem, 249
        incremental services, 249
    of PLC
        direct access point, 191–192
        interface point, 191
        management issues, 189
        repeaters, 191
        SNMP-based management, 189
categories of last mile network end users, 241
CATV (Community Antenna Television), 17
    HFC, 18
    HFC plant, 198–199
CBR (constraint-based routing), 219
CCK (Complementary Code Keying), 104–105
CCM (cumulative comprehensive model) business model, 249–250
    first mile planning process, 250–252, 257–258
    market analysis of access area, 262–263
    pricing versus CDM model, 259–262
CDM (conventional distributed model) business model, 248–249
    Classical Terminal Assignment problem, 249
    incremental services, 249
    pricing versus CCM model, 259–262
CDMA, 90–91
    DSSS, 92–93
    FHSS, 91–94
    handoff, 91
cellular networks, WCDMA, 20
channel characteristics of PLC, 185
    attenuation, 185–186
    impedance, 185
channel spacing in WDM systems, 63
characteristics of transponders, 74
characterizing OADMs in CWDM networks, 72–74
churns, 214
circuit-based voice systems, 236
Classical Terminal Assignment problem, 249
classifying PONs, 36

APONs, 36
BPONs
    advantages of, 37
    cells, 36
    data rates, 36
    EPONs, 39–40, 42
    GPONs, 38
        design characteristics, 39
        GFP, 38
        GSR, 38
CMIP (common management information protocol), 211
collision avoidance, 100
collisions
    in 802.16 networks, contention resolution, 128
    upstream communication in PON networks, 32–33
        STDM, 34–35
        TDM, 33
colored background noise, 186
commercial enterprise solutions to access network bottlenecks, 27–28
common part sublayer (IEEE 802.16 MAC layer), 124
    MAC addressing, 124–126
communication infrastructure, prohibitive development costs, 183
community networks, 233. *See also* intelligent city networks
    data services, 238
    data-centric applications, 236
    economic viability, 235
    Ethernet solution, 244
    healthcare applications, 235
    security needs, 234–235
    TDM solution, 244
    telemedicine, 240
    video services, 238–239
    visualization, 240
    voice services, 236–237
comparing
    first mile access technologies, 22–23
    performance of PON schemes, 53–54
    PLC and alternative first-mile solutions, 195
    pricing methods, CCM versus CDM business models, 259–262
concave lenses, 151
conduction state, 140
congestion, 217
contention resolution, IEEE 802.16, 128
contention transmission, SSs, 128
convex lenses, 151

core networks
    ATM, 224–225
        implementation costs, 227–228
        versus MPLS, 223, 226–227
    bandwidth management, 5
        light-trials, 6
    first mile access problems, 4
    MPLS, 225–226
CQ (custom queuing), 218
creating business case for first mile networks, 247
CSMA (carrier sense multiple access), 100, 187–188
CTS (Common Technical Specification), 36
CuPLUS, 189
CWDM (coarse wavelength division multiplexing), 14, 60
    benefits of, 67
    channel spacing, 60
    deployment standards, 66–67
    Ethernet protection
        with EtherChannel and UDLD, 71
        with STP and UDLD, 71
    interoperability with PONs, 268–269
    lightpath protection, 68
        line-switched protection, 69–70
        path-switched 1+1 protection, 69
    metro access solutions, 61
    network architecture, 72
    network design rules, 74
        examples, 76–82
        power-budget calculations, 75–76
    network elements, 63
        thin film filters, 65–66
        VCESLs, 64–65
    OADMs, characterization, 72–74
    protection bandwidth, 68
    relaxed component manufacturing tolerance, 63
    ring topology, 62
    services, 67–68
    versus DWDM, 63

D

data- oriented communication, QoS architecture, 222–223
data rates
    of BPONs, 36
    of STP, 9
data services in first mile business case, 238
data-centric applications in community network, 236
dB (decibel), 277
DBR (distributed Bragg reflector) lasers, 142
delay spread, 100
deploying
    CWDM, standards, 66–67
    DSL, low cost of, 179
    FSO networks, 156
    PLC, telco requirements, 192–193
designing
    802.16 networks, 131–132
    CWDM networks, 74
        examples, 76–82
        power-budget calculations, 75–76
devices
    CWDM, 63
        thin film filters, 65–66
        VCESLs, 64–65
    DSLAMs, 163
    lenses, 151
    modems, 9
    NF, 148
    optical receivers, 146–148
        shot noise, 147
        thermal noise, 147
    PCMCIA cards, 107
    transponders, 62
        characteristics of, 74
DFB (distributed fiber Bragg) lasers, 141
DiffServ, 219–220
digital divide, 7
direct access point in PLC first-mile networks, 191–192
disruptive technologies
    EPONs, 40, 42
    locating market niches for, 41
distance limitations of SDSL, 168
distributed feedback, 141–142
Distributed IFS, 106
DL-MAP message format, 118
DMT (Discrete Multi-Tone), 175–176
DOCSIS specification, 122
Dostert, Klaus, 185
downlink subframes (IEEE 802.16), 119–121
downstream communication in PONs, 30
DSL (Digital Subscriber Line), 11–12, 162
    ADSL, 170–171
        RADSL, 172–173
        splitterless, 171–172
    asymmetrical services, 164–165
    cost of deployment, 179
    DSLAMs, 163
    HDSL, 166–167
    HDSL2, 169–170

IDSL (ISDN-DSL), 165–166
incremental services, 179
interoperability with BPONs, 271
limitations of, 162
line encoding standards
    CAP, 176
    DMT, 175–176
LRE
    multiple-dwelling deployments, 174
    self-synchronizing scrambler mechanism, 174
poor customer service, 177
service provisioning, 178
symmetrical services, 164
    SDSL, 168
typical xDSL network, 176
USP, 162
VDSL, 173
    LRE, 174
    symmetrical DSL, 173–174
versus dialup, 179
versus PLC, 198
DSLAMs, 163
DSSS, 92–93
duplexing schemes in IEEE 802.16 standard, FDD, 117–119
DWDM (dense wavelength division multiplexing), versus CWDM, 63
dynamic rate shifting in 802.11b, 105

E

economic implications of intelligent city networks, 209
economic viability of community networks, 235
EFM (Ethernet in the First Mile), 42–43
    IPACT, 48
    P2MP topology, 43–44
    TUR, 44–46
        measuring fairness, 52
        numeric evaluation of, 49–51
        system operation, 47–48
EIRP (effective isotropic radiated power), 96–97
electrical power infrastructure, data communication, 184
electron-hole pair multiplication, Avalanche multiplication, 145–146
EMS (equipment/element management system), 210
encryption, SSL, 238

end-to-end path availability of circuit-based voice systems, 236
end-user bandwidth management
    access networks, 7, 10
        cable networks, 17–19
        DSL, 11–12
        fiber, 12–17
        MANs, 10
        wired networks, 10
    digital divide, 7
    resource provisioning, 6–7
EPONs (Ethernet PONs), 17, 39–40, 42
    EFM, P2MP topology, 43–44
    interoperability
        with ADSL, 270–271
        with WiFi, 270
    IPACT, 48
    PLC-based, fiber to the curb, 194–195
    standardization, 42–43
    TUR, 44–46
        measuring fairness, 52
        numeric evaluation of, 49–51
        system operation, 47–48
ESS (extended service set), 102
Ethernet, 244
    LRE, 174
        multiple-dwelling deployments, 174
        self-synchronizing scrambler mechanism, 174
    protecting against failure in CWDM networks
        with EtherChannel and UDLD, 71
        with STP and UDLD, 71
evolution of 802.11 standards, 100
    802.11b
        ad hoc method, 103
        dynamic rate shifting, 105
        infrastructure method, 102
        management frames, 106
        physical layer, 103–106
    mobile IP, 101–102
examples of CWDM network design, 76–82
Extended IFS, 106
EZ-DSL, 173

F

fade depth, 131
fading, 100
fast hopping, 91

FDD (frequency-division duplexing), 117–119
    downlink subframe structure, 120
FDM (frequency-division multiplexing), 89–90
features of PONs, 28
FHSS (frequency hopping spread spectrum), 91–94
fiber cut, protecting against, 68–69
fiber networks, 12–13
    attenuation, 14
    bandwidth, WDM, 59
    narrow-aperture lasers, 13
    PONs, 15–17
        ATM PONs, 16
        BPONs, 16
        EPONs, 17
        features, 15–16
        GPONs, 16
    WDM, 3, 13–14
        CWDM, 14
        Light trails, 6
fiber to the curb, PLC-EPON networks, 194–195
fiber to the home, prohibitive cost of, 194
FIFO queueing, 218
first mile access problem, 4, 9
fixed wireless networks, 21
flow control, 217
Fraunhofer theory of diffraction, 95
free space optics, 22
frequency modulation, 9
FSK (frequency shift keying), 186–187
FSO systems, 138
    deploying, 156
    factors affecting
        beam spreading, 154
        Mie scattering, 153–154
        Rayleigh scattering, 154
        weather conditions, 154–155
    lasers, 139–141
        DBR, 142
        DFB, 141–142
        safety, 155
    lenses, 151
    receivers
        photodetectors, 142–145
        optical receivers, 146–148
    tracking and acquisition techniques, 152
FTTU (fiber to the end user) through PON, 242
full duplex transmission, 12
future applications, 274–275

G-H

G.Lite, 172
gain
    antenna theory, 95–96
    dBs, 277
    processing gain of SS systems, 95
geometric path loss, 138
GFP (Generic Framing Protocol), 16, 38
GMPLS (Generalized Multiprotocol Label Switching), 211
GPC (Grant per Connection) mode, 127–128
GPONs (Gigabit PONs), 16, 38
    design characteristics, 39
    GFP, 38
    GSR, 38
GPSS (Grant per Subscriber Station) mode, 127–128
GSR (gigabit service requirement), 38

half-power beamwidth, 96
handoff, 91
HDSL (high-data-rate DSL), 11, 166–167
HDSL2 (next-generation HDSL), 168–170
healthcare, broadband network applications, 235
HFC (hybrid fiber coax)
    networks, 18
    plants, 198–199
home networking, PLC solutions, 198
HomePlug 1.0 standard, 195–197

I

IDSL (ISDN-DSL), 165–166
IEEE 802.11b, 103
    dynamic rate shifting, 105
    evolution of, 100
        Mobile IP, 101–102
    infrastructure mode, 102
    key operations, 107–108
    physical layer, 105
        CCK, 104–105
        MAC sublayer, 106
        management frames, 106
        QPSK, 103–104
IEEE 802.16
    adaptive modulation schemes, 116
    contention resolution, 128
    line-of-sight radio propagation, 131

MAC CPS, 124
    MAC addressing, 124–126
modulation schemes, 121–122
network design, 131–132
overlapping network solutions, 131–132
privacy sublayer
    key management protocol, 123
    packet data encryption, 123
    SAs, 123–124
reference model, 114
    IEs, 119
    MAC CPS, 115
    MAC layer, 115
    physical layer, 115–119
    privacy sublayer, 115, 122–123
    service-specific CS, 115
scheduling services, 126–127
    GPC mode, 127–128
    GPSS mode, 127–128
    requests, 127
typical network architecture, 114
wireless MAN-SC air interface, 113
IEEE 802.1Q VLANs, 221–222
IEs (information elements), fields, 119
IFS (Interframe space), 106
impedance of PLC channels, 185
implementing
    MPLS and ATM, cost of, 227–228
    network architectures, 222
incremental services, 249
indoor power-line networks, 184–185
infrared waves, 22
initialization of SSs, 128–130
intelligent city networks, 203–207
    backbone, 207–208
    bandwidth management, 213–214
    BoD, 204
    economic implications, 209
    managing, 210, 212
        bandwidth management, 213–214
        mobility management, 212–213
    optical layer, 208
interface point of PLC first-mile networks, 191
interoperability
    of BPONs and DSL, 271
    of CWDM and PON, 268–269
    of EPONs and ADSL, 270–271
    of EPONs and WiFi, 270
    PLC and LRE, 269–270
interwavelength distance, 60
IntServ (Integrated Services), 220–221

IPACT (Interleaved Polling with Adaptive Cycle Time), 48
    comparing performance with other PON schemes, 53–54
ISM (Industrial, Scientific, and Medical) band, 102
ITU G.694.2 recommendation, 66

J-K

jitter, 216

key management protocol (IEEE 802.16 privacy sublayer), 123

L

label stacking, 225–226
labels, 225
lasers, 139–141
    DBR, 142
    DFB, 141–142
    FSO, safety, 155
    narrow-aperture, 13
    VCSELs, 64
        line width, 65
lasing threshold, 141
last inch
    bandwidth requirements, 240
    wireless solution, 243
Legacy TDM networks, 216
legislation, Telecommunications Act of 1996, 7
lenses, 151
Light trails, 6
lightpath protection of CWDM networks, 68
    line-switched protection, 69–70
    path-switched 1+1 protection, 69
limitations of DSL, 162
line encoding standards
    CAP, 176
    DMT, 175–176
line width, 65
line-of-sight radio propagation, 802.16, 131
line-switched protection, 69–70
link-margin analysis, 152–153
load balancing, 217
LRE (long-reach Ethernet), 42–43
    interoperability with PLC, 269–270
    multiple-dwelling deployments, 174
    self-synchronizing scrambler mechanism, 174

M

MAC (media access control)
  addressing, IEEE 802.16, 124–126
  HomePlug 1.0, 196–197
  in PLC, CSMA, 187–188
MAC CPS (IEEE 802.16 standard), 115
MAC layer (IEEE 802.16 standard), 115
MAC sublayer (802.11b physical layer), 106
management frames (802.11b), 106
management systems, impact on billing, 274
managing
  intelligent city networks, 210, 212
    bandwidth management, 213–214
    mobility management, 212–213
  multiple technologies in networks, 272–274
  PLC networks, 189
MANs (metropolitan-area networks), 10,
  See also WirelessMAN standard
  intelligent city networks, 207
market analysis
  of CCM access area, 262–263
  of PLC, 197–198
market niches for new technologies, 41
maximizing optical fiber bandwidth, 60
mean center wavelength, 73
measuring
  fairness of TUR in PONs, 52
  receiver performance
    BER, 148–150
    SNR, 150–151
Mie scattering, 153–154
minislots, 119
mismatch of electronic technology and optical bandwidth, 59
MMF (multimode fiber), 272
Mobile IP, 101–102
modems, 9
modes, 272
modulation schemes, 9
  CAP, 176
  for IEEE 802.16 standard, 121–122
    physical layer, 116
  FSK, 186–187
  OFDM, 187–188
  PLC requirements, 186
  QAM, 170
    in LRE implementations, 174
  QPSK, 103–104
  SS, 187
MPCP (Multipoint Control Protocol), 44

MPLS (Multiprotocol Label Switching), 4, 211
  implementation costs, 227–228
  in core networks, 225–226
  traffic queueing, 218
  versus ATM in core networks, 223, 226–227
multiple technologies
  billing, 274
  managing, 272–274
multiple-dwelling deployments, LRE, 174
multiple-tier networks, 5
multiplexing
  CDMA, 90–91
    DSSS, 92–93
    FHSS, 91–94
  CWDM, 60
    benefits of, 67
    channel spacing, 60
    deployment standards, 66–67
    lightpath protection, 68–70
    metro access solutions, 61
    network architecture, 72
    network elements, 63
    OADMs, characterization, 72–74
    protection bandwidth, 68
    ring topology, 62
    services, 67–68
    thin film filters, 65–66
    VCESLs, 64–65
  FDM, 89–90
  multiple frequencies on power lines, 186
  OFDM, 188
  TDM, 88–89
  WDM, 63
multiplicative factor, 146
municipality-owned networks. *See* intelligent city networks

N

narrow-aperture lasers, 13
narrowband noise, 186
network management, 272–274
NF (noise figure), 148
nodal failure, protecting against in CWDM networks, 68
noise in PLC channels, 186
NP (nondeterministic polynomial time) complete, 6
nrtPS (non-real-time polling service), 127
NtPLUS, 190

O

OADMs (optical add/drop multiplexers), 62
OFDM (orthogonal FDM), 187–188
OLTUs (optical line terminal units), 17, 29–30
ONU (optical network unit), 17, 29–30
optical communications, units of power measurements, 277
optical fiber networks, 12–13
optical layer (intelligent city networks), 208
optical loss, 153
optical networks
    attenuation, 14
    CWDM, 60
        benefits of, 67
        channel spacing, 60
        deployment standards, 66–67
        design rules, 74–76
        Ethernet protection, 71
        lightpath protection, 68–70
        metro access solutions, 61
        network architecture, 72
        network design examples, 76–82
        network elements, 63
        OADMs, characterization, 72–74
        protection bandwidth, 68
        ring topology, 62
        services, 67–68
        thin film filters, 65–66
        VCESLs, 64–65
    lasers, 139–141
        DBR, 142
        DFB, 141–142
    modes, 272
    narrow-aperture lasers, 13
    PONs, 15–17
        APONs, 16, 36
        BPONs, 16, 36–37
        calculating optical power of optical splitter output, 32
        EPONs, 17, 39–40, 42–52
        features, 15–16, 28
        GPONs, 16, 38–39
        topology, 29–31
        upstream communication, 32–35
    third generation, 208
    WDM, 13–14
        interwavelength distance, 60
optical receivers, 146–148
    shot noise, 147
    thermal noise, 147
overlapping 802.16 network solutions, 131–132

P

P2MP topology, EPONs, 43–44
packet-based networks versus TDM networks, 211
PAgP (Port Aggregation Protocol), 71
paradigm shift in telecommunications business model, causes of, 233
    data services, 238
    data-centric applications, 236
    economic viability, 235
    health care applications, 235
    security needs, 234–235
    telemedicine, 240
    video services, 238–239
    visualization, 240
    voice services, 236–237
pass-band, 73
path-switched 1+1 protection, 69
PCM (pulse code modulation), 237
PCMCIA cards for WLANs, 107
peak wavelength, 73
performance
    BER, 62, 148–150
    comparitive evaluation of PON schemes, 53–54
    of 802.11b networks, 108–109
    SNR, 150–151
performance oversupply, 40
per-hop behavior, 220
periodic impulse noise, 186
phase modulation, 9
phase shift keying, QPSK, 103–104
photocurrent, 143
photodetectors, 142–143
    PIN, 143–145
photodiodes, 143
    Avalanche, 145–146
    PIN, 143–145
photonic winter, 205
physical layer
    of IEEE 802.11b standard, 105
        CCK, 104–105
        MAC sublayer, 106
        management frames, 106
        QPSK, 103–104
    of IEEE 802.16 standard, 115–116
        FDD, 117–119
        FDD downlink subframe structure, 120
        TDD downlink subframe structure, 120
        modulation schemes, 116
    of HomePlug 1.0, 197
PIN photodetectors, 143–145
PIN photodiodes, 143–145

planning first mile, CCM approach, 250–252, 257–258
PLC (power-line communication), 4, 183
    adapting to telco requirements, 192–193
    and alternative first-mile solutions, 195
    case study
        direct access point, 191–192
        interface point, 191
        management issues, 189
        repeaters, 191
        SNMP-based management, 189
    channel characteristics, 185
        attenuation, 185–186
        impedance, 185
    EPON solutions, fiber to the curb, 194–195
    HomePlug 1.0 standard, 195–196
    indoor power-line networks, 184–185
    interoperability
        with BPONs, 271
        with LRE, 269–270
    market analysis, 197–198
    modulation
        FSK, 186–187
        OFDM, 187
        SS, 187
    modulation requirements, 186
    multiplexing, 186
        OFDM, 188
    noise, 186
    signal processing, 196
    standardization, HomePlug 1.0, 196–197
    standards and frequency regulations, 189
    versus DSL, 198
PLCP (Physical Layer Convergence Protocol), 105
PLUSmms, 190
PMD (physical medium dependent), 105
point-coordination IFS, 106
PONs, 15–17
    APONs, 16, 36
    BPONs, 16
        advantages of, 37
        cells, 36
        data rates, 36
    calculating optical power of optical splitter output, 32
    comparing performance of PON schemes, 53–54
    EPONs, 17, 39–40, 42
        IPACT, 48
        P2MP topology, 43–44
        standardization, 42–43
        TUR, 44–52

    features of, 15–16, 28
    FTTU, 242
    GPONs, 16
        design characteristics, 39
        GFP, 38
        GSR, 38
    interoperability with CWDM, 268–269
    topologies, 29
        OLT, 29–30
        ONU, 29–30
        star topology, 31
    upstream communication, 32
        prohibitive cost of, 33
        STDM, 34–35
        TDM, 33
population inversion, 140
POTS splitters in ADSL implementations, 171
power measurements, 277
power-budget calculations, 75–76, 152–153
    maximum distance in point-to-point systems, 75–76
    maximum distance in ring systems, 76
PQ (priority queuing), 218
preamplifiers, 146
pricing methods, CCM versus CDM business model, 259–262
privacy sublayer (IEEE 802.16 standard), 115, 122–123
    key management protocol, 123
    packet data encryption, 123
    processing gain, 95
    SAs, 123–124
prohibitive cost of BPON deployment, 37
protecting CWDM networks from failure
    dedicated protection, 70
    equipment failure, 68
    fiber failure, 68–69
    line-switched protection, 69–70
    N protection, 70
    path-switched 1+1 protection, 69
    shared protection, 70
    with EtherChannel and UDLD, 71
    with STP and UDLD, 71
protection bandwidth, 68
PSTNs (Public Switched Telephone Networks), 19–20
    STP-based, 9
    universal connectivity, 241

Q

Q factor, 63
QAM (Quadrature Amplitude Modulation), 170
   in LRE implementations, 174
QoS (quality of service), 216–217
   architecutures
      for data-oriented communication, 222–223
      for voice-oriented communication, 222
   DiffServ, 219–220
   in small-scale networks, 216
   IntServ, 220–221
   service classes, 217–218
QPSK (Quadrature Phase Shift Keying), 11, 103–104
queuing traffic with MPLS, 218

R

RADSL (rate-adative DSL), 172
rain-fade factor, 131
Raleigh fading, 100
ranging, 36
Rayleigh scattering, 154
RBOCs (regional Bell operating companies), telco deployment requirements, 192–193
receivers
   DSSS, 93
   measuring performance
      BER, 148–150
      SNR, 150–151
   optical, 146–148
   photodetectors, 142–143
      PIN, 143–145
   shot noise, 147
   thermal noise, 147
Reciprocity theorem, 95
reference protocol model for IEEE 802.16
   standard, 114
      IEs, 119
      MAC CPS, 115
      MAC layer, 115
         CPS, 124–126
            privacy sublayer, 122–124
      physical layer, 115–116
         duplexing schemes, 116–119
         modulation schemes, 116
      privacy sublayer, 115
      service-specific CS, 115

reflective stack (thin film filters), 66
relaxed manufacturing tolerance of CWDM components, 63
repeaters in PLC first-mile networks, 191
requests, 127
residential access of HDSL, 167
resource allocation, 217
resource provisioning, 6–7
revenue matrix for service-based pricing, 259–260
ring topology, CWDM networks, 62
roaming, 108
RpPLUS, 189
RSpec, 220
RSTP, CWDM Ethernet protection, 71
RSVP, (Resource Reservation Protocol), 221
rtPS (Real-time Polling Service), 127

S

S/W (share to work) ratio, 70
safety for FSO networks, 155
SAR (segmentation and reassembly) sublayer (ATM), 225
SAs (IEEE 802.16 privacy sublayer), 123–124
scattering
   Mie scattering, 153–154
   Rayleigh scattering, 154
scheduling services, IEEE 802.16, 126–127
   GPC mode, 127–128
   GPSS mode, 127–128
   requests, 127
SDSL (symmetrical DSL), 168
security, broadband network requirements, 234–235
self-synchronizing scrambler mechanism in LRE, 174
semiconductor-based photodetectors, 143
service categories (ATM), 224
service classes, 217–218
service provisioning of DSL, 178
service-specific CS (IEEE 802.16), 115
Shannon's theory, 94
shared protection, 70
Short IFS, 106
shot noise, 147
signal processing of PLC, 196
signaling, 214
single-pair DSL, 168
single-to-multipoint splitters, 30
SLAs (service-level agreements), 217
slow-hopping, 91

small-scale networks, QoS, 216
smart antennas, 97
    major lobe, 98
    multipath phase problem, 99
    transmission strategies, 98
SNMP (Simple Network Management Protocol), PLC management, 189
SNR (signal-to-noise ratio), 150–151
    of PLC channels, 185
soft switches, 237
spacers (thin film filters), 66
spacing, 60
    in WDM systems, 63
splitterless ADSL, 171–172
    RADSL, 172–173
splitters, single-to-multipoint, 30
spontaneous emission, 140
SS (spread spectrum) systems, 187
    FHSS, 93–94
    processing gain, 95
SSL (Secure Sockets Layer), 238
SSs (subscriber stations)
    contention transmission, 128
    initialization, 128–130
standards
    EFM, 42–43
        IPACT, 48
        P2MP topology, 43–44
        TUR, 44–52
    for CWDM deployment, 66–67
    IEEE 802.16, 113
        contention resolution, 128
        line-of-sight radio propagation, 131
        MAC CPS, 124–126
        modulation schemes, 121–122
        network design considerations, 131–132
        overlapping network solutions, 131–132
        privacy sublayer, 122–124
        reference model, 114–119
        scheduling services, 126–128
        SS initialization, 128–130
        typical network architecture, 114
        wirelessMAN-SC air interface, 113
    PLC, 189
        HomePlug 1.0, 195–197
star topology of PONs, 31–32
start point indicators, 45
STDM, 34
    implementing in PONs, 34–35
    synchronization, 35
stimulated emission, 140

STP (shielded twisted-pair), 9
    data transfer rates, 9
    Ethernet protection in CWDM networks, 71
subcarriers, 196
subframes (IEEE 802.16), 119–121
subheaders of IEEE 802.16 MAC layer, 126
symmetrical DSL services, 164
    SDSL, 168
    symmetrical VDSL, 173–174
synchronization, STDM, 35

T

TDD, downlink subframe structure, 120
TDM (time-division multiplexing), 12, 88–89, 244
    implementing in PONs, 33
TE (traffic engineering), 217
technology case
    categories of last mile network end users, 241
    Ethernet solutions, 244
    FTTU, 242
    last inch solutions, 243
    TDM solutions, 244
telcos, PLC deployment requirements, 192–193
telecommunications, causes for business model paradigm shift, 233
    data services, 238
    data-centric applications, 236
    economic viability, 235
    healthcare applications, 235
    security needs, 234–235
    telemedicine, 240
    video services, 238–239
    distribution, 239
    visualization, 240
    voice services, 236–237
Telecommunications Act of 1996, 7
teleconferencing, 274
telemedicine, role in telecommunications business model, 240
thermal noise, 147
thin film filters, 65–66
third generation of optical networks, 208
topologies
    of PONs, 29
        EPONs, P2MP, 44
        OLT, 29–30
        ONU, 29–30
        star topology, 31
    ring, CWDM networks, 62

total loss, calculating, 152–153
tracking, 152
transition layer (thin film filters), 66
transponders, 62
    characteristics of, 74
TSpec, 220
TUR (Transmission Upon Reception), 44–46
    comparing performance with other PON schemes, 53–54
    measuring fairness, 52
    numeric evaluation of, 49–51
    system operation, 47–48
typical xDSL network, 176

U

UDLD (unidirectional link detection), Ethernet protection in CWDM networks, 71
UGS (Unsolicited Grant Service), 127
UL-MAP message format, 118–119
underbilling, 274
unified management of intelligent city networks, 214
units of power measurement, 277
universal connectivity of PSTN, 241
uplink subframes (IEEE 802.16), 119–121
upstream communication in PONs, 30, 32
    prohibitive cost of, 33
    STDM, 34–35
    TDM, 33
USP (user selling point) of DSL, 162
UPSR (unidirectional path-switched ring), 69

V

VCESLs (vertical cavity surface emitting lasers, 64–65
VDSL (very-high-data-rate DSL), 173
    LRE, 174
        multiple-dwelling deployments, 174
        self-synchronizing scrambler mechanism, 174
    symmetrical DSL, 173–174
video distribution, 239
video services in first mile business case, 238–239
videoconferencing, 239

visualization, role in telecommunications business model, 240
VLANs, IEEE 802.1Q, 221–222
voice oriented communication, QoS architecture, 222
voice services in first mile business case, 236–237
VoIP, 237

W-Z

WCDMA (wideband code-division multiple access), 20
WDM (wavelength division multiplexing) systems, 3, 13, 59, 63
    CWDM, 14, 60
        benefits of, 67
        channel spacing, 60
        deployment standards, 66–67
        lightpath protection, 68–70
        metro access solutions, 61
        network architecture, 72
        network design examples, 76–82
        network design rules, 74–76
        network elements, 63
        OADMs, characterization, 72–74
        protection bandwidth, 68
        relaxed component manufacturing tolerance, 63
        ring topology, 62
        services, 67–68
        thin film filters, 65–66
        VCESLs, 64–65
        versus DWDM, 63
    interwavelength distance, 60
    light trails, 6
weather conditions, effect on FSO, 154–155
WFQ (weighted fair queuing), 218
WiFi, 22
    interoperability with EPONs, 270
wired networks, 4
    as access networks, 10
        cable networks, 17–19
        DSL, 11–12
        fiber, 12–17
    PSTN, 19–20
wireless networks, 4, 20, 212
    air as medium, 87–88
    delay spread, 100

fading, 100
FHSS, 93–94
fixed wireless, 21
free space optics, 22
infrared waves, 22
mobility management, 213
multiplexing
    CDMA, 90–93
    FDM, 89–90
    TDM, 88–89
WiFi, 22
wireless solution for last inch users, 243
WirelessMAN standard, 113
    contention resolution, 128
    line-of-sight radio propagation, 131
    MAC layer
        CPS, 124–126
        privacy sublayer, 122–124
    modulation schemes, 121–122
    network design considerations, 131–132
    reference model, 114
        IEs, 119
        MAC CPS, 115
        MAC layer, 115
        physical layer, 115–119
        privacy sublayer, 115
        service-specific CS, 115
    scheduling services, 126–127
        GPC mode, 127–128
        GPSS mode, 127–128
        requests, 127
    SSs, initialization, 128–130
    typical network architecture, 114
    wirelessMAN-SC air interface, 113
WirelessMAN-SC air interface, 113
WLANs (wireless LANs), 22, 100
    active scanning, 108
    ad hoc, 101
    association, 108
    roaming, 108